Django Web 开发实例精解

[美] 爱达斯·班德拉蒂斯　等著

张华臻　译

清華大学出版社

北　京

内 容 简 介

本书详细阐述了与 Django 相关的基本解决方案，主要包括模型和数据库结构、表单和视图、模板和 JavaScript、自定义模板过滤器和标签、模型管理、安全和性能、层次结构、导入和导出数据、测试、部署、维护等内容。此外，本书还提供了相应的示例、代码，以帮助读者进一步理解相关方案的实现过程。

本书适合作为高等院校计算机及相关专业的教材和教学参考书，也可作为相关开发人员的自学用书和参考手册。

图书在版编目（CIP）数据

Django Web 开发实例精解 /（美）爱达斯·班德拉蒂斯等著；张华臻译. —北京：清华大学出版社，2023.3

书名原文：Django 3 Web DevelopmentCookbook, Fourth Edition

ISBN 978-7-302-63131-6

Ⅰ. ①D… Ⅱ. ①爱… ②张… Ⅲ. ①软件工具—程序设计—汉、英 Ⅳ. ①TP311.561

中国国家版本馆 CIP 数据核字（2023）第 047809 号

责任编辑： 贾小红
封面设计： 刘 超
版式设计： 文森时代
责任校对： 马军令
责任印制： 杨 艳

出版发行： 清华大学出版社
网　　址： http://www.tup.com.cn，http://www.wqbook.com
地　　址： 北京清华大学学研大厦 A 座　　**邮　　编：** 100084
社 总 机： 010-83470000　　**邮　　购：** 010-62786544
投稿与读者服务： 010-62776969，c-service@tup.tsinghua.edu.cn
质量反馈： 010-62772015，zhiliang@tup.tsinghua.edu.cn
印 装 者： 三河市春园印刷有限公司
经　　销： 全国新华书店
开　　本： 185mm×230mm　　**印　　张：** 34.5　　**字　　数：** 690 千字
版　　次： 2023 年 4 月第 1 版　　**印　　次：** 2023 年 4 月第 1 次印刷
定　　价： 159.00 元

产品编号：088047-01

译 者 序

Web 开发是 Python 语言应用领域的重要部分。Python 作为当前最火爆、最热门，也是最主要的 Web 开发语言之一，在其发展过程中出现了数十种 Web 框架。其中，Django 是一个功能强大的 Python Web 框架，支持快速开发过程以及简洁、实用的设计方案。Django 是高水准的 Python 编程语言驱动的一个开源模型，是一个视图、控制器风格的 Web 应用程序框架，它起源于开源社区。使用这种架构，程序员可以方便、快捷地创建高品质、易维护、数据库驱动的应用程序。另外，在 Django 框架中，还包含许多功能强大的第三方插件，使得 Django 具有较强的可扩展性。

如果读者具有一定的 Django 使用经验，并想进一步提升自己的技能，那么本书将十分适合你。本书将利用 Django 框架处理 Web 开发过程中的各个阶段，主要包括模型和数据库结构、表单和视图、模板和 JavaScript、自定义模板过滤器和标签、模型管理、安全和性能、层次结构、导入和导出数据、测试、部署、维护等内容。

在本书的翻译过程中，除张华臻外，张博、刘璋、刘祎等人也参与了部分翻译工作，在此一并表示感谢。

由于译者水平有限，难免有疏漏和不妥之处，恳请广大读者批评指正。

译 者

前 言

Django 框架是专门帮助开发人员快速、高效地构建健壮和功能强大的 Web 应用程序而设计的。其间，Django 框架负责处理大量的枯燥和重复的工作，解决诸如项目结构、数据库对象-关系映射、模板机制、表单验证、会话、身份验证、安全性、Cookie 管理、国际化、基本管理和脚本的数据访问接口等问题。Django 构建于 Python 编程语言之上，该语言自身强制执行清晰和易读的代码。除了核心框架，Django 的设计目的是让开发人员能够创建第三方模块，进而可与自己的应用程序结合使用。另外，Django 有个已经建立的充满活力的社区，读者可从中查找源代码、寻求帮助并贡献自己的意见。

本书将利用 Django 3.0 框架指导读者完成 Web 开发过程中的各个阶段。首先将讨论项目的配置和结构；随后介绍如何使用复用组件以定义数据库结构，以及如何通过项目的生命周期对其进行管理。相应地，本书将通过表单和视图访问输入和列出数据。此外，我们还将处理响应式模板和 JavaScript 以增强用户体验。接下来，我们将通过自定义过滤器和标签提升 Django 的模板系统，从而使前端开发更具灵活性。在此之后，我们还将细化管理界面，以简化内容编辑器的工作流。据此，我们将重点转移至项目的稳定性和健壮性上，进而优化应用程序，同时提升应用程序的安全性。随后将介绍如何高效地存储和管理层次结构，并展示如何从不同的数据源收集数据。其间将会发现，以某些格式向其他处提供自身的数据比想象中的要更加简单。接下来将介绍 Django 项目代码的编程和调试技巧，并通过一些有效的选择方案测试代码。在本书的最后，我们还将展示如何将项目部署至产品中，并通过设置维护操作完成开发周期。

与其他一些仅涉及框架自身的 Django 书籍相比，本书引入了一些较为重要的第三方模块和工具以帮助我们完成 Web 开发。除此之外，我们还借助 Bootstrap 前端框架和 jQuery JavaScript 库提供了一些示例，从而简化某些高级、复杂的用户界面的创建工作。

适用读者

如果读者具有一定的 Django 使用经验，并想进一步提升自己的技能，那么本书将十分适合于你。本书适用于中级和专业 Django 开发人员，旨在构建多语言、安全的、响应式的、可伸缩的健壮型项目。

本书内容

第 1 章将介绍 Django 项目所需的基本设置和配置步骤，其中涉及虚拟环境、Docker、不同环境和数据库间的项目设置。

第 2 章将解释如何编写模块构建过程中的可复用代码。这里，用新应用程序首先需要定义的是数据模块，这将构成项目的主要组成部分。随后将学习如何将多语言数据保存至数据库中。另外，本章还将学习如何利用 Django 迁移管理数据库模式变化和数据操控。

第 3 章将讨论如何构建视图和表单，用于数据显示和编辑。其间，我们将学习如何使用微格式和其他协议，以确保页面更具可读性，进而展示搜索结果和社交网络中的显示内容。另外，本章还将学习如何生成 PDF 文档和实现多语言搜索。

第 4 章将介绍模板和 JavaScript 结合使用的实际示例，其间将整合下列内容：面向用户的渲染模板表达信息，以及在现代站点中 JavaScript 提供的重要的增强内容，进而丰富用户体验。

第 5 章将介绍如何创建和使用自己的模板过滤器和标签。其间将会看到，可扩展默认的 Django 模板系统以满足模板开发人员的要求。

第 6 章将探讨默认的 Django 管理界面，并引领读者通过自己的功能扩展该管理界面。

第 7 章将深入讨论 Django 固有或外部的项目安全和优化方法。

第 8 章将介绍 Django 中树状结构的创建和操控，同时还将 django-mptt 或 treebeard 库引入工作流。本章将展示如何使用这两种库实现层次结构的显示和管理。

第 9 章将讨论不同格式的数据转换，以及不同数据源的提供者。同时，本章将采用自定义管理命令实现数据导入，并通过站点图、RSS 和 REST API 实现数据导出。

第 10 章将展示一些额外的代码片段和技巧，这对于日常的 Web 开发和调试十分有用。

第 11 章将引入不同的测试类型，并提供多个特征示例用以测试项目代码。

第 12 章将处理针对 Python Package Index 的第三方应用程序部署，以及专用服务器的 Django 项目部署。

第 13 章将解释如何创建数据库备份、为常规任务设置计划作业，以及如何生成日志事件以供进一步查看。

本书资源

采用 Django 3.0 实现本书示例时，需要安装下列内容：

- Python 3.6 或更高版本。
- 操控图像的 Pillow 库。
- MySQL 数据库和 mysqlclient 绑定库，或者基于 psycopg2-binary 绑定库的 PostgreSQL 数据库。
- Docker Desktop 或 Docker Toolbox 用以实现完整系统虚拟化效果，或者一个内建的虚拟环境，以使每个项目的 Python 模块处于分离状态。
- 针对版本控制的 Git。

本书所涉及的软件和硬件如表 1 所示。

表 1

软件和硬件	操作系统
Python 3.6 或更高版本 Django 3.0.X PostgreSQL 11.4或更高版本/MySQL 5.6或更高版本	最新的基于 UNIX 的操作系统，如 macOS 或 Linux（也有可能在 Windows 操作系统上进行开发）

除此之外，其他的特定需求将在每个示例中单独提到。

读者可采用任何代码编辑器编辑项目文件，但这里推荐使用 PyCharm（https://www.jetbrains.com/pycharm/）或 Visual Studio Code（https://code.visualstudio.com/）。

注意：

全部代码示例均通过Django 3.0测试，且应可与Django的后续版本协同工作。

下载示例代码文件

读者可通过 www.packt.com 账户下载本书的示例代码文件。如果读者购买了本书，则可访问 www.packtpub.com/support 并注册，我们将把相关文件通过电子邮件直接发送于您。

下列步骤展示了代码文件的下载过程。

- 登录 www.packt.com 并注册。
- 选择 Support 选项卡。
- 单击 Code Downloads 按钮。
- 在搜索框中输入本书名称，并遵循相应的指令。

待文件下载完毕后，确保利用下列软件的最新版本解压或析取文件夹。

- WinRAR/7-Zip（Windows 环境）。
- Zipeg/iZip/UnRarX（Mac 环境）。
- 7-Zip/PeaZip（Linux 环境）。

除此之外，本书的代码包还托管于 GitHub 上，对应网址为 https://github.com/PacktPublishing/Django-3-Web-Development-Cookbook-Fourth-Edition，且代码将与已有的 GitHub 储存库同步更新。

另外，读者还可访问 https://github.com/PacktPublishing/查看其他图书的代码包、丰富的资源分类和视频内容。

本书惯例

本书涵盖了以下文本惯例。

代码块如下所示。

```
# requirements/dev.txt
-r _base.txt
coverage
django-debug-toolbar
selenium
```

需要强调的特定代码采用粗体表示，如下所示。

```
class Idea(CreationModificationDateBase, MetaTagsBase, UrlBase):
    title = models.CharField(
        _("Title"),
        max_length=200,
    )
    content = models.TextField(
        _("Content"),
    )
```

命令行输入或输出内容如下所示。

```
(env)$ pip install -r requirements/dev.txt
```

图标表示警告或重要事项。

图标表示提示或操作技巧。

读者反馈和客户支持

欢迎读者对本书提出建议或意见并予以反馈。

对此，读者可向 customercare@packtpub.com 发送邮件，并以书名作为邮件标题。

勘误表

尽管我们希望本书做到尽善尽美，但疏漏依然在所难免。如果读者发现谬误，无论是文字错误抑或是代码错误，还望不吝赐教。对此，读者可访问 http://www.packtpub.com/submit-errata，选取对应书籍，输入并提交相关问题的详细内容。

版权须知

一直以来，互联网上的版权问题从未间断，Packt 出版社对此类问题异常重视。若读者在互联网上发现本书任意形式的副本，请告知我们网络地址或网站名称，我们将对此予以处理。关于盗版问题，读者可发送邮件至 copyright@packtpub.com。

若读者针对某项技术具有专家级的见解，抑或计划撰写书籍或完善某部著作的出版工作，则可访问 authors.packtpub.com。

问题解答

读者对本书有任何疑问，均可发送邮件至 questions@packtpub.com，我们将竭诚为您服务。

目　录

第 1 章　Django 3.0 开始之旅

本章主要涉及下列主题。

- 处理虚拟环境。
- 创建项目文件结构。
- 利用 pip 处理项目依赖关系。
- 配置设置项，用于开发、测试、预置和产品环境。
- 在设置项中定义相对路径。
- 处理敏感的设置项。
- 将外部依赖项纳入项目中。
- 以动态方式设置 STATIC_URL。
- 设置 UTF-8 作为 MySQL 配置的默认编码机制。
- 创建 Git ignore 文件。
- 删除 Python 编译文件。
- Python 文件中的导入命令。
- 创建应用程序配置。
- 定义可覆写的应用程序设置项。
- 针对 Django、Gunicorn、Nginx 和 PostgreSQL，与 Docker 容器协同工作。

1.1　简　　介

当采用 Python 3 并基于 Django 3.0 启动新的项目时，本章将介绍一些十分有用的操作方法。其中，我们将选取最有效的方式处理可伸缩的项目布局、设置项和配置，无论采用 virtualenv 或 Docker 管理项目。

此外假设读者已熟悉 Django、Git 版本控制、MySQL 和 PostgreSQL 数据库和命令行使用方面的基础知识。另外，还假设读者正在使用基于 UNIX 的操作系统，如 macOS 或 Linux。在 UNIX 平台上通过 Django 进行开发是更有意义的，因为 Django 网站很可能在 Linux 服务器上发布。这意味着，我们可以构建以相同方式工作的例程，无论是正在开发或者部署。如果采用本地方式在 Windows 环境下与 Django 协同工作，例程之间将具有相似性，而非总是保持相同。

在不考虑本地平台的情况下，针对开发环境采用 Docker 可通过部署改进应用程序的可移植性，因为 Docker 容器内的环境可准确地与部署服务器上的环境进行匹配。另外，对于本章的示例，我们假设读者已持有相应的版本控制系统，且数据库服务器已经安装在本地机器上（无论是否采用 Docker 进行开发）。

1.2 技术需求

当与本书代码协同工作时，需要安装最新稳定版本的Python，读者可访问 https://www.python.org/downloads/进行下载。在本书编写时，Python 的最新稳定版本是 3.8.X。此外，我们还需要使用 MySQL 或 PostgreSQL 数据库。相应地，读者可访问 https://dev.mysql.com/downloads/下载 MySQL 数据库服务器，并访问 https://www.postgresql.org/下载 PostgreSQL 数据库服务器。其他需求条件将在特定的示例中予以提示。

读者可访问 GitHub 储存库的 ch01 目录查看本章代码，对应网址为 https://github.com/PacktPublishing/Django-3-Web-Development-Cookbook-Fourth-Edition。

1.3 与虚拟环境协同工作

一种可能的情况是，读者将在计算机上开发多个 Django 项目。某些模块（如 virtualenv、setuptools、wheel 或 Ansible）可能已经一次性安装完毕，以供所有项目共享。其他模块（如 Django、第三方 Python 库和 Django 应用程序）则需要彼此间保持隔离。virtualenv 工具是一个实用程序，该程序负责隔离所有的 Python 项目，并将它们保存在自己的领域中。下面的示例将讨论如何使用 virtualenv 工具。

1.3.1 准备工作

当管理 Python 包时，需要使用 pip 命令。如果读者正在使用 Python 3.4+，则需要在安装过程中将该命令纳入进来。如果读者正在使用 Python 的其他版本，则需要执行 http://pip.readthedocs.org/en/stable/installing/中的安装命令安装 pip。下面更新共享 Python 模块、pip、setuptools 和 wheel。

```
$ sudo pip3 install --upgrade pip setuptools wheel
```

自 Python 3.3 起，虚拟环境已内建于 Python 中。

1.3.2　实现方式

待一切安装完毕后，生成一个目录，如主目录下的 projects，并存储所有的 Django 项目。在生成该目录后，可执行下列步骤。

（1）访问新生成的目录并创建一个使用共享系统站点包的虚拟环境。

```
$ cd ~/projects
$ mkdir myproject_website
$ cd myproject_website
$ python3 -m venv env
```

（2）当使用新创建的虚拟环境时，需要在当前 shell 中执行激活脚本，这可通过下列命令实现。

```
$ source env/bin/activate
```

（3）取决于使用的 shell，可能无法使用 source 命令。另一种获取资源文件的方法是使用下列命令，且具有相同的操作结果（注意点和 env 之间的空格）。

```
$ . env/bin/activate
```

（4）可以看到，命令行工具的提示符将获得一个当前项目名称的前缀，如下所示。

```
(env)$
```

（5）当退出虚拟环境时，可输入下列命令。

```
(env)$ deactivate
```

1.3.3　工作方式

当创建一个虚拟环境时，也会生成一些专有的目录，如 bin、include 和 lib，用于存储 Python 安装的副本，并定义一些共享 Python 路径。当激活虚拟环境时，利用 pip 或 easy_install 安装的任何内容均会被置入并由虚拟环境站点包所使用，而非 Python 安装的全局站点包。

当在虚拟环境中安装最新的 Django 3.0.x 时，可输入下列命令。

```
(env)$ pip install "Django~=3.0.0"
```

1.3.4　延伸阅读

- 创建“项目文件结构”示例。

- “针对 Django、Gunicorn、Nginx 和 PostgreSQL 与 Docker 容器协同工作”示例。
- 第 12 章中的“针对预发布（staging）环境利用 mod_wsgi 在 Apache 上部署”示例。
- 第 12 章中的“针对产品环境利用 mod_wsgi 在 Apache 上部署”示例。
- 第 12 章中的“针对预发布环境在 Nginx 和 Gunicorn 上部署”示例。
- 第 12 章中的“针对产品环境在 Nginx 和 Gunicorn 上部署”示例。

1.4 创建一个项目文件结构

一致的项目文件结构可实现良好的组织并提升生产力。一旦持有了定义好的基本工作流，即可快速进入业务逻辑并创建项目。

1.4.1 准备工作

创建一个~/projects 目录（如果不存在），并于其中放置所有的 Django 项目（参见“与虚拟环境协同工作”示例）。

随后创建一个特定的项目目录，如 myproject_website 目录，并于 env 目录中启用虚拟环境，随后激活虚拟环境并安装 Django。这里，建议针对与当前项目相关的本地 shell 脚本添加一个 commands 目录、针对数据库转储添加一个 db_backups 目录、针对网站设计文件添加一个 mockups 目录，以及针对 Django 项目添加一个重要的 src 目录。

1.4.2 实现方式

下列步骤将创建一个项目的文件结构。

（1）激活虚拟环境，访问 src 目录并启动一个新的 Django 项目，如下所示。

```
(env)$ django-admin.py startproject myproject
```

上述命令将生成一个名为 myproject 的目录，其中包含了项目文件。该目录将包含一个名为 myproject 的 Python 模块。出于简洁、方便考虑，我们将这一顶级目录重命名为 django-myproject。另外，该目录还涉及版本控制，因而其中包含了一个.git 或类似名称的子目录。

（2）在 django-myproject 目录中，创建一个新的 README.md 文件，并将项目表述为新的 developdjango-admin.py startproject myprojecters。

（3）django-myproject 项目包含下列内容。

- 名为 myproject 的项目的 Python 包。
- 基于 Django 框架和其他外部依赖项（参见“利用 pip 处理项目的依赖项”）的项目的 pip 需求条件。
- LICENSE 文件中的项目许可证。如果项目开源，可从 https://choosealicense.com 中选择较为常见的许可证之一。

（4）在项目的根目录 django-myproject 中，创建下列目录。

- 项目上传的 media 目录。
- 收集静态文件的 static 目录。
- 项目翻译的 locale 目录。
- 外部依赖项的 externals 目录，当无法使用 pip 需求条件时，这些外部依赖项将纳入当前项目中。

（5）myproject 目录应包含下列目录和文件。

- apps 目录放置项目中所有的内部 Django 应用程序。此处建议包含一个名为 core 或 utils 的应用程序，用以实现项目的共享功能。
- 用于项目设置项的 settings 目录（参见“部署、测试、预发布和产品环境的配置设置项”示例）。
- 特定于项目静态文件的 site_static 目录。
- 项目 HTML 模板的 templates 目录。
- 项目 URL 配置的 urls.py 文件。
- 项目 Web 服务器配置的 wsgi.py 文件。

（6）在 site_static 目录中，创建 site 目录并作为特定站点静态文件的命名空间。随后将在其中的分类子目录之间划分静态文件，如下所示。

- Sass 文件的 scss 子目录（可选）。
- 生成的最小的层叠样式表（Cascading Style Sheets，CSS）的 css 子目录。
- 样式图像、收藏夹图表和 Logo 的 img 子目录。
- JavaScriptdjango-admin.py 启动项目 myproject 的 js 子目录。
- 第三方模块的 vendor 子目录（整合了所有的文件类型，如 TinyMCE 富文本编辑器）。

（7）除了 site 目录，site_static 可能也包含第三方应用程序的覆写静态目录。例如，site_static 目录可能包含 cms，该目录覆写 Django CMS 的静态文件。当从 Sass 中生成 CSS 文件并最小化 JavaScript 文件时，可通过图形用户界面使用 CodeKit（https://codekitapp.com/）或 Prepros（https://prepros.io/）应用程序。

（8）将由应用程序分隔的模板置于 templates 目录中。如果模板文件表示为一个页面（如 change_item.html 或 item_list.html），那么可将该文件直接置于应用程序的模板目录中。如果该模板包含在另一个模板（如 similar_items.html）中，则可将其置于 includes 子目录中。除此之外，模板目录还可包含一个名为 utils 的目录，供全局复用代码片段使用，如分页和语言选择器。

1.4.3　工作方式

一个完整项目的整体文件结构如下所示。

```
myproject_website/
├── commands/
├── db_backups/
├── mockups/
├── src/
│   └── django-myproject/
│       ├── externals/
│       │   ├── apps/
│       │   │   └── README.md
│       │   └── libs/
│       │       └── README.md
│       ├── locale/
│       ├── media/
│       ├── myproject/
│       │   ├── apps/
│       │   │   ├── core/
│       │   │   │   ├── __init__.py
│       │   │   │   └── versioning.py
│       │   │   └── __init__.py
│       │   ├── settings/
│       │   │   ├── __init__.py
│       │   │   ├── _base.py
│       │   │   ├── dev.py
│       │   │   ├── production.py
│       │   │   ├── sample_secrets.json
│       │   │   ├── secrets.json
│       │   │   ├── staging.py
│       │   │   └── test.py
│       │   ├── site_static/
│       │   │   └── site/
│       │   │   django-admin.py startproject myproject ├── css/
```

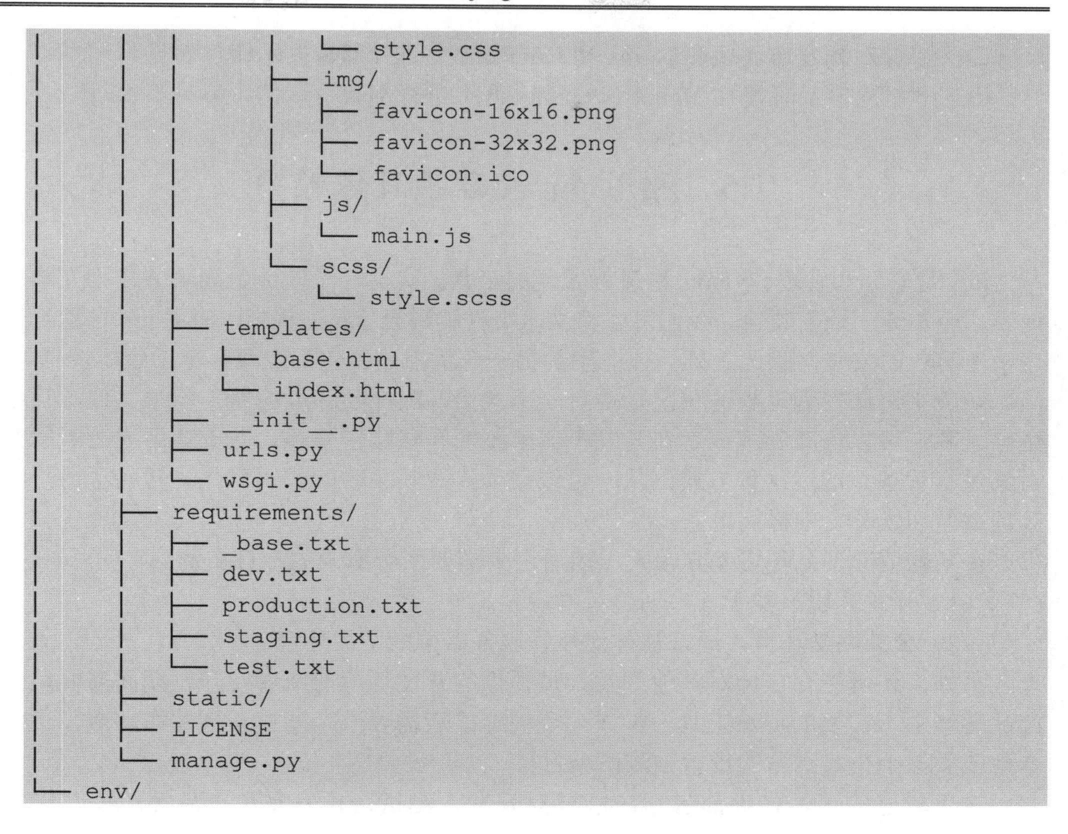

```
│   │   │   │   └── style.css
│   │   │   ├── img/
│   │   │   │   ├── favicon-16x16.png
│   │   │   │   ├── favicon-32x32.png
│   │   │   │   └── favicon.ico
│   │   │   ├── js/
│   │   │   │   └── main.js
│   │   │   └── scss/
│   │   │       └── style.scss
│   │   ├── templates/
│   │   │   ├── base.html
│   │   │   └── index.html
│   │   ├── __init__.py
│   │   ├── urls.py
│   │   └── wsgi.py
│   ├── requirements/
│   │   ├── _base.txt
│   │   ├── dev.txt
│   │   ├── production.txt
│   │   ├── staging.txt
│   │   └── test.txt
│   ├── static/
│   ├── LICENSE
│   └── manage.py
└── env/
```

1.4.4　更多内容

为了加快项目的创建速度，可采用来自 https://github.com/archatas/django-myproject 中的项目样板文件。在下载了相关代码后，可执行全局搜索并利用你的项目名称替换 myproject。

1.4.5　延伸阅读

- “利用 pip 处理项目依赖项”示例。
- “在项目中包含外部依赖项”示例。
- “针对开发、测试、预发布和产品环境，配置设置项”示例。
- “针对预发布环境利用 mod_wsgi 在 Apache 上进行部署”示例，参见第 12 章。
- “针对产品环境利用 mod_wsgi 在 Apache 上进行部署”示例，参见第 12 章。

- “针对预发布环境在 Nginx 和 Gunicorn 上进行部署”示例，参见第 12 章。
- “针对产品环境在 Nginx 和 Gunicorn 上进行部署”示例，参见第 12 章。

1.5 利用 pip 处理项目依赖项

pip 是安装和管理 Python 包最为方便的工具。定义一个安装包列表并作为文本文件内容，而非逐一地安装包。我们可以将文本文件传递至 pip 工具中，随后 pip 自动处理列表中所有包的安装。除此之外，该方案的另一个优点是，包列表可存储于版本控制中。

一般来讲，较为理想的状况是持有一个直接匹配产品环境的单一需求文件，且已然足够。我们可以修改版本或者在开发机器上添加和移除依赖项，并于随后通过版本控制对其进行管理。通过这种方式，从一组依赖项（及其关联的代码变化）转移至另一组依赖项就像切换分支一样简单。

在某些情况下，环境差异较大，因而至少需要两个不同的项目实例。

（1）开发环境，并于其中创建新的特性。

（2）公共网站环境，通常被称作托管服务器中的产品环境。

另外，还可能存在针对其他开发人员提供的开发环境，或者在开发期间需要，但在产品环境中不需要的特殊工具。抑或，我们可能还需要一个测试和预发布环境，进而在本地或类似于公共网站的设置中测试项目。

对于较好的可维护性而言，应能够对开发、测试、预发布和产品环境安装所需的 Python 模块。其中，一些模块将被共享，而某些模块则特定于具体的环境子集。在当前示例中，我们将学习如何组织多重环境的项目依赖项，并通过 pip 对其进行管理。

1.5.1 准备工作

在应用当前示例之前，Django 项目应处于就绪状态，且 pip 安装完毕，同时虚拟环境处于激活状态。更多内容可参见“与虚拟环境协同工作”示例。

1.5.2 实现方式

逐一执行下列步骤，以对虚拟环境 Django 项目准备 pip 需求。

（1）访问版本控制下所持有的 Django 项目，并利用下列文本文件创建 requirements 目录。

- 针对共享模块的_base.txt 文件。

- 针对开发环境的 dev.txt 文件。
- 针对测试环境的 test.txt 文件。
- 针对预发布环境的 staging.txt。
- 针对产品的 production.txt 文件。

（2）编辑_base.txt，并添加在全部环境中共享的 Python 模块。

```
# requirements/_base.txt
Django~=3.0.4
djangorestframework
-e git://github.com/omab/python-socialauth.
git@6b1e301c79#egg=python-social-auth
```

（3）如果特定环境的需求条件等同于_base.txt 文件中的内容，则可在该环境的需求条件中添加包含_base.txt 的代码行，如下所示。

```
# requirements/production.txt
-r _base.txt
```

（4）如果对某一环境存在特殊要求，那么可在_base.txt 后进行添加，如下所示。

```
# requirements/dev.txt
-r _base.txt
coverage
django-debug-toolbar
selenium
```

（5）可在虚拟环境中运行下列命令，并安装开发环境（或其他环境的类似命令）所需的全部依赖项。

```
(env)$ pip install -r requirements/dev.txt
```

1.5.3　工作方式

无论在虚拟环境中还是在全局级别上以显式方式执行，上述 pip install 命令都会从 requirements/_base.txt 和 requirements/dev.txt 中下载并安装所有的项目依赖项。可以看到，我们可以指定一个 Django 框架所需的模块版本，甚至还可直接从 Git 储存库的特定提交中安装所需的模块版本，正如当前示例中 python-social-auth 所做的那样。

当在项目中包含许多依赖项时，较好的做法是将发布版本限制在一定的范围内。据此，项目的完整性不会因依赖项更新而遭到破坏。这些更新的依赖项可能会包含冲突或无法向后兼容。在部署项目或将项目移交给新开发人员时，这一点尤为重要。

如果已经通过手动方式和 pip 逐一地安装了项目的所需条件，随后即可在虚拟环境中通过下列命令生成 requirements/_base.txt 文件。

```
(env)$ pip freeze > requirements/_base.txt
```

1.5.4　更多内容

为了简化事物并确保所有环境均使用相同的依赖项，那么可针对相应的需求使用一个名为 requirements.txt 的文件，该文件由定义生成，如下所示。

```
(env)$ pip freeze > requirements.txt
```

当在新的虚拟环境中安装模块时，可简单地使用下列命令。

```
(env)$ pip install -r requirements.txt
```

如果需要从另一个版本控制系统或者本地路径中安装一个 Python 库，则可查看官方文档以了解与 pip 相关的更多内容，对应网址为 https://pip.pypa.io/en/stable/user_guide/。

另一种管理 Python 依赖项的方法是 Pipenv，且越发变得流行。对此，可访问 https://github.com/pypa/pipenv 以了解更多内容。

1.5.5　延伸阅读

- “与虚拟环境协同工作”示例。
- “针对 Django、Gunicorn、Nginx 和 PostgreSQL，与 Docker 容器协同工作”示例。
- “在项目中包含外部依赖项”示例。
- “针对开发、测试、预发布和产品环境，配置设置项”示例。

1.6　针对开发、测试、预发布和产品环境，配置设置项

如前所述，我们将在开发环境中创建新特性，在测试环境中对其进行测试，并将站点置入预发布的服务器上，以便其他人尝试使用新特性。接下来，将网站部署至产品服务器上以供公共访问。其中，每一个环境均包含特定的设置项，当前示例将学习如何对其进行组织。

1.6.1　准备工作

在 Django 项目中，我们将针对每种环境创建设置项，其中包括开发、测试、预发布

和产品阶段。

1.6.2　实现方式

下列步骤将配置设置项。

（1）在 myproject 目录中，利用下列文件创建一个 settings Python 模块。

- __init__.py 文件生成 Python 模块的设置项目录。
- _base.py 文件用于共享设置项。
- dev.py 文件用于开发设置项。
- test.py 文件用于测试设置项。
- staging.py 文件用于预发布设置项。
- production.py 文件用于产品设置项。

（2）将 settings.py 文件中的内容（启动新的 Django 项目时自动生成）复制至 settings/_base.py 文件中。随后删除 settings.py 文件。

（3）修改 settings.py 文件中 BASE_DIR，并指向上一级，如下所示。

```
BASE_DIR =
os.path.dirname(os.path.dirname(os.path.abspath(__file__)))
```

修改完毕后，最终结果如下所示。

```
BASE_DIR = os.path.dirname(
    os.path.dirname(os.path.dirname(os.path.abspath(__file__)))
)
```

（4）如果某个环境设置项等同于共享设置项，则仅需从_base.py 文件中导入每项内容即可，如下所示。

```
# myproject/settings/production.py
from ._base import *
```

（5）在其他文件中应用需要绑定或覆写特定环境的设置项。例如，开发环境应访问 dev.py 文件，如下列代码片段所示。

```
# myproject/settings/dev.py
from ._base import *
EMAIL_BACKEND = "django.core.mail.backends.console.EmailBackend"
```

（6）修改 manage.py 和 myproject/wsgi.py 文件，并通过修改下列代码行使用默认状态下的环境设置项之一。

```
os.environ.setdefault('DJANGO_SETTINGS_MODULE',
'myproject.settings')
```

（7）修改结果如下所示。

```
os.environ.setdefault('DJANGO_SETTINGS_MODULE',
'myproject.settings.production')
```

1.6.3　工作方式

默认状态下，Django 管理命令使用 myproject/settings.py 文件中的设置项。当使用定义于当前示例中的方法时，我们可以将所有环境中所需的非敏感设置项保存在 config 目录中的版本控制下。另一方面，settings.py 文件自身将会被版本控制忽略，且仅包含当前开发、测试、预发布和产品环境所需的设置项。

提示：

对于每个环境，建议单独设置 PyCharm（env/bin/activate 脚本）或.bash_profile 中的 DJANGO_SETTINGS_MODULE 环境变量。

1.6.4　延伸阅读

- “针对 Django、Gunicorn、Nginx 和 PostgreSQL，与 Docker 容器协同工作”示例。
- “处理敏感设置项”示例。
- “在设置项中定义相对路径”示例。
- “创建 Git 忽略文件”示例。

1.7　在设置项中定义相对路径

Django 需要我们在设置项中定义不同的文件路径，如媒体的根路径、静态文件的根路径、模板路径，以及翻译文件的路径。对于项目的每一位开发人员，路径可能有所不同，因为虚拟环境可能在各处予以设置，并且用户可能在 macOS、Linux 或 Windows 环境下工作。即使项目封装在 Docker 容器中，定义绝对路径也将会降低可维护性和可移植性。尽管如此，仍存在一种方法并以动态方式定义这些路径，以便形成相对于 Django 项目目录的路径。

1.7.1　准备工作

启动 Django 项目并打开 settings/_base.py 文件。

1.7.2　实现方式

相应地，修改相对路径设置项，而非硬编码本地目录的路径，如下所示。

```
# settings/_base.py
import os
BASE_DIR = os.path.dirname(
    os.path.dirname(os.path.dirname(os.path.abspath(__file__)))
)
# ...
TEMPLATES = [{
    # ...
    DIRS: [
        os.path.join(BASE_DIR, 'myproject', 'templates'),
    ],
    # ...
}]
# ...
LOCALE_PATHS = [
    os.path.join(BASE_DIR, 'locale'),
]
#
STATICFILES_DIRS = [
    os.path.join(BASE_DIR, 'myproject', 'site_static'),
]
STATIC_ROOT = os.path.join(BASE_DIR, 'static')
MEDIA_ROOT = os.path.join(BASE_DIR, 'media')
```

1.7.3　工作方式

默认状态下，Django 设置项包含 BASE_DIR 值，该值表示为包含 manage.py 文件（该文件通常高于 settings.py 文件一级，或者高于 settings/_base.py 文件两级）的目录的绝对路径。随后，可利用 os.path.join()函数设置相对于 BASE_DIR 的全部路径。

基于所设置的目录布局（参见“创建一个项目文件结构”示例），在前面的示例中，我们会插入'myproject'作为中间路径段，因为相关的文件夹是在这个路径段中创建的。

1.7.4 延伸阅读

- “创建一个项目文件结构”示例。
- “针对 Django、Gunicorn、Nginx 和 PostgreSQL，与 Docker 容器协同工作”示例。
- “在项目中包含外部依赖项”示例。

1.8 处理敏感设置项

当配置一个 Django 项目时，一般会处理一些敏感信息，如密码和 API 密钥。这里，不建议将这一类敏感信息置于版本控制下。对此，存在两种主要的方法存储此类信息，即环境变量和独立的未跟踪（untracked）文件中。当前示例将探讨这两种情况。

1.8.1 准备工作

大多数项目设置项将在所有的环境中共享，并存储于版本控制中。这些内容可直接定义于设置项文件中。尽管如此，仍存在一些特定于项目实例环境的设置项，以及较为敏感和提升安全级别的设置项，如数据库或电子邮件设置项。我们将通过环境变量使用这些设置项。

1.8.2 实现方式

当从环境变量中读取敏感设置项时，可执行下列步骤。

（1）在 settings/_base.py 文件开始处，定义 get_secret()函数，如下所示。

```
# settings/_base.py
import os
from django.core.exceptions import ImproperlyConfigured

def get_secret(setting):
    """Get the secret variable or return explicit exception."""
    try:
        return os.environ[setting]
    except KeyError:
        error_msg = f'Set the {setting} environment variable'
        raise ImproperlyConfigured(error_msg)
```

（2）当定义一个敏感值时，可使用 get_secret()函数，如下所示。

```
SECRET_KEY = get_secret('DJANGO_SECRET_KEY')

DATABASES = {
    'default': {
        'ENGINE': 'django.db.backends.postgresql_psycopg2',
        'NAME': get_secret('DATABASE_NAME'),
        'USER': get_secret('DATABASE_USER'),
        'PASSWORD': get_secret('DATABASE_PASSWORD'),
        'HOST': 'db',
        'PORT': '5432',
    }
}
```

1.8.3　工作方式

如果在缺少环境变量集的情况下运行 Django 管理命令，我们将会看到一条错误消息，如 Set the DJANGO_SECRET_KEY environment variable。

我们可在 PyCharm 配置中、远程服务器配置控制台中、env/bin/activate 脚本、.bash_profile 中，或直接在终端中设置环境变量。

```
$ export DJANGO_SECRET_KEY="change-this-to-50-characters-long-
 random-string"
$ export DATABASE_NAME="myproject"
$ export DATABASE_USER="myproject"
$ export DATABASE_PASSWORD="change-this-to-database-password"
```

注意，应对全部密码、API 密钥，以及 Django 项目配置中所需的任何敏感信息使用 get_secret()函数。

1.8.4　更多内容

如果不采用环境变量，还可使用包含敏感信息的文本文件，这些敏感信息在版本控制下不会被跟踪。这些文件可以是置于硬盘某处 YAML、INI、CSV 或 JSON 文件。例如，对于 JSON 文件，可以使用 get_secret()函数，如下所示。

```
# settings/_base.py
import os
import json
```

```
with open(os.path.join(os.path.dirname(__file__), 'secrets.json'), 'r')
as f:
    secrets = json.loads(f.read())

def get_secret(setting):
    """Get the secret variable or return explicit exception."""
    try:
        return secrets[setting]
    except KeyError:
        error_msg = f'Set the {setting} secret variable'
        raise ImproperlyConfigured(error_msg)
```

这将从设置项目录中读取 secrets.json 文件，并期望该文件应至少涵盖下列结构。

```
{
  "DATABASE_NAME": "myproject",
  "DATABASE_USER": "myproject",
  "DATABASE_PASSWORD":"change-this-to-database-password",
  "DJANGO_SECRET_KEY":"change-this-to-50-characters-long-random-string"
}
```

此外，应确保 secrets.json 文件被版本控制所忽略。但出于方便考虑，可创建一个包含空值的 sample_secrets.json 文件，并将其置于版本控制之下。

```
{
  "DATABASE_NAME": "",
  "DATABASE_USER": "",
  "DATABASE_PASSWORD":"",
  "DJANGO_SECRET_KEY":"change-this-to-50-characters-long-random-string"
}
```

1.8.5　延伸阅读

- “创建一个项目文件结构”示例。
- “针对 Django、Gunicorn、Nginx 和 PostgreSQL，与 Docker 容器协同工作”示例。

1.9　在项目中包含外部依赖项

有时，我们无法利用 pip 安装外部依赖项，且需要将依赖项直接包含至项目中，如在以下情况下。

- 当拥有一个打补丁的第三方应用程序，其间需要修复一个 Bug 或添加了一个未被项目所有者接受的特性。
- 当需要使用 Python Package Index（PyPI）或公共版本控制库无法访问的私有应用程序时。
- 当使用 PyPI 中无效的遗留依赖项版本时。

在项目中包含外部依赖项时，可确保开发人员在任何时刻更新依赖模块；其他开发人员都将在下次版本控制系统的更新中获得升级版本。

1.9.1　准备工作

在虚拟环境中启动一个 Django 项目。

1.9.2　实现方式

在虚拟环境项目中逐个执行下列步骤。

（1）在 Django 项目目录 django-myproject 下创建一个 externals 目录（如果不存在）。

（2）随后在 externals 目录下创建 libs 和 apps 目录。其中，libs 目录用于项目所需的 Python 模块，如 Boto、Requests、Twython 和 Whoosh。另外，apps 目录则用于第三方 Django 应用程序，如 Django CMS、Django Haystack 和 Django 存储。

这里强烈建议在 libs 和 apps 目录中创建 README.md 文件，并于其中指出每个模块的内容、所采用的版本及其来源。

（3）目录结构如下所示。

```
externals/
  ├── apps/
  │   ├── cms/
  │   ├── haystack/
  │   ├── storages/
  │   └── README.md
  └── libs/
      ├── boto/
      ├── requests/
      ├── twython/
      └── README.md
```

（4）接下来将外部库和应用程序置于 Python 路径下，以便按照安装方式进行识别。

这可通过在设置项中添加下列代码实现。

```
# settings/_base.py
import os
import sys
BASE_DIR = os.path.dirname(
    os.path.dirname(os.path.dirname(os.path.abspath(__file__)))
)
EXTERNAL_BASE = os.path.join(BASE_DIR, "externals")
EXTERNAL_LIBS_PATH = os.path.join(EXTERNAL_BASE, "libs")
EXTERNAL_APPS_PATH = os.path.join(EXTERNAL_BASE, "apps")
sys.path = ["", EXTERNAL_LIBS_PATH, EXTERNAL_APPS_PATH] + sys.path
```

1.9.3 工作方式

如果可以运行 Python 并导入模块，那么该模块应位于 Python 路径下。相应地，将某个模块置于 Python 路径之下的方法之一是调整 sys.path 变量，随后导入非常规位置处的模块。设置文件项指定的 sys.path 值是一个目录列表，该列表始于一个表示当前目录的空字符串，随后是项目中的目录，最后是 Python 安装的全局共享目录。我们可以在 Python shell 中查看 sys.path 值，如下所示。

```
(env)$ python manage.py shell
>>> import sys
>>> sys.path
```

当尝试导入一个模块时，Python 将搜索上述列表中的模块，并返回所找到的第一个结果。

因此，首先定义 BASE_DIR 变量，该变量表示为 django-myproject 的绝对路径，或者比 myproject/settings/_base.py 高 3 个级别。随后定义 EXTERNAL_LIBS_PATH 和 EXTERNAL_APPS_PATH 变量（相对于 BASE_DIR）。最后调整 sys.path 属性，并向列表起始处添加新路径。需要注意的是，我们还添加了一个空字符串作为搜索的第一个路径，这意味着，任何模块的当前目录一般会被首先检查，随后检查其他 Python 路径。

这种包含外部库的方式无法与基于 C 语言绑定的 Python 包实现跨平台操作，如 lxml。针对这种依赖项，建议采用之前介绍的 pip 命令。

1.9.4 延伸阅读

- “创建一个项目文件结构”示例。

- “针对 Django、Gunicorn、Nginx 和 PostgreSQL，与 Docker 容器协同工作”示例。
- “利用 pip 处理项目依赖项”示例。
- “在设置项中定义相对路径”示例。
- 第 10 章中的“使用 Django shell”示例。

1.10　以动态方式设置 STATIC_URL

如果将 STATIC_URL 设置为静态值，那么每次更新一个 CSS 文件、JavaScript 文件或图像时，网站访问者需要清除浏览器缓存以查看变化内容，这里，存在一个浏览器缓存清除技巧，即在 STATIC_URL 中最近变化内容上设置时间戳。当代码更新后，访问者的浏览器将强制加载全部新的静态文件。

当前示例将探讨 Git 用户如何在 STATIC_URL 中设置时间戳。

1.10.1　准备工作

确保当前项目在 Git 版本控制下，且设置项中已经定义了 BASE_DIR。参见“在设置项中定义相对路径”示例。

1.10.2　实现方式

在 STATIC_URL 设置项中置入时间戳包含下列两个步骤。

（1）在 Django 项目中创建 myproject.apps.core 应用程序（若不存在），同时还应于其中创建一个 versioning.py 文件。

```
# versioning.py
import subprocess
from datetime import datetime

def get_git_changeset_timestamp(absolute_path):
    repo_dir = absolute_path
    git_log = subprocess.Popen(
        "git log --pretty=format:%ct --quiet -1 HEAD",
        stdout=subprocess.PIPE,
        stderr=subprocess.PIPE,
        shell=True,
        cwd=repo_dir,
```

```
        universal_newlines=True,
    )

    timestamp = git_log.communicate()[0]
    try:
        timestamp = datetime.utcfromtimestamp(int(timestamp))
    except ValueError:
        # Fallback to current timestamp
        return datetime.now().strftime('%Y%m%d%H%M%S')
    changeset_timestamp = timestamp.strftime('%Y%m%d%H%M%S')
    return changeset_timestamp
```

（2）将新创建的 get_git_changeset_timestamp()函数导入设置项中，并用于 STATIC_URL 路径，如下所示。

```
# settings/_base.py
from myproject.apps.core.versioning import
get_git_changeset_timestamp
# ...
timestamp = get_git_changeset_timestamp(BASE_DIR)
STATIC_URL = f'/static/{timestamp}/'
```

1.10.3　工作方式

get_git_changeset_timestamp()函数作为参数接收 absolute_path 目录，并利用参数调用 git log shell 命令，进而显示目录中 HEAD 修订的 UNIX 时间戳。这里，我们向函数传递 BASE_DIR，因为我们已经确保它已处于版本控制下。时间戳被解析后将转换为包含年、月、日、小时、分钟和秒的字符串，随后就被包含在 STATIC_URL 的定义中。

1.10.4　更多内容

上述方法仅适用于每个环境包含完整的项目 Git 储存库。某些情况下，当使用 Heroku 或 Docker 进行部署时，我们无须访问 Git 储存库和远程服务器中的 git log 命令。为了使 STATIC_URL 具有动态片段，需要从文本文件（如 myproject/settings/lastmodified.txt）中读取时间戳，该文件应随着每次提交而更新。

在这种情况下，设置项将包含下列代码行。

```
# settings/_base.py
with open(os.path.join(BASE_DIR, 'myproject', 'settings',
'last-update.txt'), 'r') as f:
```

```
    timestamp = f.readline().strip()

STATIC_URL = f'/static/{timestamp}/'
```

我们可采用一个预提交钩子使 Git 储存库更新 last-modified.txt。这是一个可执行的 bash 脚本并被称作预提交且置于 django-myproject/.git/hooks/下。

```
# django-myproject/.git/hooks/pre-commit
#!/usr/bin/env python
from subprocess import check_output, CalledProcessError
import os
from datetime import datetime

def root():
    ''' returns the absolute path of the repository root '''
    try:
        base = check_output(['git', 'rev-parse', '--show-toplevel'])
    except CalledProcessError:
        raise IOError('Current working directory is not a git repository')
    return base.decode('utf-8').strip()

def abspath(relpath):
    ''' returns the absolute path for a path given relative to the
        root of the git repository
    '''
    return os.path.join(root(), relpath)

def add_to_git(file_path):
    ''' adds a file to git '''
    try:
        base = check_output(['git', 'add', file_path])
    except CalledProcessError:
        raise IOError('Current working directory is not a git repository')
    return base.decode('utf-8').strip()

def main():
    file_path = abspath("myproject/settings/last-update.txt")

    with open(file_path, 'w') as f:
        f.write(datetime.now().strftime("%Y%m%d%H%M%S"))
    add_to_git(file_path)

if __name__ == '__main__':
    main()
```

当提交至 Git 储存库并将该文件添加至 Git 索引时，该脚本将更新 last-modified.txt。

1.10.5　延伸阅读

读者可参考“创建 Git 忽略文件”示例。

1.11　将 UTF-8 设置为 MySQL 配置的默认编码机制

MySQL 将自身描述为一个最受欢迎的开源数据库。在当前示例中，我们将讨论如何将 UTF-8 设置为 MySQL 的默认编码机制。注意，如果未在数据库配置中设置该编码机制，可能会出现的情况是，默认状态下，LATIN1 与 UTF-8 编码数据协同使用。当使用诸如“€”这种符号时，将会导致数据库错误。另外，当持有采用 LATIN1 或其他 UTF-8 格式编码的表时，当前示例可在 LATIN1 和 UTF-8 之间简单地转换数据。

1.11.1　准备工作

确保安装了 MySQL 数据库管理系统和 mysqlclient Python 模块，并在读取的项目设置中使用 MySQL 引擎。

1.11.2　实现方式

在编辑器中打开 MySQL 配置文件/etc/mysql/my.cnf，确保在[client]、[mysql]和[mysqld]部分中设置了下列内容。

```
# /etc/mysql/my.cnf
[client]
default-character-set = utf8

[mysql]
default-character-set = utf8

[mysqld]
collation-server = utf8_unicode_ci
init-connect = 'SET NAMES utf8'
character-set-server = utf8
```

如果缺少上述任何一部分内容，即可在文件/etc/mysql/my.cnf 中进行创建。如果相关部分已经存在，则可将这些设置项添加至现有的配置中，随后在命令行工具中重新启动

MySQL，如下所示。

```
$ /etc/init.d/mysql restart
```

1.11.3　工作方式

当创建新的 MySQL 数据库时，默认状态下，数据库及其全部表均设置为 UTF-8 编码机制。对此，不要忘记在开发或发布项目的所有计算机上对此加以设置。

1.11.4　更多内容

在 PostgreSQL 中，默认的服务器编码机制已经设置为 UTF-8。如果希望利用 UTF-8 编码机制显式地创建一个 PostgreSQL 数据库，可使用下列命令。

```
$ createdb --encoding=UTF8 --locale=en_US.UTF-8 --template=template0
myproject
```

1.11.5　延伸阅读

- “创建一个项目文件结构”示例。
- “针对 Django、Gunicorn、Nginx 和 PostgreSQL，与 Docker 容器协同工作”示例。

1.12　创建 Git 忽略文件

Git 是较为流行的分布式版本控制系统，读者很可能已在项目中使用过 Git。虽然我们一直在跟踪大多数文件的变化，但建议保持某些特殊文件和文件夹不受版本控制。通常情况下，缓存、编译后的代码、日志文件和隐藏的系统文件不应在 Git 储存库中被跟踪。

1.12.1　准备工作

确保 Django 项目处于 Git 版本控制之下。

1.12.2　实现方式

利用文本编辑器在 Django 项目的根目录中创建一个.gitignore 文件，并将下列文件和目录置于其中。

```
# .gitignore
### Python template
# Byte-compiled / optimized / DLL files
__pycache__/
*.py[cod]
*$py.class

# Installer logs
pip-log.txt
pip-delete-this-directory.txt

# Unit test / coverage reports
htmlcov/
.tox/
.nox/
.coverage
.coverage.*
.cache
nosetests.xml
coverage.xml
*.cover
.hypothesis/
.pytest_cache/

# Translations
*.mo
*.pot

# Django stuff:
*.log
db.sqlite3

# Sphinx documentation
docs/_build/

# IPython
profile_default/
ipython_config.py

# Environments
env/

# Media and Static directories
/media/
```

```
!/media/.gitkeep

/static/
!/static/.gitkeep

# Secrets
secrets.json
```

1.12.3　工作方式

.gitignore 文件指定了 Git 版本控制系统的非跟踪模式。在当前示例中，我们所创建的.gitignore 文件将忽略 Python 编译文件、本地设置项、收集的静态文件和包含上传文件的多媒体目录。

注意，这里针对多媒体和静态文件采用了带有感叹号的特殊语法。

```
/media/
!/media/.gitkeep
```

这将通知 Git 忽略/media/目录，但对/media/.gitkeep 文件则保持版本系统下的跟踪。因为 Git 版本控制跟踪文件，但不涉及目录。这里，我们使用.gitkeep 确保 media 目录在每个环境下被创建，但不会被跟踪。

1.12.4　延伸阅读

- ❑ “创建一个项目文件结构”示例。
- ❑ “针对 Django、Gunicorn、Nginx 和 PostgreSQL，与 Docker 容器协同工作”示例。

1.13　删除 Python 编译文件

当首次运行项目时，Python 将编译字节码编译文件*.pyc 中的全部*.py 代码，稍后用于执行。通常情况下，当修改*.py 文件时，*.pyc 将被重新编译。然而，某些时候，在切换分支或移动目录时，需要手动清理已编译的文件。

1.13.1　准备工作

利用编辑器在 home 目录中编辑或创建一个.bash_profile 文件。

1.13.2　实现方式

（1）在.bash_profile 结尾处添加别名，如下所示。

```
# ~/.bash_profile
alias delpyc='
find . -name "*.py[co]" -delete
find . -type d -name "__pycache__" -delete'
```

（2）当清除 Python 编译文件时，访问项目目录，并在命令行中输入下列命令。

```
(env)$ delpyc
```

1.13.3　工作方式

首先创建一个 UNIX 别名，用于搜索*.pyc 和*.pyo 文件以及__pycache__目录，并在当前目录及其子目录中删除它们。当在命令行工具中启动一个新会话时，.bash_profile 文件将被执行。

1.13.4　更多内容

如果希望避免创建 Python 编译文件，则可在.bash_profile、env/bin/activate 脚本或 PyCharm 配置中将环境变量 PYTHONDONTWRITEBYTECODE 设置为 1。

1.13.5　延伸阅读

除此之外，读者还可参考“创建 Git 忽略文件”示例。

1.14　遵循 Python 文件中的导入顺序

当创建 Python 模块时，较好的做法是与文件中的结构保持一致，从而提升代码的可读性。当前示例将展示如何结构化导入内容。

1.14.1　准备工作

创建一个虚拟环境并于其中创建一个 Django 项目。

1.14.2　实现方式

针对创建的每个 Python 文件使用下列结构，进而将导入内容划分为多个部分。

```
# System libraries
import os
import re
from datetime import datetime

# Third-party libraries
import boto
from PIL import Image

# Django modules
from django.db import models
from django.conf import settings

# Django apps
from cms.models import Page

# Current-app modules
from .models import NewsArticle
from . import app_settings
```

1.14.3　工作方式

此处共计 5 个导入分类，如下所示。

- ❑ Python 默认安装中的包的系统库。
- ❑ 附加安装的 Python 包的第三方库。
- ❑ 有别于 Django 框架的不同模块的 Django 模块。
- ❑ 第三方和本地应用程序的 Django 应用程序。
- ❑ 从当前应用中进行相对导入的当前应用程序模块。

1.14.4　更多内容

当在 Python 和 Django 中编码时，可使用 Python 代码的官方风格指南 PEP8，具体内容可参考 https://www.python.org/dev/peps/pep-0008/。

1.14.5 延伸阅读

- “利用 pip 处理项目依赖项”示例。
- “在项目中包含外部依赖项”示例。

1.15 创建应用程序配置

Django 项目由多个称之为应用程序（常称作应用）的 Python 模块构成，这些模块组合了不同的模块化功能。其中，每个应用程序可包含模型、视图、表单、URL 配置、管理命令、迁移、信号、测试、上下文处理器、中间件等。Django 框架包含一个应用程序注册表，其中，所有的应用程序和模型均被收集起来，随后用于配置和自省。自 Django 1.7 以来，与应用程序相关的元信息可保存在每个应用程序的 AppConfig 实例中。接下来将创建一个 magazine 示例应用程序，以探讨如何使用应用程序配置。

1.15.1 准备工作

通过调用 startapp 管理命令，或者采用手动方式构建应用程序模块，我们可以创建一个 Django 应用程序。

```
(env)$ cd myproject/apps/
(env)$ django-admin.py startapp magazine
```

利用已创建的 magazine 应用程序，可分别向 models.py 中添加一个 NewsArticle 模型、在 admin.py 中创建该模型的管理，并将"myproject.apps.magazine"置于设置项的 INSTALLED_APPS 中。如果读者尚不熟悉上述各项任务，则可参考 Django 官方教程，对应网址为 https://docs.djangoproject.com/en/3.0/intro/tutorial01/。

1.15.2 实现方式

下列步骤可以创建和使用应用程序配置。

（1）修改 apps.py 文件并向其中添加下列内容。

```
# myproject/apps/magazine/apps.py
from django.apps import AppConfig
from django.utils.translation import gettext_lazy as _
```

```
class MagazineAppConfig(AppConfig):
    name = "myproject.apps.magazine"
    verbose_name = _("Magazine")

    def ready(self):
        from . import signals
```

（2）编辑 magazine 模块中的__init__.py 文件并涵盖下列内容。

```
# myproject/apps/magazine/__init__.py
default_app_config =
"myproject.apps.magazine.apps.MagazineAppConfig"
```

（3）创建一个 signals.py 文件，并添加下列信号处理程序。

```
# myproject/apps/magazine/signals.py
from django.db.models.signals import post_save, post_delete
from django.dispatch import receiver
from django.conf import settings

from .models import NewsArticle

@receiver(post_save, sender=NewsArticle)
def news_save_handler(sender, **kwargs):
    if settings.DEBUG:
        print(f"{kwargs['instance']} saved.")

@receiver(post_delete, sender=NewsArticle)
def news_delete_handler(sender, **kwargs):
    if settings.DEBUG:
        print(f"{kwargs['instance']} deleted.")
```

1.15.3 工作方式

当运行一个 HTTP 服务器或调用一个管理命令时，django.setup()方法将被调用。该方法加载设置项、设置日志机制并准备应用程序注册表。其间，注册表将分 3 个步骤被初始化。首先，Django 针对设置项中的 INSTALLED_APPS 的每个条目导入配置。这些条目可直接指向应用程序的名称或配置，如"myproject.apps.magazine"或"myproject.apps.magazine.apps.MagazineAppConfig"。

随后，Django 尝试从 INSTALLED_APPS 中的每个应用程序中导入 models.py，并收集全部模型。

最后，Django 运行每个应用程序配置的 ready()方法。该方法为开发过程提供了一个较为适宜之处，用于注册信号处理程序（如果存在）。另外，ready()方法则是一个可选项。

在当前示例中，MagazineAppConfig 类负责设置 magazine 应用程序的配置。name 参数则定义当前应用程序的模块。verbose_name 参数定义 Django 模型管理中的人名。其中，模型按照应用程序分组呈现。ready()方法导入并激活信号处理程序，当处于 DEBUG 模式时，信号处理程序将在终端中输出，进而保存或删除一个 NewsArticle 对象。

1.15.4　更多内容

在调用了 django.setup()方法后，即可加载应用程序配置和注册表中的模型，如下所示。

```
>>> from django.apps import apps as django_apps
>>> magazine_app_config = django_apps.get_app_config("magazine")
>>> magazine_app_config
<MagazineAppConfig: magazine>
>>> magazine_app_config.models_module
<module 'magazine.models' from
'/path/to/myproject/apps/magazine/models.py'>
>>> NewsArticle = django_apps.get_model("magazine", "NewsArticle")
>>> NewsArticle
<class 'magazine.models.NewsArticle'>
```

关于应用程序配置的更多内容，读者可参考 Django 官方文档，对应网址为 https://docs.djangoproject.com/en/2.2/ref/applications/。

1.15.5　延伸阅读

- "与虚拟环境协同工作"示例。
- "针对 Django、Gunicorn、Nginx 和 PostgreSQL，与 Docker 容器协同工作"示例。
- "定义可覆写的应用程序设置项"示例。
- 第 6 章。

1.16　定义可覆写的应用程序设置项

当前示例将讨论如何定义应用程序的设置项，这些设置项随后将在项目的设置项文件中被覆写。通过添加一项配置，这对于自定义的可复用应用程序来说十分有用。

1.16.1　准备工作

遵循“创建应用程序配置”示例中的步骤构建一个 Django 应用程序。

1.16.2　实现方式

（1）在 models.py 文件（如果仅包含一个或两个设置项）或 app_settings.py 文件（如果设置项可扩展且打算自定义设置项）中，通过 getattr()模式定义应用程序设置项。

```
# myproject/apps/magazine/app_settings.py
from django.conf import settings
from django.utils.translation import gettext_lazy as _

# Example:
SETTING_1 = getattr(settings, "MAGAZINE_SETTING_1", "default value")

MEANING_OF_LIFE = getattr(settings, "MAGAZINE_MEANING_OF_LIFE", 42)

ARTICLE_THEME_CHOICES = getattr(
    settings,
    "MAGAZINE_ARTICLE_THEME_CHOICES",
     [
        ('futurism', _("Futurism")),
        ('nostalgia', _("Nostalgia")),
        ('sustainability', _("Sustainability")),
        ('wonder', _("Wonder")),
    ]
)
```

（2）models.py 将包含 NewsArticle 模型，如下所示。

```
# myproject/apps/magazine/models.py
from django.db import models
from django.utils.translation import gettext_lazy as _

class NewsArticle(models.Model):
    created_at = models.DateTimeField(_("Created at"),
     auto_now_add=True)
    title = models.CharField(_("Title"), max_length=255)
    body = models.TextField(_("Body"))
    theme = models.CharField(_("Theme"), max_length=20)
```

```
    class Meta:
        verbose_name = _("News Article")
        verbose_name_plural = _("News Articles")

    def __str__(self):
        return self.title
```

（3）在 admin.py 中，导入并使用 app_settings.py 中的设置项，如下所示。

```
# myproject/apps/magazine/admin.py
from django import forms
from django.contrib import admin

from .models import NewsArticle

from .app_settings import ARTICLE_THEME_CHOICES

class NewsArticleModelForm(forms.ModelForm):
    theme = forms.ChoiceField(
        label=NewsArticle._meta.get_field("theme").verbose_name,
        choices=ARTICLE_THEME_CHOICES,
        required=not NewsArticle._meta.get_field("theme").blank,
    )
    class Meta:
        fields = "__all__"

@admin.register(NewsArticle)
class NewsArticleAdmin(admin.ModelAdmin):
form = NewsArticleModelForm
```

（4）如果打算覆写给定项目中的 ARTICLE_THEME_CHOICES 设置项，则应在项目设置项中添加 MAGAZINE_ARTICLE_THEME_CHOICES，如下所示。

```
# myproject/settings/_base.py
from django.utils.translation import gettext_lazy as _
# ...
MAGAZINE_ARTICLE_THEME_CHOICES = [
     ('futurism', _("Futurism")),
     ('nostalgia', _("Nostalgia")),
     ('sustainability', _("Sustainability")),
     ('wonder', _("Wonder")),
     ('positivity', _("Positivity")),
     ('solutions', _("Solutions")),
```

```
    ('science', _("Science")),
]
```

1.16.3　工作方式

getattr(object, attribute_name[, default_value]) Python 函数尝试获取 object 中的 attribute_name 属性，如果没有找到则返回 default_value。这里，我们尝试读取 Django 项目设置项模块中的不同设置项，如果不存在，则采用默认值。

注意，我们本可以在 models.py 中为 theme 字段定义 choices，但我们在管理中创建了一个自定义 ModelForm 并于此处设置了选项。这样做是为了避免在 ARTICLE_THEME_CHOICES 更改时创建新的数据库迁移。

1.16.4　延伸阅读

- ❑ “创建应用程序配置”示例。
- ❑ 第 6 章。

1.17　针对 Django、Gunicorn、Nginx 和 PostgreSQL，与 Docker 容器协同工作

Django 项目不仅依赖于 Python 需求条件，还依赖于许多系统需求，如 Web 服务器、数据库、服务器为缓存和邮件服务器。当开发一个 Django 项目时，应确保所有环境和开发人员持有完全相同的安装需求条件。对此，保持依赖项同步的方法之一是使用 Docker。当使用 Docker 时，对于每个项目，可持有不同的数据库版本、Web 或者其他服务器。

Docker（称作容器）系统用于生成配置完毕的自定义虚拟机，进而可精确地复制任何产品环境的设置。Docker 容器从称之为 Docker 镜像中创建。其中，镜像包含关于容器构建方式的多个层次（或指令）。相应地，可存在 PostgreSQL 镜像、Redis 镜像、Memcached 镜像以及 Django 项目的自定义镜像。所有这些镜像均可通过 Docker Compose 整合至伴随容器中。

当前示例将采用项目样板构建一个基于 PostgreSQL 数据库的 Django 项目，由 Nginx 和 Gunicorn 提供服务，并使用 Docker Compose 对其进行管理。

1.17.1　准备工作

首先需要安装 Docker Engine，并遵循 https://www.docker.com/get-started 中的各项指令。这通常会包含 Compose 工具，从而可管理基于多个容器的系统，这也是完全隔离的 Django 项目的理想选择方案。如果需要单独使用，关于 Compose 的安装细节，读者可访问 https://docs.docker.com/compose/install/以了解更多内容。

1.17.2　实现方式

下面探讨 Django 和 Docker 的样板代码。

（1）访问 https://github.com/archatas/django_docker，并将代码下载至计算机的~/projects/django_docker 目录中。

注意：

如果打算选择另一个目录，如 myproject_docker，则需要执行全局搜索，并利用 myproject_docker 替换 django_docker。

（2）打开 docker-compose.yml 文件。此处需要创建 3 个容器，即 nginx、gunicorn 和 db。稍后将对此进行深入讨论。

```
# docker-compose.yml
version: "3.7"

services:
  nginx:
    image: nginx:latest
    ports:
      - "80:80"
    volumes:
      - ./config/nginx/conf.d:/etc/nginx/conf.d
      - static_volume:/home/myproject/static
      - media_volume:/home/myproject/media
    depends_on:
      - gunicorn

  gunicorn:
    build:
      context: .
```

```
    args:
      PIP_REQUIREMENTS: "${PIP_REQUIREMENTS}"

    command: bash -c "/home/myproject/env/bin/gunicorn --workers 3
    --bind 0.0.0.0:8000 myproject.wsgi:application"
    depends_on:
      - db
    volumes:
      - static_volume:/home/myproject/static
      - media_volume:/home/myproject/media
    expose:
      - "8000"
    environment:
      DJANGO_SETTINGS_MODULE: "${DJANGO_SETTINGS_MODULE}"
      DJANGO_SECRET_KEY: "${DJANGO_SECRET_KEY}"
      DATABASE_NAME: "${DATABASE_NAME}"
      DATABASE_USER: "${DATABASE_USER}"
      DATABASE_PASSWORD: "${DATABASE_PASSWORD}"
      EMAIL_HOST: "${EMAIL_HOST}"
      EMAIL_PORT: "${EMAIL_PORT}"
      EMAIL_HOST_USER: "${EMAIL_HOST_USER}"
      EMAIL_HOST_PASSWORD: "${EMAIL_HOST_PASSWORD}"

  db:
    image: postgres:latest
    restart: always
    environment:
      POSTGRES_DB: "${DATABASE_NAME}"
      POSTGRES_USER: "${DATABASE_USER}"
      POSTGRES_PASSWORD: "${DATABASE_PASSWORD}"
    ports:
      - 5432
    volumes:
      - postgres_data:/var/lib/postgresql/data/

volumes:
  postgres_data:
  static_volume:
  media_volume:
```

（3）打开并读取 Dockerfile 文件。创建 gunicorn 容器需要多个层次（指令）。

```
# Dockerfile
```

```
# pull official base image
FROM python:3.8

# accept arguments
ARG PIP_REQUIREMENTS=production.txt

# set environment variables
ENV PYTHONDONTWRITEBYTECODE 1
ENV PYTHONUNBUFFERED 1

# install dependencies
RUN pip install --upgrade pip setuptools

# create user for the Django project
RUN useradd -ms /bin/bash myproject

# set current user
USER myproject

# set work directory
WORKDIR /home/myproject

# create and activate virtual environment
RUN python3 -m venv env

# copy and install pip requirements
COPY --chown=myproject ./src/myproject/requirements
/home/myproject/requirements/
RUN ./env/bin/pip3 install -r
/home/myproject/requirements/${PIP_REQUIREMENTS}

# copy Django project files
COPY --chown=myproject ./src/myproject /home/myproject/
```

（4）将 build_dev_example.sh 脚本复制至 build_dev.sh 中并编辑其内容。下列内容显示了传递至 docker-compose 脚本的环境变量。

```
# build_dev.sh
#!/usr/bin/env bash
DJANGO_SETTINGS_MODULE=myproject.settings.dev \
DJANGO_SECRET_KEY="change-this-to-50-characters-long-random-
  string" \
DATABASE_NAME=myproject \
```

```
DATABASE_USER=myproject \
DATABASE_PASSWORD="change-this-too" \
PIP_REQUIREMENTS=dev.txt \
docker-compose up --detach --build
```

（5）在命令行工具中，向 build_dev.sh 添加运行许可，运行并构建容器。

```
$ chmod +x build_dev.sh
$ ./build_dev.sh
```

（6）当访问 http://0.0.0.0/en/时，可看到“Hello, World!”页面。
当访问 http://0.0.0.0/en/admin/时，可看到下列内容。

```
OperationalError at /en/admin/
 FATAL: role "myproject" does not exist
```

这意味着，需要在 Docker 容器内创建数据库用户和数据库。
（7）SSH 至 db 容器，并在 Docker 容器内创建数据库用户、密码和数据库自身。

```
$ docker exec -it django_docker_db_1 bash
/# su - postgres
/$ createuser --createdb --password myproject
/$ createdb --username myproject myproject
```

当询问时，输入与 build_dev.sh 脚本中相同的数据库密码。
按 Ctrl+D 快捷键两次退出 PostgreSQL 用户和 Docker 容器。
当访问 http://0.0.0.0/en/admin/时，即可看到下列内容。

```
ProgrammingError at /en/admin/ relation "django_session" does not
exist LINE 1: ...ession_data", "django_session"."expire_date" FROM
"django_se...
```

这意味着，需要运行迁移以生成数据库模式。
（8）SSH 至 gunicorn 容器并运行所需的 Django 管理命令。

```
$ docker exec -it django_docker_gunicorn_1 bash
$ source env/bin/activate
(env)$ python manage.py migrate
(env)$ python manage.py collectstatic
(env)$ python manage.py createsuperuser
```

回答管理命令询问的全部问题。
按 Ctrl+D 快捷键两次退出 Docker 容器。
当访问 http://0.0.0.0/en/admin/时，应该可以看到 Django 管理页面，并利用已生成的超级用户证书登录。

（9）创建类似的脚本 build_test.sh、build_staging.sh 和 build_production.sh，其中，仅环境变量有所不同。

1.17.3　工作方式

样板中的代码结构类似于虚拟环境中的结构。项目源文件位于 src 目录中。另外，用于预提交钩子的 git-hooks 目录用于跟踪最近一次修改日期；此外还有一个容器中使用的服务配置 config 目录。

```
django_docker
├── config/
│   └── nginx/
│       └── conf.d/
│           └── myproject.conf
├── git-hooks/
│   ├── install_hooks.sh
│   └── pre-commit
├── src/
│   └── myproject/
│       ├── locale/
│       ├── media/
│       ├── myproject/
│       │   ├── apps/
│       │   │   └── __init__.py
│       │   ├── settings/
│       │   │   ├── __init__.py
│       │   │   ├── _base.py
│       │   │   ├── dev.py
│       │   │   ├── last-update.txt
│       │   │   ├── production.py
│       │   │   ├── staging.py
│       │   │   └── test.py
│       │   ├── site_static/
│       │   │   └── site/
│       │   │       ├── css/
│       │   │       ├── img/
│       │   │       ├── js/
│       │   │       └── scss/
│       │   ├── templates/
│       │   │   ├── base.html
```

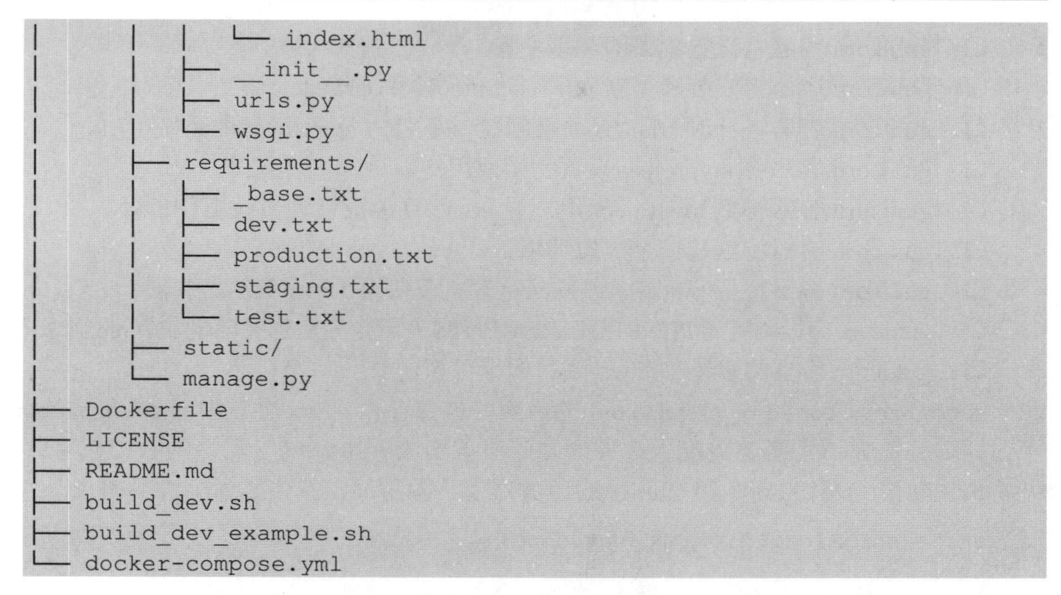

```
│       │   │   └── index.html
│       │   ├── __init__.py
│       │   ├── urls.py
│       │   └── wsgi.py
│       ├── requirements/
│       │   ├── _base.txt
│       │   ├── dev.txt
│       │   ├── production.txt
│       │   ├── staging.txt
│       │   └── test.txt
│       ├── static/
│       └── manage.py
├── Dockerfile
├── LICENSE
├── README.md
├── build_dev.sh
├── build_dev_example.sh
└── docker-compose.yml
```

其中，与主 Docker 相关的配置位于 docker-compose.yml 和 Dockerfile 中。Docker Compose 是一个围绕 Docker 命令行 API 的封装器。build_dev.sh 脚本在 Gunicorn WSGI HTTP 服务器下构建和运行 Django 项目，对应的端口为 8000；Nginx 的对应端口为 80（服务静态和多媒体文件并代理针对 Gunicorn 的其他请求）；PostgreSQL 数据库的对应端口为 5432。

在 docker-compose.yml 文件中，需要创建 3 个容器。

（1）Nginx Web 服务器的 nginx。

（2）基于 Gunicorn Web 服务器并针对 Django 项目的 gunicorn。

（3）PostgreSQL 数据库的 db。

其中，nginx 和 db 容器将从官方镜像（https://hub.docker.com）中被创建，二者均包含特定的配置参数，如运行的端口、环境变量、其他容器上的依赖项和卷。

Docker 卷是在重建 Docker 容器时保持不变的特定目录。相应地，需要为数据库数据文件、多媒体、静态等定义卷。

gunicorn 容器则根据 Dockerfile 处的指令（由 docker-compose.yml 文件中的构建上下文定义）构建。接下来将查看各个层（或指令）。

- gunicorn 容器基于 python:3.7 镜像。
- gunicorn 容器将 PIP_REQUIREMENTS 视为 docker-compose.yml 文件中的一个参数。

- gunicorn 容器将设置容器的环境变量。
- gunicorn 容器将安装和更新 pip、setuptools 和 virtualenv。
- gunicorn 容器将针对 Django 项目创建一个名为 myproject 的系统用户。
- gunicorn 容器将 myproject 设置为当前用户。
- gunicorn 容器将把 myproject 用户的 home 目录设置为当前工作目录。
- gunicorn 容器将创建一个虚拟环境。
- gunicorn 容器将把 pip 需求条件从计算机复制至 Docker 容器中。
- gunicorn 容器将针对 PIP_REQUIREMENTS 变量定义的当前环境安装 pip 需求条件。
- gunicorn 容器将复制 Django 项目的全部资源。

config/nginx/conf.d/myproject.conf 的内容将被保存在 nginx 容器中的/etc/nginx/conf.d/下，即 Nginx Web 服务器的配置，并通知该服务器监听端口 80（默认的 HTTP 端口），并将请求转发到端口 8000 的 Gunicorn 服务器上，除了静态或多媒体内容的请求。

```
#/etc/nginx/conf.d/myproject.conf
upstream myproject {
    server django_docker_gunicorn_1:8000;
}

server {
    listen 80;

    location / {
        proxy_pass http://myproject;
        proxy_set_header X-Forwarded-For $proxy_add_x_forwarded_for;
        proxy_set_header Host $host;
        proxy_redirect off;
    }

    rewrite "/static/\d+/(.*)" /static/$1 last;

    location /static/ {
        alias /home/myproject/static/;
    }

    location /media/ {
        alias /home/myproject/media/;
    }
}
```

关于 Nginx 和 Gunicorn 配置的更多内容，读者可参考第 12 章。

1.17.4　更多内容

我们可利用 docker-compose down 命令销毁 Docker 容器，并通过构建脚本对其进行重建。

```
$ docker-compose down
$ ./build_dev.sh
```

如果过程未按预期进行，则可利用 docker-compose logs 命令查看日志。

```
$ docker-compose logs nginx
$ docker-compose logs gunicorn
$ docker-compose logs db
```

当通过 SSH 连接至容器上时，可使用下列命令之一。

```
$ docker exec -it django_docker_gunicorn_1 bash
$ docker exec -it django_docker_nginx_1 bash
$ docker exec -it django_docker_db_1 bash
```

使用 docker cp 命令，可在 Docker 容器上的卷之间复制文件和目录。

```
$ docker cp ~/avatar.png django_docker_gunicorn_1:/home/myproject/media/
$ docker cp django_docker_gunicorn_1:/home/myproject/media ~/Desktop/
```

关于 Docker 和 Docker Compose 的更多内容，读者可参考官方文档，对应网址为 https://docs.docker.com/和 https://docs.docker.com/compose/。

1.17.5　延伸阅读

- “创建一个项目文件结构”示例。
- 第 12 章。

第 2 章　模型和数据库结构

本章主要涉及下列主题。

- ❑ 使用模型混入（mixin）。
- ❑ 利用与 URL 相关的方法创建一个模型混入。
- ❑ 创建一个模型混入以处理日期的生成和修改。
- ❑ 创建一个模型混入以关注元标签。
- ❑ 创建一个模型混入以处理泛型关系。
- ❑ 处理多语言字段。
- ❑ 处理模型翻译表。
- ❑ 避免环状依赖项。
- ❑ 添加数据库约束。
- ❑ 使用迁移。
- ❑ 将外键修改为多对多字段。

2.1　简　　介

当开始一个新的应用程序时，首先需要创建表示数据库结构的模型。此外假设已经创建了 Django 应用程序，或者至少已经阅读并理解了 Django 官方教程。本章将介绍多种有趣的技术，以使数据库结构在不同的项目应用程序之间保持一致。随后，我们还将讨论如何处理数据库中数据的国际化问题。接下来，本章将学习如何避免模型中的环状依赖，以及如何设置数据库约束。在本章的最后，我们将讨论如何在开发过程中利用迁移修改数据库结构。

2.2　技 术 需 求

当处理本章的代码时，至少需要安装 Python 的稳定版本、MySQL 或 PostgreSQL，以及基于虚拟环境的 Django 项目。

读者可访问 GitHub 储存库中 ch02 目录查看本章的全部代码，对应网址为 https://

github.com/PacktPublishing/Django-3-Web-Development-Cookbook-Fourth-Edition。

2.3　使用模型混入

在面向对象的语言（如 Python）中，混入类可视为一个包含实现特性的接口。当某个模型扩展一个混入时，需实现接口，同时包含其所有的字段、属性、特性和方法。当需要多次复用不同模型中的通用功能时，Django 模型中的混入将十分有用。Django 中的模型混入定义为抽象的模型基类，稍后将对此予以介绍。

2.3.1　准备工作

首先需要创建可复用的混入。保存模型混入的较好的地方是 myproject.apps.core 应用程序。如果创建了一个可与他人分享的可复用的应用程序，那么应该将模型混入保存至这一个可复用的应用程序自身中，可能是在一个 base.py 文件中。

2.3.2　实现方式

打开使用混入的 Django 应用程序的 models.py 文件并输入下列代码。

```
# myproject/apps/ideas/models.py
from django.db import models
from django.urls import reverse
from django.utils.translation import gettext_lazy as _

from myproject.apps.core.models import (
    CreationModificationDateBase,
    MetaTagsBase,
    UrlBase,
)

class Idea(CreationModificationDateBase, MetaTagsBase, UrlBase):
    title = models.CharField(
        _("Title"),
        max_length=200,
    )
    content = models.TextField(
        _("Content"),
    )
```

```
    # other fields…

    class Meta:
        verbose_name = _("Idea")
        verbose_name_plural = _("Ideas")

    def __str__(self):
        return self.title

    def get_url_path(self):
        return reverse("idea_details", kwargs={
            "idea_id": str(self.pk),
        })
```

2.3.3　工作方式

Django 的模型继承机制支持 3 种继承类型，即抽象基类、多表继承和代理模型。模型混入表示为抽象模型类，因为我们通过使用一个抽象的 Meta 类对其加以定义。Meta 类包含指定的字段、属性和方法。

当创建诸如 Idea 这一类模型时，该模型将继承 CreationModificationDateMixin、MetaTagsMixin 和 UrlMixin 中的全部特性。这些抽象类的所有字段与扩展模型的字段都保存在同一个数据库表中。在当前示例中，我们将学习如何定义模型混入。

2.3.4　更多内容

在常规的 Python 类继承中，如果存在多个基类，且全部基类均实现了特定的方法，并在子类的实例上调用该方法。那么，仅来自第 1 个父类的方法会被调用，如下所示。

```
>>> class A(object):
... def test(self):
...     print("A.test() called")
...

>>> class B(object):
... def test(self):
...     print("B.test() called")
...

>>> class C(object):
... def test(self):
```

```
...     print("C.test() called")
...

>>> class D(A, B, C):
... def test(self):
...     super().test()
...     print("D.test() called")

>>> d = D()
>>> d.test()
A.test() called
D.test() called
```

Django 模型基类也是如此，但存在一种特殊情况。

注意：

Django 框架使用元类从每个基类中调用 save()和 delete()方法。

这意味着，针对混入中定义的特殊字段（通过覆写 save()和 delete()方法），可执行前保存、后保存、前删除和后删除操作。

关于模型继承的不同类型，读者可参考 Django 官方文档，对应网址为 https://docs.djangoproject.com/en/2.2/topics/db/models/#model-inheritance。

2.3.5　延伸阅读

- “利用与 URL 相关的方法创建一个模型混入”示例。
- “创建一个模型混入以处理日期的创建和修改”示例。
- “创建一个模型混入以关注元标签”示例。

2.4　利用与 URL 相关的方法创建一个模型混入

对于包含自身独特详细页面的每个模型，较好的做法是定义 get_absolute_url()方法。该方法可被用于模板和 Django 管理站点中，进而预览保存的对象。由于 get_absolute_url()方法返回 URL 路径而非完整的 URL，因而该方法是不确定的。

在当前示例中，我们将研究如何创建一个模型混入，并针对特定模型的 URL 提供简单的支持。该混入可执行下列操作。

（1）定义模型中的 URL 或完整的 URL。

（2）根据定义的 URL 自动生成其他 URL。

（3）定义幕后的 get_absolute_url()方法。

2.4.1　准备工作

创建 myproject.apps.core 应用程序（如果不存在），并于其中保存模型混入。接下来，在 core 包中创建 models.py 文件。另外，如果创建了一个可复用的应用程序，可将混入置于该应用程序的 base.py 文件中。

2.4.2　实现方式

接下来逐一执行下列步骤。

（1）将下列内容添加至 core 应用程序的 models.py 文件中。

```
# myproject/apps/core/models.py
from urllib.parse import urlparse, urlunparse
from django.conf import settings
from django.db import models

class UrlBase(models.Model):
    """
    A replacement for get_absolute_url()
    Models extending this mixin should have either get_url or
     get_url_path implemented.
    """
    class Meta:
        abstract = True

    def get_url(self):
        if hasattr(self.get_url_path, "dont_recurse"):
            raise NotImplementedError
        try:
            path = self.get_url_path()
        except NotImplementedError:
            raise
        return settings.WEBSITE_URL + path
    get_url.dont_recurse = True

    def get_url_path(self):
        if hasattr(self.get_url, "dont_recurse"):
```

```
            raise NotImplementedError
        try:
            url = self.get_url()
        except NotImplementedError:
            raise
        bits = urlparse(url)
        return urlunparse(("", "") + bits[2:])
    get_url_path.dont_recurse = True

    def get_absolute_url(self):
        return self.get_url()
```

（2）将 WEBSITE_URL 设置项（不包含尾斜杠）添加至 dev、test、staging 和 production 设置项中。例如，在开发环境下，该过程如下所示。

```
# myproject/settings/dev.py
from ._base import *

DEBUG = True
WEBSITE_URL = "http://127.0.0.1:8000" # without trailing slash
```

（3）当在应用程序中使用混入时，需要导入 core 应用程序中的混入、继承模型类中的混入，并定义 get_url_path()方法，如下所示。

```
# myproject/apps/ideas/models.py
from django.db import models
from django.urls import reverse
from django.utils.translation import gettext_lazy as _

from myproject.apps.core.models import UrlBase

class Idea(UrlBase):
    # fields, attributes, properties and methods…

    def get_url_path(self):
        return reverse("idea_details", kwargs={
                "idea_id": str(self.pk),
        })
```

2.4.3　工作方式

UrlBase 类是一个抽象模型，其中包含 3 个方法。

（1）get_url()方法检索对象的完整 URL。

（2）get_url_path()方法检索对象的绝对路径。

（3）get_absolute_url()方法模拟 get_url_path()方法。

get_url()和 get_url_path()方法需要在扩展后的类（如 Idea）中被覆写。我们可以定义 get_url()，而 get_url_path()将把它“剥离”至路径中。另外，还可定义 get_url_path()，而 get_url()将网站 URL 添加至路径的开始处。

注意：

经验法则是，总是覆写 get_url_path()方法。

在模板中，当需要一个同一网站上对象的链接时，可使用 get_url_path()方法，如下所示。

```
<a href="{{ idea.get_url_path }}">{{ idea.title }}</a>
```

在外部通信中针对链接（如电子邮件、RSS 摘要或 API）使用 get_url()方法时，具体操作如下所示。

```
<a href="{{ idea.get_url }}">{{ idea.title }}</a>
```

默认的 get_absolute_url()方法将在 Django 模型管理中用于 View on site 功能，也可能被一些第三方 Django 应用程序使用。

提示：

不要在 URL 中使用增量主键，因为将这些主键暴露给最终用户是不安全的：此时条目的总数将是可见的，而且只需要更改 URL 路径就可以轻易地在不同的条目中进行导航。

只有当详细页面是通用唯一标识符（Universal Unique Identifier，UUID）或生成的随机字符串时，才可以使用 URL 中的主键。否则，创建并使用一个 slug 字段，如下所示。

```
class Idea(UrlBase):
    slug = models.SlugField(_("Slug for URLs"), max_length=50)
```

2.4.4　延伸阅读

- “使用模型混入”示例。
- “创建一个模型混入以处理日期的创建和修改”示例。
- “创建一个模型混入以关注元标签”示例。
- “创建一个模型混入以处理 Generic Relation”示例。

- 第 1 章中的“针对开发、测试、预发布和产品环境，配置设置项”示例。

2.5 创建一个模型混入以处理日期的创建和修改

在模型中包含时间戳用以创建和修改模型实例是较为常见的做法。在当前示例中，我们将学习如何创建一个简单的模型混入，以保存模型的创建和修改日期与时间。使用这一种混入将确保所有的模型使用相同的时间戳字段名，并包含相同的行为。

2.5.1 准备工作

创建 myproject.apps.core 包（如果不存在）以保存混入。随后在 core 包中创建 models.py 文件。

2.5.2 实现方式

打开 myprojects.apps.core 包中的 models.py 文件，并添加下列内容。

```
# myproject/apps/core/models.py
from django.db import models
from django.utils.translation import gettext_lazy as _

class CreationModificationDateBase(models.Model):
    """
    Abstract base class with a creation and modification date and time
    """

    created = models.DateTimeField(
        _("Creation Date and Time"),
        auto_now_add=True,
    )

    modified = models.DateTimeField(
        _("Modification Date and Time"),
        auto_now=True,
    )

    class Meta:
        abstract = True
```

2.5.3　工作方式

CreationModificationDateMixin 类是一个抽象模型，这意味着，扩展模型类将在相同的数据库表中生成全部字段。也就是说，不存在增加表处理难度的一对一关系。

混入包含两个日期—时间字段，即 created 和 modified。当使用 auto_now_add 和 auto_now 属性时，保存一个模型实例时，时间戳将被自动保存。该字段将自动获取 editable = False 属性，因而将以管理表单形式被隐藏。如果 USE_TZ 在设置项中被设置为 True（这也是默认和推荐行为），那么将使用时区时间戳，否则将使用非时区时间戳。这里，时区时间戳将保存在数据库的（Coordinated Universal Time，UTC）时区；并在读写时转换为项目默认的时区。非时区时间戳则被保存在数据库中的本地项目时区中，且通常不具备实际操作意义，因为这使得时区之间的时间管理变得更加复杂。

当使用这一混入时，我们仅需导入该混入并扩展模型，如下所示。

```
# myproject/apps/ideas/models.py
from django.db import models

from myproject.apps.core.models import CreationModificationDateBase

class Idea(CreationModificationDateBase):
    # other fields, attributes, properties, and methods…
```

2.5.4　延伸阅读

- “使用模型混入”示例。
- “创建一个模型混入以关注元标签”示例。
- “创建一个模型混入以处理 Generic Relation”示例。

2.6　创建一个模型混入以关注元标签

当针对搜索引擎优化站点时，不仅需要针对每个页面使用语义标记，还需要包含相应的元标签。出于灵活性考虑，这将有助于定义通用元标签的内容，并特定于网站上包含自身详情页面的对象。在当前示例中，我们将研究如何为与关键字、描述、作者和版权元标签相关的字段和方法创建模型混入。

2.6.1　准备工作

如前所述，确保已经安装了混入的 myproject.apps.core 包。除此之外，还应在该包中创建 templates/utils/includes/目录结构，并于其中生成 meta.html 文件以存储基本的元标签标记。

2.6.2　实现方式

下面创建模型混入。

（1）确保将"myproject.apps.core"添加至设置项中的 INSTALLED_APPS 中，因为我们需要考虑当前模块的 templates 目录。

（2）将下列基本的元标签标记添加至 meta_field.html 中。

```
{# templates/core/includes/meta_field.html #}
<meta name="{{ name }}" content="{{ content }}" />
```

（3）在编辑器中打开 core 包中的 models.py 文件，并添加下列内容。

```
# myproject/apps/core/models.py
from django.conf import settings
from django.db import models
from django.utils.translation import gettext_lazy as _
from django.utils.safestring import mark_safe
from django.template.loader import render_to_string

class MetaTagsBase(models.Model):
    """
    Abstract base class for generating meta tags
    """
    meta_keywords = models.CharField(
        _("Keywords"),
        max_length=255,
        blank=True,
        help_text=_("Separate keywords with commas."),
    )
    meta_description = models.CharField(
        _("Description"),
        max_length=255,
        blank=True,
    )
```

```
meta_author = models.CharField(
    _("Author"),
    max_length=255,
    blank=True,
)
meta_copyright = models.CharField(
    _("Copyright"),
    max_length=255,
    blank=True,
)

class Meta:
    abstract = True

def get_meta_field(self, name, content):
    tag = ""
    if name and content:
        tag = render_to_string("core/includes/meta_field.html",
        {
            "name": name,
            "content": content,
        })
    return mark_safe(tag)

def get_meta_keywords(self):
    return self.get_meta_field("keywords", self.meta_keywords)

def get_meta_description(self):
    return self.get_meta_field("description",self.meta_description)

def get_meta_author(self):
    return self.get_meta_field("author", self.meta_author)

def get_meta_copyright(self):
    return self.get_meta_field("copyright",self.meta_copyright)

def get_meta_tags(self):
    return mark_safe("\n".join((
        self.get_meta_keywords(),
        self.get_meta_description(),
        self.get_meta_author(),
        self.get_meta_copyright(),
    )))
```

2.6.3 工作方式

该混入向扩展模型中添加了 4 个字段，即 meta_keywords、meta_description、meta_author 和 meta_copyright。此外还添加了用于渲染关联元标签的对应的 get_*()方法，并将名称和相应的字段内容传递至 get_meta_field()方法中，该方法使用这些输入并根据 meta_field.html 模板返回渲染后的标记。最后，get_meta_tags()方法一次性地生成全部有效元数据的组合标记。

如果在模型中使用该混入，如 Idea，则可将下列内容置于 detail 页面模板的 HEAD 部分，进而一次性地渲染全部元标签。

```
{% block meta_tags %}
{{ block.super }}
{{ idea.get_meta_tags }}
{% endblock %}
```

这里，meta_tags 块定义于父模板中，且上述代码片段展示了子模板如何再次定义这一代码块，包含父模板中的内容首先作为 block.super，并利用 idea 对象中的附加标签对其进行扩展。此外，还可使用{{idea.get_meta_description}}仅渲染特定的元标签。

相信读者已经从 models.py 中的代码看到，渲染后的元标签被标记为 safe。也就是说，这些标签未被转义（escaped），且无须使用 safe 模板过滤器。相应地，仅来自数据库中的值才会被转义，以确保最终的 HTML 具有良好的格式。当调用 meta_field.html 模板的 render_to_string()方法时，meta_keywords 中的数据库数据和其他字段将自动被转义，因为该模板未在其内容中指定{% autoescape off %}。

2.6.4 延伸阅读

- “使用模型混入”示例。
- “创建一个模型混入以处理日期的创建和修改”示例。
- “创建一个模型混入以处理 Generic Relation”示例。
- 第 4 章中的“安排 base.html 模板”示例。

2.7 创建一个模型混入以处理 Generic Relation

除标准的数据库关系之外，如外键关系或多对多关系，Django 还涵盖一种机制可将某一个模型与任何其他模型关联。这一概念被称作 Generic Relation。对于每一个 Generic Relation，可保存关联模型的内容类型和该模型实例的 ID。

在当前示例中，我们将研究如何在模型混入中抽象 Generic Relation 的创建过程。

2.7.1 准备工作

针对当前示例，我们需要安装 contenttypes 应用程序。该应用程序默认状态下位于设置项的 INSTALLED_APPS 列表中，如下代码所示。

```
# myproject/settings/_base.py

INSTALLED_APPS = [
    # contributed
    "django.contrib.admin",
    "django.contrib.auth",
    "django.contrib.contenttypes",
    "django.contrib.sessions",
    "django.contrib.messages",
    "django.contrib.staticfiles",
    # third-party
    # ...
    # local
    "myproject.apps.core",
    "myproject.apps.categories",
    "myproject.apps.ideas",
]
```

再次强调，应确保已经创建了模型混入的 myproject.apps.core 应用程序。

2.7.2 实现方式

接下来创建并使用 Generic Relation 的混入，如下所示。

（1）在文本编辑器中打开 core 包中的 models.py 文件，并添加下列内容。

```
# myproject/apps/core/models.py
from django.db import models
from django.utils.translation import gettext_lazy as _
from django.contrib.contenttypes.models import ContentType
from django.contrib.contenttypes.fields import GenericForeignKey
from django.core.exceptions import FieldError
```

```
def object_relation_base_factory(
        prefix=None,
        prefix_verbose=None,
        add_related_name=False,
        limit_content_type_choices_to=None,
        is_required=False):
    """
    Returns a mixin class for generic foreign keys using
    "Content type - object ID" with dynamic field names.
    This function is just a class generator.

    Parameters:
    prefix:           a prefix, which is added in front of
                      the fields
    prefix_verbose:   a verbose name of the prefix, used to
                      generate a title for the field column
                      of the content object in the Admin
    add_related_name:a boolean value indicating, that a
                      related name for the generated content
                      type foreign key should be added. This
                      value should be true, if you use more
                      than one ObjectRelationBase in your
                      model.

    The model fields are created using this naming scheme:
        <<prefix>>_content_type
        <<prefix>>_object_id
        <<prefix>>_content_object
    """
    p = ""
    if prefix:
        p = f"{prefix}_"
    prefix_verbose = prefix_verbose or _("Related object")
    limit_content_type_choices_to = limit_content_type_choices_to
     or {}

    content_type_field = f"{p}content_type"
    object_id_field = f"{p}object_id"
    content_object_field = f"{p}content_object"

    class TheClass(models.Model):
        class Meta:
            abstract = True
```

```
if add_related_name:
    if not prefix:
        raise FieldError("if add_related_name is set to "
                         "True, a prefix must be given")
    related_name = prefix
else:
    related_name = None

optional = not is_required

ct_verbose_name = _(f"{prefix_verbose}'s type (model)")

content_type = models.ForeignKey(
    ContentType,
    verbose_name=ct_verbose_name,
    related_name=related_name,
    blank=optional,
    null=optional,
    help_text=_("Please select the type (model) "
                "for the relation, you want to build."),
    limit_choices_to=limit_content_type_choices_to,
    on_delete=models.CASCADE)

fk_verbose_name = prefix_verbose

object_id = models.CharField(
    fk_verbose_name,
    blank=optional,
    null=False,
    help_text=_("Please enter the ID of the related object."),
    max_length=255,
    default="") # for migrations

content_object = GenericForeignKey(
    ct_field=content_type_field,
    fk_field=object_id_field)

TheClass.add_to_class(content_type_field, content_type)
TheClass.add_to_class(object_id_field, object_id)
TheClass.add_to_class(content_object_field, content_object)

return TheClass
```

（2）下列代码片段展示了如何在应用程序中使用两个 Generic Relation（将代码置于 ideas/models.py 文件中）的示例。

```
# myproject/apps/ideas/models.py
from django.db import models
from django.utils.translation import gettext_lazy as _

from myproject.apps.core.models import (
    object_relation_base_factory as generic_relation,
)

FavoriteObjectBase = generic_relation(
    is_required=True,
)

OwnerBase = generic_relation(
    prefix="owner",
    prefix_verbose=_("Owner"),
    is_required=True,
    add_related_name=True,
    limit_content_type_choices_to={
        "model__in": (
            "user",
            "group",
        )
    }
)

class Like(FavoriteObjectBase, OwnerBase):
    class Meta:
        verbose_name = _("Like")
        verbose_name_plural = _("Likes")

    def __str__(self):
        return _("{owner} likes {object}").format(
            owner=self.owner_content_object,
            object=self.content_object
        )
```

2.7.3　工作方式

不难发现，上述代码片段稍显复杂。

object_relation_base_factory 函数（导入中的别名为 generic_relation）并不是混入自身，它是一个生成模型混入的函数，即一个要扩展的抽象模型类。动态创建的混入添加了 content_type 和 object_id 字段以及指向相关实例的 content_object 通用外键。

为什么不能用这 3 个属性定义一个简单的模型混入呢？动态生成的抽象类允许我们为每个字段名称添加前缀。因此，在同一个模型中，我们可以持有多个 Generic Relation。例如，之前出现过的 Link 模型将包含所喜欢对象的 content_type、object_id 和 content_object 字段，以及喜欢该对象的用户或组的 owner_content_type、owner_object_id 和 owner_content_object 字段。

object_relation_base_factory（别名简写为 generic_relation）通过 limit_content_type_choices_to 参数添加了限制内容类型选择的能力。上述示例将 owner_content_type 的选择结果限制为 User 和 Group 模型的内容类型。

2.7.4　延伸阅读

- “利用与 URL 相关的方法创建一个模型混入”示例。
- “创建一个模型混入以处理日期的创建和修改”示例。
- “创建一个模型混入以关注元标签”示例。
- 第 4 章中的“实现 Like 微件”示例。

2.8　处理多语言字段

Django 采用国际化机制翻译代码和模板中的详细的字符串，但这取决于开发人员确定如何实现模型中的多语言内容。对此，我们将展示多种方式以在项目中直接实现多语言模型。其中，第一个方案是在模型中使用特定于语言的字段。

该方案包含下列特性。

- 直接在模型中定义多语言字段。
- 在数据库查询中可简单地使用多语言字段。
- 可使用贡献管理编辑包含多语言字段的模型，且无须进行额外的修改。
- 必要时，可方便地在同一个模板中显示一个对象的全部翻译结果。
- 在设置项中修改了语言数量后，将需要针对全部多语言模型创建并运行迁移。

2.8.1　准备工作

在前述示例中，我们已经创建了 myproject.apps.core 包。当前，我们需要在 core 应用程序中创建一个新的 model_fields.py 文件，用于自定义模型字段。

2.8.2　实现方式

执行下列步骤将定义多语言字符字段和多语言文本字段。

（1）打开 model_fields.py 文件，并创建多语言基本字段，如下所示。

```
# myproject/apps/core/model_fields.py
from django.conf import settings
from django.db import models
from django.utils.translation import get_language
from django.utils import translation

class MultilingualField(models.Field):
    SUPPORTED_FIELD_TYPES = [models.CharField, models.TextField]

    def __init__(self, verbose_name=None, **kwargs):
        self.localized_field_model = None
        for model in MultilingualField.SUPPORTED_FIELD_TYPES:
            if issubclass(self.__class__, model):
                self.localized_field_model = model
        self._blank = kwargs.get("blank", False)
        self._editable = kwargs.get("editable", True)
        super().__init__(verbose_name, **kwargs)
    @staticmethod
    def localized_field_name(name, lang_code):
        lang_code_safe = lang_code.replace("-", "_")
        return f"{name}_{lang_code_safe}"

    def get_localized_field(self, lang_code, lang_name):
        _blank = (self._blank
                  if lang_code == settings.LANGUAGE_CODE
                  else True)
        localized_field = self.localized_field_model(
            f"{self.verbose_name} ({lang_name})",
            name=self.name,
            primary_key=self.primary_key,
            max_length=self.max_length,
            unique=self.unique,
            blank=_blank,
            null=False, # we ignore the null argument!
            db_index=self.db_index,
```

```
            default=self.default or "",
            editable=self._editable,
            serialize=self.serialize,
            choices=self.choices,
            help_text=self.help_text,
            db_column=None,
            db_tablespace=self.db_tablespace)
        return localized_field

    def contribute_to_class(self, cls, name,
                            private_only=False,
                            virtual_only=False):
        def translated_value(self):
            language = get_language()
            val = self.__dict__.get(
                MultilingualField.localized_field_name(
                        name, language))
            if not val:
                val = self.__dict__.get(
                    MultilingualField.localized_field_name(
                        name, settings.LANGUAGE_CODE))
            return val

        # generate language-specific fields dynamically
        if not cls._meta.abstract:
            if self.localized_field_model:
                for lang_code, lang_name in settings.LANGUAGES:
                    localized_field = self.get_localized_field(
                        lang_code, lang_name)
                    localized_field.contribute_to_class(
                        cls,
                        MultilingualField.localized_field_name(
                            name, lang_code))

                setattr(cls, name, property(translated_value))
            else:
                super().contribute_to_class(
                    cls, name, private_only, virtual_only)
```

（2）在同一个文件中，子类化字符基本字段和文本字段表单，如下所示。

```
class MultilingualCharField(models.CharField, MultilingualField):
    pass
```

```
class MultilingualTextField(models.TextField, MultilingualField):
    pass
```

（3）在 core 应用程序中创建 admin.py 文件，并添加下列内容。

```
# myproject/apps/core/admin.py
from django.conf import settings

def get_multilingual_field_names(field_name):
    lang_code_underscored = settings.LANGUAGE_CODE.replace("-",
    "_")
    field_names = [f"{field_name}_{lang_code_underscored}"]
    for lang_code, lang_name in settings.LANGUAGES:
        if lang_code != settings.LANGUAGE_CODE:
            lang_code_underscored = lang_code.replace("-", "_")
            field_names.append(
                f"{field_name}_{lang_code_underscored}"
            )
    return field_names
```

接下来介绍如何在应用程序中使用多语言字段，如下所示。

（1）在项目的设置项中设置多语言。假设网站支持欧盟的所有官方语言，且默认语言为英语。

```
# myproject/settings/_base.py
LANGUAGE_CODE = "en"

# All official languages of European Union
LANGUAGES = [
    ("bg", "Bulgarian"),    ("hr", "Croatian"),
    ("cs", "Czech"),        ("da", "Danish"),
    ("nl", "Dutch"),        ("en", "English"),
    ("et", "Estonian"),     ("fi", "Finnish"),
    ("fr", "French"),       ("de", "German"),
    ("el", "Greek"),        ("hu", "Hungarian"),
    ("ga", "Irish"),        ("it", "Italian"),
    ("lv", "Latvian"),      ("lt", "Lithuanian"),
    ("mt", "Maltese"),      ("pl", "Polish"),
    ("pt", "Portuguese"),   ("ro", "Romanian"),
    ("sk", "Slovak"),       ("sl", "Slovene"),
    ("es", "Spanish"),      ("sv", "Swedish"),
]
```

（2）打开 myproject.apps.ideas 应用程序中的 models.py 文件，创建 Idea 模型的多语言字段，如下所示。

```
# myproject/apps/ideas/models.py
from django.db import models
from django.utils.translation import gettext_lazy as _

from myproject.apps.core.model_fields import (
    MultilingualCharField,
    MultilingualTextField,
)

class Idea(models.Model):
    title = MultilingualCharField(
        _("Title"),
        max_length=200,
    )
    content = MultilingualTextField(
        _("Content"),
    )

    class Meta:
        verbose_name = _("Idea")
        verbose_name_plural = _("Ideas")

    def __str__(self):
        return self.title
```

（3）创建 ideas 应用程序的 admin.py 文件。

```
# myproject/apps/ideas/admin.py
from django.contrib import admin
from django.utils.translation import gettext_lazy as _

from myproject.apps.core.admin import get_multilingual_field_names

from .models import Idea

@admin.register(Idea)
class IdeaAdmin(admin.ModelAdmin):
    fieldsets = [
        (_("Title and Content"), {
            "fields": get_multilingual_field_names("title") +
```

```
                get_multilingual_field_names("content")
        }),
    ]
```

2.8.3 工作方式

Idea 示例将生成如下所示的模型。

```
class Idea(models.Model):
    title_bg = models.CharField(
        _("Title (Bulgarian)"),
        max_length=200,
    )
    title_hr = models.CharField(
        _("Title (Croatian)"),
        max_length=200,
    )
    # titles for other languages…
    title_sv = models.CharField(
        _("Title (Swedish)"),
        max_length=200,
    )

    content_bg = MultilingualTextField(
        _("Content (Bulgarian)"),
    )
    content_hr = MultilingualTextField(
        _("Content (Croatian)"),
    )
    # content for other languages…
    content_sv = MultilingualTextField(
        _("Content (Swedish)"),
    )

    class Meta:
        verbose_name = _("Idea")
        verbose_name_plural = _("Ideas")

    def __str__(self):
        return self.title
```

对于带有连接号的语言代码，如瑞士德语中的“de-ch”，那么这些语言的字段将被下画线所替换，如 title_de_ch 和 content_de_ch。

除了生成的特定语言的字段，还存在两个属性，即 title 和 content，这将返回当前活动语言的对应字段。如果不存在有效的本地化字段内容，那么将回退至默认的语言。

根据 LANGUAGES 设置项，MultilingualCharField 和 MultilingualTextField 字段将动态地调整模型字段。二者将在 Django 框架创建模型类时覆写所用的 contribute_to_class()方法。这些多语言字段将以动态方式为项目中的每种语言添加字符或文本字段。我们需要创建一个数据库迁移，并向数据库中添加相应的字段。另外，默认状态下，还应创建相应的属性以返回当前活动语言或主语言的翻译值。

在管理阶段，get_multilingual_field_names()将返回一个特定于语言字段名称的列表，从一种默认语言开始，然后是 LANGUAGES 设置项中的其他语言。

关于如何在模板和视图中使用多语言字段，下面将给出一些示例。

如果在模板中持有下列代码，这将返回当前处于激活状态下的语言中的文本，如立陶宛语；如果对应的翻译结果不存在，则回退至英文。

```
<h1>{{ idea.title }}</h1>
<div>{{ idea.content|urlize|linebreaks }}</div>
```

如果需要包含以翻译标题排序的 QuerySet，则可按照下列方式对其加以定义。

```
>>> lang_code = input("Enter language code: ")
>>> lang_code_underscored = lang_code.replace("-", "_")
>>> qs = Idea.objects.order_by(f"title_{lang_code_underscored}")
```

2.8.4 延伸阅读

- “与模型翻译表协同工作”示例。
- “使用迁移”示例。
- 第 6 章。

2.9 与模型翻译表协同工作

处理数据库中多语言内容的第二种方案是针对每个多语言模型采用模型翻译表。

该方案的特征如下所示。

- 可使用贡献管理将翻译编辑为内联。
- 在修改了设置项中的语言数量后，无须迁移或其他进一步的操作。

- ❑ 可方便地在模板中显示当前语言的翻译结果，但难以在同一页面中的特定语言下显示多个翻译结果。
- ❑ 需要了解和使用当前示例中所描述的某种特定模式，以创建模型翻译。
- ❑ 对于数据库查询，当前方案虽不简单但仍为可行。

2.9.1　准备工作

再次说明，这里将启动 myprojects.apps.core 应用程序。

2.9.2　实现方式

执行下列步骤并为多语言模型做好准备。

（1）在 core 应用程序中，利用下列内容创建 model_fields.py 文件。

```
# myproject/apps/core/model_fields.py
from django.conf import settings
from django.utils.translation import get_language
from django.utils import translation

class TranslatedField(object):
    def __init__(self, field_name):
        self.field_name = field_name

    def __get__(self, instance, owner):
        lang_code = translation.get_language()
        if lang_code == settings.LANGUAGE_CODE:
            # The fields of the default language are in the main model
            return getattr(instance, self.field_name)
        else:
            # The fields of the other languages are in the translation
            # model, but falls back to the main model
            translations = instance.translations.filter(
                language=lang_code,
            ).first() or instance
            return getattr(translations, self.field_name)
```

（2）利用下列内容将 admin.py 文件添加至 core 应用程序中。

```
# myproject/apps/core/admin.py
from django import forms
```

```
from django.conf import settings
from django.utils.translation import gettext_lazy as _

class LanguageChoicesForm(forms.ModelForm):
    def __init__(self, *args, **kwargs):
        LANGUAGES_EXCEPT_THE_DEFAULT = [
            (lang_code, lang_name)
            for lang_code, lang_name in settings.LANGUAGES
            if lang_code != settings.LANGUAGE_CODE
        ]
        super().__init__(*args, **kwargs)
        self.fields["language"] = forms.ChoiceField(
            label=_("Language"),
            choices=LANGUAGES_EXCEPT_THE_DEFAULT,
            required=True,
        )
```

接下来实现多语言模型。

（1）在项目的设置项中设置多个语言。也就是说，站点支持欧盟的全部官方语言，且默认语言为英语。

```
# myproject/settings/_base.py
LANGUAGE_CODE = "en"

# All official languages of European Union
LANGUAGES = [
    ("bg", "Bulgarian"),    ("hr", "Croatian"),
    ("cs", "Czech"),        ("da", "Danish"),
    ("nl", "Dutch"),        ("en", "English"),
    ("et", "Estonian"),     ("fi", "Finnish"),
    ("fr", "French"),       ("de", "German"),
    ("el", "Greek"),        ("hu", "Hungarian"),
    ("ga", "Irish"),        ("it", "Italian"),
    ("lv", "Latvian"),      ("lt", "Lithuanian"),
    ("mt", "Maltese"),      ("pl", "Polish"),
    ("pt", "Portuguese"),   ("ro", "Romanian"),
    ("sk", "Slovak"),       ("sl", "Slovene"),
    ("es", "Spanish"),      ("sv", "Swedish"),
]
```

（2）创建 Idea 和 IdeaTranslations 模型。

```
# myproject/apps/ideas/models.py
```

```
from django.db import models
from django.conf import settings
from django.utils.translation import gettext_lazy as _

from myproject.apps.core.model_fields import TranslatedField

class Idea(models.Model):
    title = models.CharField(
        _("Title"),
        max_length=200,
    )
    content = models.TextField(
        _("Content"),
    )
    translated_title = TranslatedField("title")
    translated_content = TranslatedField("content")

    class Meta:
        verbose_name = _("Idea")
        verbose_name_plural = _("Ideas")

    def __str__(self):
        return self.title

class IdeaTranslations(models.Model):
    idea = models.ForeignKey(
        Idea,
        verbose_name=_("Idea"),
        on_delete=models.CASCADE,
        related_name="translations",
    )
    language = models.CharField(_("Language"), max_length=7)

    title = models.CharField(
        _("Title"),
        max_length=200,
    )
    content = models.TextField(
        _("Content"),
    )

    class Meta:
```

```
        verbose_name = _("Idea Translations")
        verbose_name_plural = _("Idea Translations")
        ordering = ["language"]
        unique_together = [["idea", "language"]]

    def __str__(self):
        return self.title
```

（3）创建 ideas 应用程序的 admin.py 文件，如下所示。

```
# myproject/apps/ideas/admin.py
from django.contrib import admin
from django.utils.translation import gettext_lazy as _

from myproject.apps.core.admin import LanguageChoicesForm

from .models import Idea, IdeaTranslations

class IdeaTranslationsForm(LanguageChoicesForm):
    class Meta:
        model = IdeaTranslations
        fields = "__all__"

class IdeaTranslationsInline(admin.StackedInline):
    form = IdeaTranslationsForm
    model = IdeaTranslations
    extra = 0

@admin.register(Idea)
class IdeaAdmin(admin.ModelAdmin):
    inlines = [IdeaTranslationsInline]

    fieldsets = [
        (_("Title and Content"), {
            "fields": ["title", "content"]
        }),
    ]
```

2.9.3　工作方式

在 Idea 模型中，我们保存了默认语言的特定语言字段。每种语言的翻译内容均位于 IdeaTranslations 模型中，该模型将作为内联翻译内容在管理中被列出。这里，IdeaTranslation

模型在模型中不包含语言选择是有原因的——我们不希望每次添加新语言或删除某些语言时都创建迁移。相反，语言选择在管理表单中设置，并确保跳过默认语言，或者在列表中无法选取默认语言。另外，使用 LanguageChoicesForm 类将对语言选择有所限制。

要获取当前语言中的特定字段时，可使用定义为 TranslatedField 的字段。在模板中，如下所示。

```
<h1>{{ idea.translated_title }}</h1>
<div>{{ idea.translated_content|urlize|linebreaks }}</div>
```

当在特定语言中以翻译标题排序条目时，可使用 annotate()方法，如下所示。

```
>>> from django.conf import settings
>>> from django.db import models
>>> lang_code = input("Enter language code: ")

>>> if lang_code == settings.LANGUAGE_CODE:
...     qs = Idea.objects.annotate(
...         title_translation=models.F("title"),
...         content_translation=models.F("content"),
...     )
... else:
...     qs = Idea.objects.filter(
...         translations__language=lang_code,
...     ).annotate(
...         title_translation=models.F("translations__title"),
...         content_translation=models.F("translations__content"),
...     )
>>> qs = qs.order_by("title_translation")

>>> for idea in qs:
...     print(idea.title_translation)
```

在当前示例中，我们在 Django shell 中提示输入语言代码。如果对应语言为默认语言，那么将 title 和 content 作为 title_translation 和 content_translation 存储在 Idea 模型中。如果存在另一种可选的语言，我们将利用可选语言并作为 IdeaTranslations 模型中的 title_translation 和 content_translation 读取 title 和 content。

随后根据 title_translation 或 content_translation 过滤或排序 QuerySet。

2.9.4 延伸阅读

- “处理多语言字段”示例。

- 第 6 章。

2.10　避免环状依赖项

当开发 Django 模型时，需要注意的是，应该避免 models.py 文件中的环状依赖项。环状依赖项是指不同 Python 模块彼此间的导入行为。我们不应在不同的 models.py 文件间进行交叉导入，这将导致严重的稳定性问题。相反，对于相互依赖项，我们应使用当前示例所描述的操作。

2.10.1　准备工作

接下来将处理 categories 和 ideas 应用程序，以进一步展示如何处理交叉依赖项问题。

2.10.2　实现方式

当处理使用其他应用程序中的模型时，应执行下列步骤。

（1）对于其他应用程序中的模型的外键和多对多关系，可使用"<app_label>.<model>"进行声明，而非导入模型。在 Django 中，这常与 ForeignKey、OneToOneField 和 ManyToManyField 协同工作，如下所示。

```
# myproject/apps/ideas/models.py
from django.db import models
from django.conf import settings
from django.utils.translation import gettext_lazy as _

class Idea(models.Model):
    author = models.ForeignKey(
        settings.AUTH_USER_MODEL,
        verbose_name=_("Author"),
        on_delete=models.SET_NULL,
        blank=True,
        null=True,
    )
    category = models.ForeignKey(
        "categories.Category",
        verbose_name=_("Category"),
        blank=True,
```

```
        null=True,
        on_delete=models.SET_NULL,
    )
    # other fields, attributes, properties and methods…
```

这里，settings.AUTH_USER_MODEL 可视为包含诸如"auth.User"这一类值的设置项。

（2）如果需要访问某个方法中另一个应用程序中的一个模型，可在该方法中导入模型，而非模块级别，如下所示。

```
# myproject/apps/categories/models.py
from django.db import models
from django.utils.translation import gettext_lazy as _

class Category(models.Model):
    # fields, attributes, properties, and methods…

    def get_ideas_without_this_category(self):
        from myproject.apps.ideas.models import Idea
        return Idea.objects.exclude(category=self)
```

（3）当针对模型混入采用模型继承时，可将基类保存至一个独立的应用程序中，同时置于 INSTALLED_APPS 中使用这些基类的其他应用程序之前，如下所示。

```
# myproject/settings/_base.py

INSTALLED_APPS = [
    # contributed
    "django.contrib.admin",
    "django.contrib.auth",
    "django.contrib.contenttypes",
    "django.contrib.sessions",
    "django.contrib.messages",
    "django.contrib.staticfiles",
    # third-party
    # ...
    # local
    "myproject.apps.core",
    "myproject.apps.categories",
    "myproject.apps.ideas",
]
```

这里，ideas 应用程序将使用来自 core 应用程序中的模型混入，如下所示。

```
# myproject/apps/ideas/models.py
```

```
from django.db import models
from django.conf import settings
from django.utils.translation import gettext_lazy as _

from myproject.apps.core.models import (
    CreationModificationDateBase,
    MetaTagsBase,
    UrlBase,
)

class Idea(CreationModificationDateBase, MetaTagsBase, UrlBase):
    # fields, attributes, properties, and methods…
```

2.10.3　延伸阅读

- 第 1 章中的“针对开发、测试、预发布和产品环境，配置设置项”示例。
- 第 1 章中的“遵循 Python 文件中的导入顺序”示例。
- “使用模型混入”示例。
- “将外键修改为多对多字段”示例。

2.11　添加数据库约束

对于较好的数据库完整性，常见的做法是定义数据库约束、通知某些字段与其他数据库表的字段进行绑定、使某些字段保持唯一或非 null。对于高级的数据库约束，如通过相关条件使字段唯一，或者针对某些字段值设置特定的条件。Django 定义了特殊的类，即 UniqueConstraint 和 CheckConstraint。在当前示例中，我们将介绍具体的应用示例。

2.11.1　准备工作

下面将启用 ideas 应用程序，以及至少包含 title 和 author 字段的 Idea 模型。

2.11.2　实现方式

接下来在 Idea 模型的 Meta 类中设置数据库约束，如下所示。

```
# myproject/apps/ideas/models.py
from django.db import models
```

```
from django.utils.translation import gettext_lazy as _

class Idea(models.Model):
    author = models.ForeignKey(
        settings.AUTH_USER_MODEL,
        verbose_name=_("Author"),
        on_delete=models.SET_NULL,
        blank=True,
        null=True,
        related_name="authored_ideas",
    )
    title = models.CharField(
        _("Title"),
        max_length=200,
    )

    class Meta:
        verbose_name = _("Idea")
        verbose_name_plural = _("Ideas")
        constraints = [
            models.UniqueConstraint(
                fields=["title"],
                condition=~models.Q(author=None),
                name="unique_titles_for_each_author",
            ),
            models.CheckConstraint(
                check=models.Q(
                    title__iregex=r"^\S.*\S$"
                    # starts with non-whitespace,
                    # ends with non-whitespace,
                    # anything in the middle
                ),
                name="title_has_no_leading_and_trailing_whitespaces",
            )
        ]
```

2.11.3　工作方式

如前所述，我们在数据库中定义了两个约束。

其中，第一个约束 UniqueConstraint 通知标题对于每个作者都是唯一的。如果未设置作者，标题则可以重复。当检查作者是否被设置时，我们采用了逆查找~models.Q(author=

None)。注意，在 Django 中，查找的“~”操作符等价于 QuerySet 的 exclude()方法。因此，这些 QuerySet 等价于：

```
ideas_with_authors = Idea.objects.exclude(author=None)
ideas_with_authors2 = Idea.objects.filter(~models.Q(author=None))
```

第 2 个约束则检查标题是否没有以空格开始和结束。对此，我们使用正则表达式查找。

2.11.4　更多内容

数据库约束并不会对表单验证产生任何影响。在将条目保存至数据库时，如果任何数据未通过其条件，则会引发 django.db.utils.IntegrityError。

如果需要执行表单验证，则需要亲自实现验证，如在模型的 clean()方法中。对于 Idea 模型，对应操作如下所示。

```
# myproject/apps/ideas/models.py
from django.db import models
from django.conf import settings
from django.core.exceptions import ValidationError
from django.utils.translation import gettext_lazy as _

class Idea(models.Model):
    author = models.ForeignKey(
        settings.AUTH_USER_MODEL,
        verbose_name=_("Author"),
        on_delete=models.SET_NULL,
        blank=True,
        null=True,
        related_name="authored_ideas2",
    )
    title = models.CharField(
        _("Title"),
        max_length=200,
    )

    # other fields and attributes…

    class Meta:
        verbose_name = _("Idea")
        verbose_name_plural = _("Ideas")
        constraints = [
            models.UniqueConstraint(
```

```
                fields=["title"],
                condition=~models.Q(author=None),
                name="unique_titles_for_each_author2",
            ),
            models.CheckConstraint(
                check=models.Q(
                    title__iregex=r"^\S.*\S$"
                    # starts with non-whitespace,
                    # ends with non-whitespace,
                    # anything in the middle
                ),
                name="title_has_no_leading_and_trailing_whitespaces2",
            )
        ]

    def clean(self):
        import re
        if self.author and Idea.objects.exclude(pk=self.pk).filter(
            author=self.author,
            title=self.title,
        ).exists():
            raise ValidationError(
                _("Each idea of the same user should have a unique title.")
            )
        if not re.match(r"^\S.*\S$", self.title):
            raise ValidationError(
                _("The title cannot start or end with a whitespace.")
            )

  # other properties and methods…
```

2.11.5 延伸阅读

- 第 3 章。
- 第 10 章中的“使用数据库查询表达式”示例。

2.12 使用迁移

在敏捷软件开发中，项目的需求在开发过程中不时地演化和更新。当开发过程以迭代方式呈现时，其间需要执行数据库模式变化。利用 Django 迁移，我们无须以手动方式修改数据库表和字段，因为大多数内容通过命令行接口自动完成。

2.12.1　准备工作

在命令行工具中激活虚拟环境，并将活动目录修改为项目目录。

2.12.2　实现方式

当创建数据库迁移时，请查看下列步骤。

（1）当在新的 categories 或 ideas 应用程序中创建模型时，需要生成一个初始迁移，进而创建应用程序的数据库表。这可通过下列命令完成。

```
(env)$ python manage.py makemigrations ideas
```

（2）当首次需要创建项目的全部表时，可运行下列命令。

```
(env)$ python manage.py migrate
```

当需要执行全部应用程序的新迁移时，可运行上述命令。

（3）如果针对特定的应用程序执行迁移，则可运行下列命令。

```
(env)$ python manage.py migrate ideas
```

（4）如果在数据库模式中稍作修改，则需要针对该模式创建一个迁移。例如，如果向 idea 模型中添加一个新的子标题字段，则需要通过下列命令创建对应的迁移。

```
(env)$ python manage.py makemigrations --name=subtitle_added ideas
```

然而，--name=subtitle_added 字段将会被跳过，因为大多数情况下，Django 将会生成具有自解释性的默认名称。

（5）有时，我们可能需要添加或修改现有模式中的数据，可通过数据迁移完成，而不是模式迁移。当创建修改数据库表中的数据迁移时，可使用下列命令。

```
(env)$ python manage.py makemigrations --name=populate_subtitle \
> --empty ideas
```

--empty 参数通知 Django 创建一个框架性质的数据迁移，且在应用之前需要调整并执行所需的数据操作。对于数据迁移，建议设置对应的名称。

（6）当列出所有有效的已应用和未应用迁移，可运行下列命令。

```
(env)$ python manage.py showmigrations
```

其中，已应用的迁移采用“[X]”前缀列出，而未应用的迁移则采用“[]”前缀列出。

（7）针对特定的应用程序列出所有的有效迁移，可运行相同的命令，但需要传递应用程序名称，如下所示。

```
(env)$ python manage.py showmigrations ideas
```

2.12.3　工作方式

Django 迁移表示为数据库迁移的机制的指令性文件。这些指令性文件通知我们创建或移除哪些数据库表、添加或移除哪些字段，以及插入、更新或删除哪些数据。除此之外，这些文件还定义了迁移之间的依赖关系。

在 Django 中，存在两种迁移类型。一种是模式迁移，另一种是数据迁移。其中，模式迁移应在添加新模型、添加或移除字段时创建。数据模型则应在利用某些值填充数据库，或大规模地从数据库中删除值时创建。相应地，数据迁移应在命令行工具中通过命令创建，并于随后在迁移文件中进行编码。

每个应用程序的迁移应保存在 migrations 目录中。其中，第一次迁移通常称作 0001_initial.py，示例中的其他迁移则称作 0002_subtitle_added.py 和 0003_populate_subtitle.py。每次迁移将得到一个数字前缀并自动递增。对于所执行的每次迁移，存在一个保存于 django_migrations 数据库表中的入口。

通过指定迁移的数量，还可实现往复迁移，如下所示。

```
(env)$ python manage.py migrate ideas 0002
```

执行下列命令，可取消应用程序的所有迁移，包括初始迁移。

```
(env)$ python manage.py migrate ideas zero
```

这里，取消迁移要求每个迁移同时具有前向和后向操作。理想情况下，后向操作将取消前向操作做出的更改。但是，在某些情况下，此类更改是不可能恢复的。例如，当前向操作移除了模式中的一个列——这将销毁数据。在这种情况下，后向操作可能会恢复模式，但数据将永远丢失；否则可能根本不存在后向操作。

提示：

在测试前向和后向迁移过程并确保二者在其他开发环境和公共站点环境下运行良好之前，不要将迁移提交至版本控制中。

2.12.4　更多内容

关于如何编写数据库迁移，读者可参考官方的 *How To* 教程，对应网址为 https://docs.

djangoproject.com/en/2.2/howto/writing-migrations/。

2.12.5　延伸阅读

- 第 1 章中的“与虚拟环境协同工作”示例。
- 第 1 章中的“针对 Django、Gunicorn、Nginx 和 PostgreSQL，与 Docker 容器协同工作”示例。
- 第 1 章中的“利用 pip 处理项目依赖项”示例。
- 第 1 章中的“在项目中包含外部依赖项”示例。
- “将外键修改为多对多字段”示例。

2.13　将外键修改为多对多字段

当前示例将介绍如何将多对一关系修改为多对多关系，同时保存现有的数据；在这种情况下，我们将同时使用模式和数据迁移。

2.13.1　准备工作

假设已持有一个 Idea 模型，且外键指向 Category 模型。

（1）在 Category 应用程序中定义 Category 模型，如下所示。

```
# myproject/apps/categories/models.py
from django.db import models
from django.utils.translation import gettext_lazy as _

from myproject.apps.core.model_fields import MultilingualCharField

class Category(models.Model):
    title = MultilingualCharField(
        _("Title"),
        max_length=200,
    )

    class Meta:
        verbose_name = _("Category")
        verbose_name_plural = _("Categories")
```

```
    def __str__(self):
        return self.title
```

（2）在 ideas 应用程序中定义 Idea 模型，如下所示。

```
# myproject/apps/ideas/models.py
from django.db import models
from django.conf import settings
from django.utils.translation import gettext_lazy as _

from myproject.apps.core.model_fields import (
    MultilingualCharField,
    MultilingualTextField,
)

class Idea(models.Model):
    title = MultilingualCharField(
        _("Title"),
        max_length=200,
    )
    content = MultilingualTextField(
        _("Content"),
    )
    category = models.ForeignKey(
        "categories.Category",
        verbose_name=_("Category"),
        blank=True,
        null=True,
        on_delete=models.SET_NULL,
        related_name="category_ideas",
    )

    class Meta:
        verbose_name = _("Idea")
        verbose_name_plural = _("Ideas")

    def __str__(self):
        return self.title
```

（3）运行下列命令并创建和执行初始迁移。

```
(env)$ python manage.py makemigrations categories
(env)$ python manage.py makemigrations ideas
(env)$ python manage.py migrate
```

2.13.2　实现方式

下列步骤展示了从外键关系切换到多对多关系，同时保存已有的数据。

（1）添加名为 categories 的多对多字段，如下所示。

```
# myproject/apps/ideas/models.py
from django.db import models
from django.conf import settings
from django.utils.translation import gettext_lazy as _

from myproject.apps.core.model_fields import (
    MultilingualCharField,
    MultilingualTextField,
)

class Idea(models.Model):
    title = MultilingualCharField(
        _("Title"),
        max_length=200,
    )
    content = MultilingualTextField(
        _("Content"),
    )
    category = models.ForeignKey(
        "categories.Category",
        verbose_name=_("Category"),
        blank=True,
        null=True,
        on_delete=models.SET_NULL,
        related_name="category_ideas",
    )
    categories = models.ManyToManyField(
        "categories.Category",
        verbose_name=_("Categories"),
        blank=True,
        related_name="ideas",
    )

    class Meta:
        verbose_name = _("Idea")
        verbose_name_plural = _("Ideas")
```

```
    def __str__(self):
        return self.title
```

（2）创建并运行模式迁移以便向数据库中添加新关系，如下所示。

```
(env)$ python manage.py makemigrations ideas
(env)$ python manage.py migrate ideas
```

（3）创建一个数据迁移，并将类别从外键复制到多对多字段，如下所示。

```
(env)$ python manage.py makemigrations --empty \
> --name=copy_categories ideas
```

（4）打开新创建的迁移文件（0003_copy_categories.py），并定义前向迁移指令，如下所示。

```
# myproject/apps/ideas/migrations/0003_copy_categories.py
from django.db import migrations

def copy_categories(apps, schema_editor):
    Idea = apps.get_model("ideas", "Idea")
    for idea in Idea.objects.all():
        if idea.category:
            idea.categories.add(idea.category)

class Migration(migrations.Migration):

    dependencies = [
        ('ideas', '0002_idea_categories'),
    ]

    operations = [
        migrations.RunPython(copy_categories),
    ]
```

（5）运行新的数据迁移，如下所示。

```
(env)$ python manage.py migrate ideas
```

（6）删除 models.py 文件中的外键 category 字段，仅保留新的 categories 多对多字段，如下所示。

```
# myproject/apps/ideas/models.py
from django.db import models
```

```
from django.conf import settings
from django.utils.translation import gettext_lazy as _

from myproject.apps.core.model_fields import (
    MultilingualCharField,
    MultilingualTextField,
)

class Idea(models.Model):
    title = MultilingualCharField(
        _("Title"),
        max_length=200,
    )
    content = MultilingualTextField(
        _("Content"),
    )

    categories = models.ManyToManyField(
        "categories.Category",
        verbose_name=_("Categories"),
        blank=True,
        related_name="ideas",
    )

    class Meta:
        verbose_name = _("Idea")
        verbose_name_plural = _("Ideas")

    def __str__(self):
        return self.title
```

（7）创建并运行模式迁移，以便删除数据库表中的 Categories 字段，如下所示。

```
(env)$ python manage.py makemigrations ideas
(env)$ python manage.py migrate ideas
```

2.13.3　工作方式

首先，我们向 Idea 模型添加了一个新的多对多字段，并生成了一个迁移以相应地更新数据库。随后创建一个数据迁移，并将现有的关系从外键 category 复制至多对多 categories 中。最后，我们从模型中移除了外键字段，并再次更新数据库。

2.13.4　更多内容

数据迁移目前只包含前向操作，并作为第 1 个关联条目将外键 category 复制至新的 categories 关系中。虽然这里没有详细说明，但在真实场景中最好也包含后向操作。这可通过将第 1 个关联条目复制回 category 外键来完成。但是，包含多个 categories 的 Idea 将会丢失额外数据。

2.13.5　延伸阅读

- “使用迁移”示例。
- “处理多语言字段”示例。
- “与模型翻译表协同工作”示例。
- “避免环状依赖项”示例。

第 3 章　表单和视图

本章主要涉及下列主题。

- 利用 CRUDL 函数创建一个应用程序。
- 保存模型实例的作者。
- 上传图像。
- 利用自定义模板创建一个表单布局。
- 利用 django-crispy-forms 创建一个表单布局。
- 与表单集协同工作。
- 过滤对象列表。
- 管理分页列表。
- 构成基于类的视图。
- 提供 Open Graph 和 Twitter Card 数据。
- 提供 schema.org 的词汇表。
- 创建 PDF 文档。
- 利用 Haystack and Whoosh 实现多语言搜索。
- 利用 Elasticsearch DSL 实现多语言搜索。

3.1　简　　介

虽然数据库结构在模型中被定义，但视图提供了所需的端点，进而向用户展示内容，或者令用户输入新的和更新后的内容。本章将重点介绍管理表单的视图、列表视图，以及向 HTML 生成可替代输出的视图。在最简单的示例中，我们将 URL 规则和模板的创建留给读者。

3.2　技术需求

当与本章代码协同工作时，需要安装 Python、MySQL 或 PostgreSQL 数据库的最新稳定版本，以及包含虚拟环境的 Django 项目。其中，一些示例需要特定的 Python 依赖项。

此外，当生成 PDF 文档时，还需要使用 cairo、pango、gdk-pixbuf 和 libffi 库。对于搜索而言，我们还需要使用 Elasticsearch 服务器，稍后将对此加以详细讨论。

本章中的大多数模板将使用具有较好观感的 Bootstrap 4 CSS 框架。

本章中的所有代码位于 GitHub 储存库的 ch03 目录中，对应网址为 https://github.com/PacktPublishing/Django-3-Web-Development-Cookbook-Fourth-Edition。

3.3　利用 CRUDL 函数创建一个应用程序

在计算机科学中，CRUDL 是指创建、读取、更新、删除和列表函数。基于交互式功能的 Django 项目需要实现上述全部函数，以管理站点上的数据。在当前示例中，我们将讨论如何创建这些基本函数的 URL 和视图。

3.3.1　准备工作

下面创建一个名为 ideas 的新应用程序，并将其置于设置项的 INSTALLED_APPS 中。利用 IdeaTranslations 模型创建下列 Idea 模型，并在该应用程序中实现翻译功能。

```
# myproject/apps/idea/models.py
import uuid

from django.db import models
from django.urls import reverse
from django.conf import settings
from django.utils.translation import gettext_lazy as _

from myproject.apps.core.model_fields import TranslatedField
from myproject.apps.core.models import (
   CreationModificationDateBase, UrlBase
)

RATING_CHOICES = (
    (1, "★☆☆☆☆"),
    (2, "★★☆☆☆"),
    (3, "★★★☆☆"),
    (4, "★★★★☆"),
    (5, "★★★★★"),
)
```

```
class Idea(CreationModificationDateBase, UrlBase):
    uuid = models.UUIDField(
        primary_key=True, default=uuid.uuid4, editable=False
    )
    author = models.ForeignKey(
        settings.AUTH_USER_MODEL,
        verbose_name=_("Author"),
        on_delete=models.SET_NULL,
        blank=True,
        null=True,
        related_name="authored_ideas",
    )
    title = models.CharField(_("Title"), max_length=200)
    content = models.TextField(_("Content"))

    categories = models.ManyToManyField(
        "categories.Category",
        verbose_name=_("Categories"),
        related_name="category_ideas",
    )
    rating = models.PositiveIntegerField(
        _("Rating"), choices=RATING_CHOICES, blank=True, null=True
    )
    translated_title = TranslatedField("title")
    translated_content = TranslatedField("content")

    class Meta:
        verbose_name = _("Idea")
        verbose_name_plural = _("Ideas")

    def __str__(self):
        return self.title

    def get_url_path(self):
        return reverse("ideas:idea_detail", kwargs={"pk": self.pk})

class IdeaTranslations(models.Model):
    idea = models.ForeignKey(
        Idea,
        verbose_name=_("Idea"),
        on_delete=models.CASCADE,
        related_name="translations",
```

```
    )
    language = models.CharField(_("Language"), max_length=7)

    title = models.CharField(_("Title"), max_length=200)
    content = models.TextField(_("Content"))

    class Meta:
        verbose_name = _("Idea Translations")
        verbose_name_plural = _("Idea Translations")
        ordering = ["language"]
        unique_together = [["idea", "language"]]

    def __str__(self):
        return self.title
```

这里使用了前述章节中介绍的多个概念。具体来说，我们继承了模型混入，并采用了模型翻译表。对此，读者可参考第 2 章中的“使用模型混入”示例和“与模型翻译表协同工作”示例。对于本章中的全部示例，我们将使用 ideas 应用程序和这些模型。

除此之外，我们还将利用 Category 和 CategoryTranslations 模型构建一个类似的 Categories 应用程序。

```
# myproject/apps/categories/models.py
from django.db import models
from django.utils.translation import gettext_lazy as _

from myproject.apps.core.model_fields import TranslatedField

class Category(models.Model):
    title = models.CharField(_("Title"), max_length=200)

    translated_title = TranslatedField("title")

    class Meta:
        verbose_name = _("Category")
        verbose_name_plural = _("Categories")

    def __str__(self):
        return self.title

class CategoryTranslations(models.Model):
    category = models.ForeignKey(
        Category,
```

```
        verbose_name=_("Category"),
        on_delete=models.CASCADE,
        related_name="translations",
    )
    language = models.CharField(_("Language"), max_length=7)

    title = models.CharField(_("Title"), max_length=200)

    class Meta:
        verbose_name = _("Category Translations")
        verbose_name_plural = _("Category Translations")
        ordering = ["language"]
        unique_together = [["category", "language"]]

    def __str__(self):
        return self.title
```

3.3.2 实现方式

Django 中的 CRUDL 功能包含表单、视图和 URL 规则，下面对其予以逐一创建。

（1）利用模型表单向 ideas 应用程序中添加一个新的 forms.py 文件，以添加和修改 Idea 模型实例。

```
# myprojects/apps/ideas/forms.py
from django import forms
from .models import Idea

class IdeaForm(forms.ModelForm):
    class Meta:
        model = Idea
        fields = "__all__"
```

（2）利用视图向 ideas 应用程序中添加新的 views.py 文件，以操控 Idea 模型。

```
# myproject/apps/ideas/views.py
from django.contrib.auth.decorators import login_required
from django.shortcuts import render, redirect, get_object_or_404
from django.views.generic import ListView, DetailView

from .forms import IdeaForm
from .models import Idea
```

```
class IdeaList(ListView):
    model = Idea

class IdeaDetail(DetailView):
    model = Idea
    context_object_name = "idea"

@login_required
def add_or_change_idea(request, pk=None):
    idea = None
    if pk:
        idea = get_object_or_404(Idea, pk=pk)

    if request.method == "POST":
        form = IdeaForm(
            data=request.POST,
            files=request.FILES,
            instance=idea
        )

        if form.is_valid():
            idea = form.save()
            return redirect("ideas:idea_detail", pk=idea.pk)
    else:
        form = IdeaForm(instance=idea)

    context = {"idea": idea, "form": form}
    return render(request, "ideas/idea_form.html", context)

@login_required
def delete_idea(request, pk):
    idea = get_object_or_404(Idea, pk=pk)
    if request.method == "POST":
        idea.delete()
        return redirect("ideas:idea_list")
    context = {"idea": idea}
    return render(request,"ideas/idea_deleting_confirmation.html",context)
```

（3）利用 URL 规则在 ideas 应用程序中创建 urls.py 文件。

```
# myproject/apps/ideas/urls.py
from django.urls import path
```

```
from .views import (
    IdeaList,
    IdeaDetail,
    add_or_change_idea,
    delete_idea,
)

urlpatterns = [
    path("", IdeaList.as_view(), name="idea_list"),
    path("add/", add_or_change_idea, name="add_idea"),
    path("<uuid:pk>/", IdeaDetail.as_view(), name="idea_detail"),
    path("<uuid:pk>/change/", add_or_change_idea, name="change_idea"),
    path("<uuid:pk>/delete/", delete_idea, name="delete_idea"),
]
```

（4）下面向项目的 URL 配置中插入这些 URL 规则。另外，我们还将包含来自 Django 贡献的 auth 应用程序中的账户的 URL 规则，以便@login_required 装饰器能够正常工作。

```
# myproject/urls.py
from django.contrib import admin
from django.conf.urls.i18n import i18n_patterns
from django.urls import include, path
from django.conf import settings
from django.conf.urls.static import static
from django.shortcuts import redirect

urlpatterns = i18n_patterns(
    path("", lambda request: redirect("ideas:idea_list")),
    path("admin/", admin.site.urls),
    path("accounts/", include("django.contrib.auth.urls")),
    path("ideas/", include(("myproject.apps.ideas.urls", "ideas"),
      namespace="ideas")),
)
urlpatterns += static(settings.STATIC_URL,
document_root=settings.STATIC_ROOT)
urlpatterns += static("/media/", document_root=settings.MEDIA_ROOT)
```

（5）创建下列模板。

- 基于登录表单的 registration/login.html。
- 包含 ideas 列表的 ideas/idea_list.html。
- 包含 idea 细节内容的 ideas/idea_detail.html。
- 包含表单的 ideas/idea_form.html，以添加或修改一个 idea。

❑ 包含一个空表单的 ideas/idea_deleting_confirmation.html，以确认 idea 删除操作。

在模板中，我们可以通过命名空间和路径名称处理 ideas 应用程序的 URL，如下所示。

```
{% load i18n %}
<a href="{% url 'ideas:change_idea' pk=idea.pk %}">{% trans "Change this
idea" %}</a>
<a href="{% url 'ideas:add_idea' %}">{% trans "Add idea" %}</a>
```

注意：

如果读者遇到问题或者希望节省时间，则可查看本书代码文件中对应的模板，对应网址为 https://github.com/PacktPublishing/Django-3-Web-Development-Cookbook-Fourth-Edition/tree/master/ch03/myproject_virtualenv/src/djangomyproject/myproject/templates/ideas。

3.3.3 工作方式

在当前示例中，我们针对 Idea 模型的主键使用了一个 UUID 字段。根据该 ID，每个 idea 包含一个无法猜测的、唯一的 URL。或者，还可使用 URL 的 slug 字段，但随后需要确保每个 slug 被填写且在站点中保持唯一。

提示：

出于安全性考虑，此处并不建议对 URL 使用默认的增量 ID。用户可计算数据库中条目的数量，并尝试访问下一个或上一个条目（虽然这些用户不具备执行此类操作的权限）。

在当前示例中，我们采用了基于类的通用视图列出和读取 idea，并使用基于函数的视图创建、更新和删除 idea。修改数据库中记录的视图需要使用基于@login_required 装饰器的授权用户。针对全部 CRUDL 函数，使用基于类的视图或基于函数的视图可视为一种较好的做法。

在成功地添加或修改 idea 后，用户将被重定向至详细视图。在删除了某个 idea 后，用户将被重定向至列表视图。

3.3.4 更多内容

此外，我们还可使用 Django 的消息框架，并在每次成功地添加、修改或删除后在页面顶部显示成功消息。

关于 Django 消息框架的更多内容，读者可访问 https://docs.djangoproject.com/en/2.2/ref/contrib/messages/。

3.3.5　延伸阅读

- 第 2 章中的“使用模型混入”示例。
- 第 2 章中的“与模型翻译表协同工作”示例。
- “保存模型实例的作者”示例。
- 第 4 章中的“安排 base.html 模板”示例。

3.4　保存模型实例的作者

每个 Django 视图的第 1 个参数为 HttpRequest 对象，该对象根据规则被称作 request，并包含与发送自浏览器或其他客户端请求相关的元数据，如当前语言代码、用户信息、cookie 和会话等条目。默认状态下，视图所使用的表单接收 GET 或 POST 数据、文件、初始数据和其他参数，但无法访问 HttpRequest 对象。在某些情况下，可以向表单额外地传递 HttpRequest，特别是想要根据其他请求数据过滤表单字段的选择结果，或者处理保存表单中的当前用户或 IP 等内容时。

在当前示例中，我们将探讨一个表单示例（添加或修改 idea），其间，当前用户将被保存为作者。

3.4.1　准备工作

我们将在前述示例的基础上构建一个新示例。

3.4.2　实现方式

当前示例若要完成，请执行下列两个步骤。

（1）调整 IdeaForm 模型表单，如下所示。

```
# myprojects/apps/ideas/forms.py
from django import forms
from .models import Idea

class IdeaForm(forms.ModelForm):
    class Meta:
        model = Idea
        exclude = ["author"]
```

```
    def __init__(self, request, *args, **kwargs):
        self.request = request
        super().__init__(*args, **kwargs)

    def save(self, commit=True):
        instance = super().save(commit=False)
        instance.author = self.request.user
        if commit:
            instance.save()
            self.save_m2m()
        return instance
```

（2）调整视图并添加或修改 idea。

```
# myproject/apps/ideas/views.py
from django.contrib.auth.decorators import login_required
from django.shortcuts import render, redirect, get_object_or_404

from .forms import IdeaForm
from .models import Idea

@login_required
def add_or_change_idea(request, pk=None):
    idea = None
    if pk:
        idea = get_object_or_404(Idea, pk=pk)
    if request.method == "POST":
        form = IdeaForm(request, data=request.POST,
          files=request.FILES, instance=idea)
        if form.is_valid():
            idea = form.save()
            return redirect("ideas:idea_detail", pk=idea.pk)
    else:
        form = IdeaForm(request, instance=idea)

    context = {"idea": idea, "form": form}
    return render(request, "ideas/idea_form.html", context)
```

3.4.3 工作方式

接下来介绍表单。首先从表单中排除了 author 字段，因为我们打算通过编程方式处理 author 字段。另外，我们覆写了__init__()方法，并作为第 1 个参数接收 HttpRequest，随后将其存储至表单中。模型表单的 save()方法处理模型的保存过程。commit 参数通知模型表单即刻保存实例，否则创建和填写该实例，但不执行保存操作。在当前示例中，我们在不保存实例的情况下获取该实例，随后赋予当前用户中的作者。最后，如果 commit 为 True，则保存实例。我们将以动态方式添加表单的 save_m2m()方法，以保存多对多关系，如 Categories。

在当前视图中，我们仅将 request 变量作为第 1 个参数传递至表单中。

3.4.4 延伸阅读

- “利用 CRUDL 函数创建一个应用程序”示例。
- “上传图像”示例。

3.5 上 传 图 像

在当前示例中，我们将采用最简单的方式处理图像上传问题。其间，我们将把 picture 字段添加至 Idea 模型中，并创建不同尺寸、不同功能的多个图像版本。

3.5.1 准备工作

对于包含图像版本的图像，我们将使用 Pillow 和 django-imagekit 库。下面在虚拟环境中使用 pip 安装这两个库（并将其包含在 requirements/_base.txt 中）。

```
(env)$ pip install Pillow
(env)$ pip install django-imagekit==4.0.2
```

随后在设置项中将"imagekit"添加至 INSTALLED_APPS 中。

3.5.2 实现方式

执行以下步骤以完成示例。

（1）调整 Idea 模型并添加 picture 字段和图像版本规范。

```
# myproject/apps/ideas/models.py
import contextlib
import os

from imagekit.models import ImageSpecField
```

```
from pilkit.processors import ResizeToFill

from django.db import models
from django.utils.translation import gettext_lazy as _
from django.utils.timezone import now as timezone_now

from myproject.apps.core.models import
(CreationModificationDateBase, UrlBase)

def upload_to(instance, filename):
    now = timezone_now()
    base, extension = os.path.splitext(filename)
    extension = extension.lower()
    return f"ideas/{now:%Y/%m}/{instance.pk}{extension}"

class Idea(CreationModificationDateBase, UrlBase):
    # attributes and fields…
    picture = models.ImageField(
        _("Picture"), upload_to=upload_to
    )
    picture_social = ImageSpecField(
        source="picture",
        processors=[ResizeToFill(1024, 512)],
        format="JPEG",
        options={"quality": 100},
    )
    picture_large = ImageSpecField(
        source="picture",
        processors=[ResizeToFill(800, 400)],
        format="PNG"
    )
    picture_thumbnail = ImageSpecField(
        source="picture",
        processors=[ResizeToFill(728, 250)],
        format="PNG"
    )
    # other fields, properties, and methods…

    def delete(self, *args, **kwargs):
        from django.core.files.storage import default_storage
        if self.picture:
            with contextlib.suppress(FileNotFoundError):
                default_storage.delete(
```

```
                self.picture_social.path
            )
            default_storage.delete(
                self.picture_large.path
            )
            default_storage.delete(
                self.picture_thumbnail.path
            )
        self.picture.delete()
    super().delete(*args, **kwargs)
```

（2）针对 forms.py 中的 Idea 模型，创建一个模型表单 IdeaForm。

（3）在添加或修改 idea 的视图中，确保将 request.POST 旁边的 request.FILES 发送至表单中。

```
# myproject/apps/ideas/views.py
from django.contrib.auth.decorators import login_required
from django.shortcuts import (render, redirect, get_object_or_404)
from django.conf import settings

from .forms import IdeaForm
from .models import Idea

@login_required
def add_or_change_idea(request, pk=None):
    idea = None
    if pk:
        idea = get_object_or_404(Idea, pk=pk)
    if request.method == "POST":
        form = IdeaForm(
            request,
            data=request.POST,
            files=request.FILES,
            instance=idea,
        )
        if form.is_valid():
            idea = form.save()
            return redirect("ideas:idea_detail", pk=idea.pk)
    else:
        form = IdeaForm(request, instance=idea)

    context = {"idea": idea, "form": form}
    return render(request, "ideas/idea_form.html", context)
```

（4）在模板中，确保编码类型设置为"multipart/form-data"，如下所示。

```
<form action="{{ request.path }}" method="post"
enctype="multipart/form-data">{% csrf_token %}
{{ form.as_p }}
<button type="submit">{% trans "Save" %}</button>
</form>
```

注意：

如果正在使用 django-crispy-form，那么 enctype 属性将自动添加至表单中。

3.5.3 工作方式

Django 模型表单自动从模型中创建，同时还提供了模型中的特定字段，因而无须在表单中通过手动方式重新定义它们。在前述示例中，我们创建了一个 Idea 模型的模型表单。当保存表单时，表单知晓如何保存数据库中的每个字段，以及如何上传文件并将其保存至 media 目录中。

当前示例中的 upload_to()函数用于将图像保存于特定的目录，并定义其名称以避免与其他模型实例的文件名发生冲突。具体来说，每个文件将存储于诸如 ideas/2020/01/0422c6fe-b725-4576-8703-e2a9d9270986.jpg 这一类路径下，其中包含了上传的年份和月份，以及 Idea 实例的主键。

提示：

某些文件系统（如 FAT32 和 NTFS）在每个目录中包含有限数量的可用文件。因此，较好的做法是按照上传日期、字母或其他标准将文件划分至目录中。

通过 django-imagekit 中的 ImageSpecField，我们创建了 3 个图像版本。

（1）picture_social 用于社交分享。

（2）picture_large 用于详细视图。

（3）picture_thumbnail 用于列表视图。

图像版本并未链接至数据库，而是保存在某个文件路径下的默认的文件存储中，如 CACHE/images/ideas/2020/01/0422c6fe-b725-4576-8703-e2a9d9270986/。

在模板中，可以使用最初或特定的图像版本，如下所示。

```
<img src="{{ idea.picture.url }}" alt="" />
<img src="{{ idea.picture_large.url }}" alt="" />
```

在 Idea 模型定义的最后，我们覆写了 delete()方法，并在删除 Idea 自身实例之前删除了图像版本和硬盘中的图像。

3.5.4 延伸阅读

- ❑ “利用 django-crispy-forms 创建一个表单布局”示例。
- ❑ 第 4 章中的“安排 base.html 模板”示例。
- ❑ 第 4 章中的“提供响应式图像”示例。

3.6 利用自定义模板创建一个表单布局

在 Django 的早期版本中，全部表单渲染均在 Python 代码中进行专门处理。自 Django 1.11 起，引入了基于模板的表单微件渲染机制。在当前示例中，我们将介绍如何使用表单微件的自定义模板，并采用 Django 管理表单展示自定义微件模板如何改进字段的可用性。

3.6.1 准备工作

下面创建 Idea 模型及其翻译的默认 Django 管理。

```
# myproject/apps/ideas/admin.py
from django import forms
from django.contrib import admin
from django.utils.translation import gettext_lazy as _

from myproject.apps.core.admin import LanguageChoicesForm

from .models import Idea, IdeaTranslations

class IdeaTranslationsForm(LanguageChoicesForm):
    class Meta:
        model = IdeaTranslations
        fields = "__all__"

class IdeaTranslationsInline(admin.StackedInline):
    form = IdeaTranslationsForm
    model = IdeaTranslations
    extra = 0

@admin.register(Idea)
```

```
class IdeaAdmin(admin.ModelAdmin):
    inlines = [IdeaTranslationsInline]

    fieldsets = [
        (_("Author and Category"), {"fields": ["author", "categories"]}),
        (_("Title and Content"), {"fields": ["title", "content",
         "picture"]}),
        (_("Ratings"), {"fields": ["rating"]}),
    ]
```

当访问 idea 的管理表单时，对应结果如图 3.1 所示。

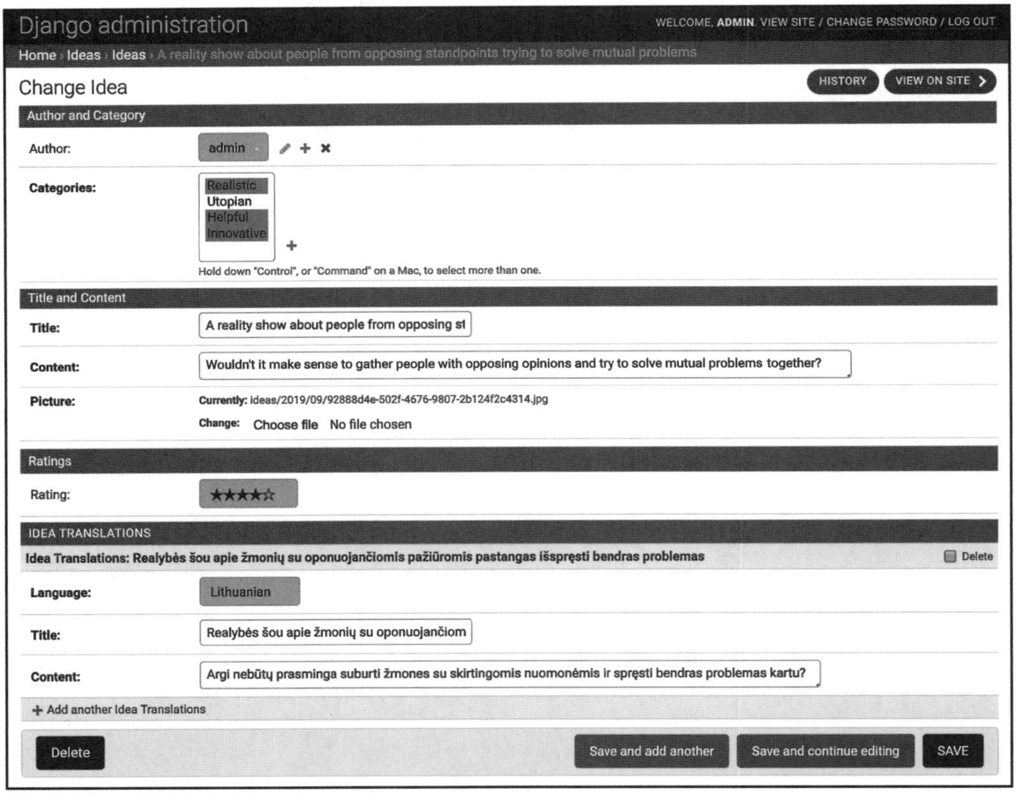

Django administration	WELCOME, ADMIN. VIEW SITE / CHANGE PASSWORD / LOG OUT
Home › Ideas › Ideas › A reality show about people from opposing standpoints trying to solve mutual problems	
Change Idea	HISTORY · VIEW ON SITE
Author and Category	
Author:	admin
Categories:	Realistic / Utopian / Helpful / Innovative — Hold down "Control", or "Command" on a Mac, to select more than one.
Title and Content	
Title:	A reality show about people from opposing st
Content:	Wouldn't it make sense to gather people with opposing opinions and try to solve mutual problems together?
Picture:	Currently: ideas/2019/09/92888d4e-502f-4676-9807-2b124f2c4314.jpg — Change: Choose file No file chosen
Ratings	
Rating:	★★★★☆
IDEA TRANSLATIONS	
Idea Translations: Realybės šou apie žmonių su oponuojančiomis pažiūromis pastangas išspręsti bendras problemas	Delete
Language:	Lithuanian
Title:	Realybės šou apie žmonių su oponuojančiom
Content:	Argi nebūtų prasminga suburti žmones su skirtingomis nuomonėmis ir spręsti bendras problemas kartu?
+ Add another Idea Translations	
Delete	Save and add another · Save and continue editing · SAVE

图 3.1

3.6.2　实现方式

具体实现方式包含下列步骤。

（1）通过在 INSTALLED_APPS 中添加"django.forms"确保模板系统能够查找到自定义模板，包括在模板配置中将 APP_DIRS 标记为 True，并使用"TemplatesSetting"表单渲染器。

```
# myproject/settings/_base.py
INSTALLED_APPS = [
    "django.contrib.admin",
    "django.contrib.auth",
    "django.contrib.contenttypes",
    "django.contrib.sessions",
    "django.contrib.messages",
    "django.contrib.staticfiles",
    "django.forms",
    # other apps…
]

TEMPLATES = [
    {
        "BACKEND":
        "django.template.backends.django.DjangoTemplates",
        "DIRS": [os.path.join(BASE_DIR, "myproject", "templates")],
        "APP_DIRS": True,
        "OPTIONS": {
            "context_processors": [
                "django.template.context_processors.debug",
                "django.template.context_processors.request",
                "django.contrib.auth.context_processors.auth",
                "django.contrib.messages.context_processors.messages",
                "django.template.context_processors.media",
                "django.template.context_processors.static",
                "myproject.apps.core.context_processors.website_url",
            ]
        },
    }
]

FORM_RENDERER = "django.forms.renderers.TemplatesSetting"
```

（2）编辑 admin.py 文件，如下所示。

```
# myproject/apps/ideas/admin.py
from django import forms
from django.contrib import admin
```

```
from django.utils.translation import gettext_lazy as _

from myproject.apps.core.admin import LanguageChoicesForm

from myproject.apps.categories.models import Category
from .models import Idea, IdeaTranslations

class IdeaTranslationsForm(LanguageChoicesForm):
    class Meta:
        model = IdeaTranslations
        fields = "__all__"

class IdeaTranslationsInline(admin.StackedInline):
    form = IdeaTranslationsForm
    model = IdeaTranslations
    extra = 0

class IdeaForm(forms.ModelForm):
    categories = forms.ModelMultipleChoiceField(
        label=_("Categories"),
        queryset=Category.objects.all(),
        widget=forms.CheckboxSelectMultiple(),
        required=True,
    )

    class Meta:
        model = Idea
        fields = "__all__"

    def __init__(self, *args, **kwargs):
        super().__init__(*args, **kwargs)

        self.fields[
            "picture"
        ].widget.template_name = "core/widgets/image.html"

@admin.register(Idea)
class IdeaAdmin(admin.ModelAdmin):
    form = IdeaForm
    inlines = [IdeaTranslationsInline]

    fieldsets = [
        (_("Author and Category"), {"fields": ["author", "categories"]}),
        (_("Title and Content"), {"fields": ["title", "content",
```

```
        "picture"]}),
        (_("Ratings"), {"fields": ["rating"]}),
    ]
```

（3）创建 picture 字段的模板。

```
{# core/widgets/image.html #}
{% load i18n %}

<div style="margin-left: 160px; padding-left: 10px;">
    {% if widget.is_initial %}
        <a href="{{ widget.value.url }}">
            <img src="{{ widget.value.url }}" width="624"
                height="auto" alt="" />
        </a>
        {% if not widget.required %}<br />
            {{ widget.clear_checkbox_label }}:
            <input type="checkbox" name="{{ widget.checkbox_name
                }}" id="{{ widget.checkbox_id }}">
        {% endif %}<br />
        {{ widget.input_text }}:
    {% endif %}
    <input type="{{ widget.type }}" name="{{ widget.name }}"{%
     include "django/forms/widgets/attrs.html" %}>
</div>
<div class="help">
    {% trans "Available formats are JPG, GIF, and PNG." %}
    {% trans "Minimal size is 800 x 800 px." %}
</div>
```

3.6.3　工作方式

当前，若查看管理表单中的 idea，则对应结果如图 3.2 所示。

此处存在两项变化，如下所示。

（1）Categories 选择采用了包含多个复选框的微件。

（2）Picture 字段利用特定的模板进行渲染，并通过首选文件类型和尺寸预览了图像和帮助文本。

此处覆写了 idea 的模型表单，并调整了类别微件和 Picture 字段的模板。

在 Django 中，默认的表单渲染器是"django.forms.renderers.DjangoTemplates"，且仅查找 app 目录中的模板。这里，我们将其修改为"django.forms.renderers.TemplatesSetting"，并查看了 DIRS 路径下的模板。

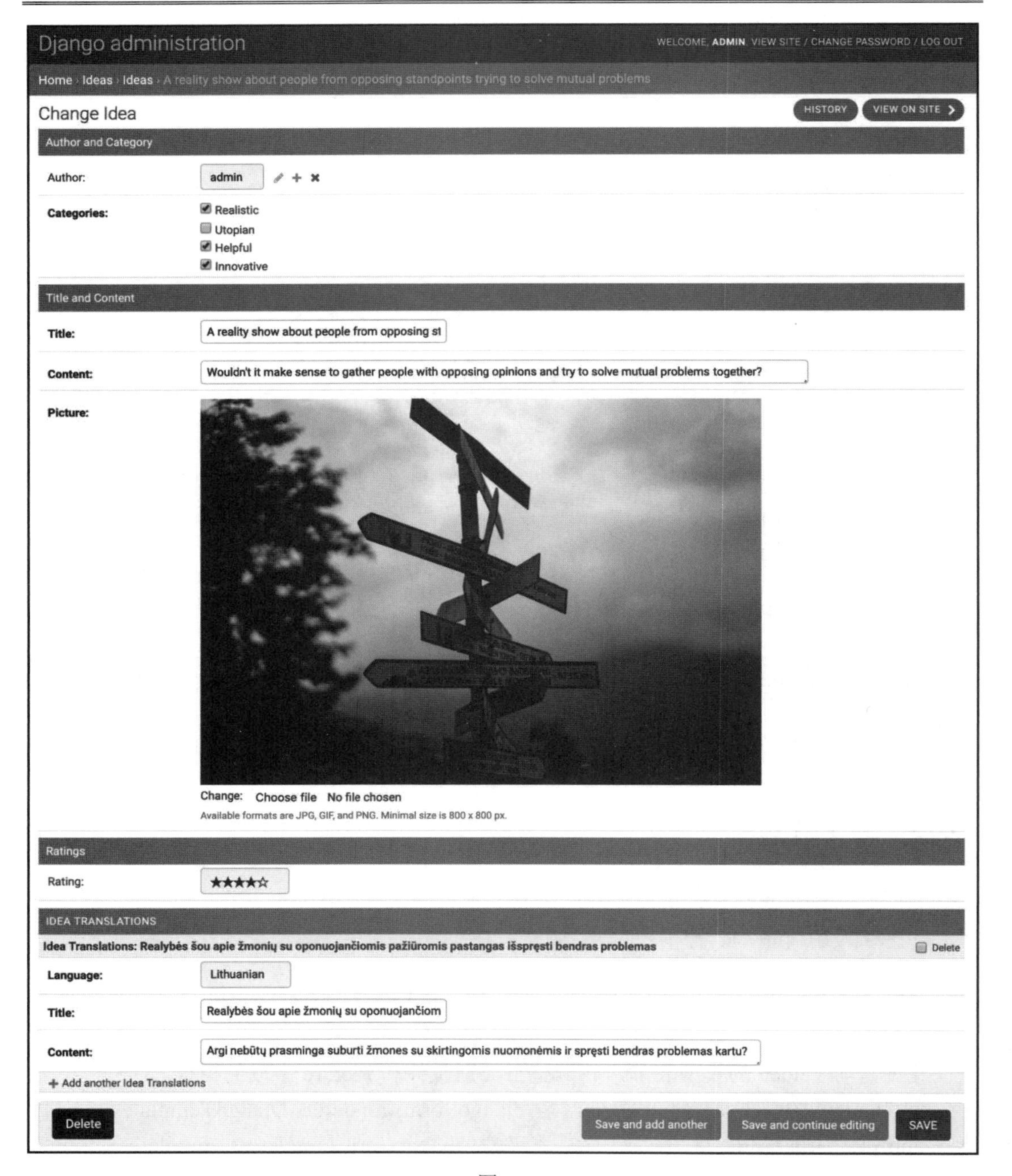

Django administration
WELCOME, ADMIN. VIEW SITE / CHANGE PASSWORD / LOG OUT
Home › Ideas › Ideas › A reality show about people from opposing standpoints trying to solve mutual problems
Change Idea
HISTORY
VIEW ON SITE
Author and Category
Author:
admin
Categories:
Realistic
Utopian
Helpful
Innovative
Title and Content
Title:
A reality show about people from opposing st
Content:
Wouldn't it make sense to gather people with opposing opinions and try to solve mutual problems together?
Picture:
Change: Choose file No file chosen
Available formats are JPG, GIF, and PNG. Minimal size is 800 x 800 px.
Ratings
Rating:
IDEA TRANSLATIONS
Idea Translations: Realybės šou apie žmonių su oponuojančiomis pažiūromis pastangas išspręsti bendras problemas
Delete
Language:
Lithuanian
Title:
Realybės šou apie žmonių su oponuojančiom
Content:
Argi nebūtų prasminga suburti žmones su skirtingomis nuomonėmis ir spręsti bendras problemas kartu?
Add another Idea Translations
Delete
Save and add another
Save and continue editing
SAVE

图 3.2

3.6.4　延伸阅读

- 第 2 章中的“与模型翻译表协同工作”示例。
- “上传图像”示例。
- “利用 django-crispy-forms 创建一个表单布局”示例。

3.7　利用 django-crispy-forms 创建一个表单布局

django-crispy-forms Django 应用程序可通过下列 CSS 框架之一构建、自定义和复用表单，即 Uni-Form、Bootstrap 3、Bootstrap 4 或 Foundation。django-crispy-forms 的使用有些类似于 Django 贡献管理中的字段集，但 django-crispy-forms 的使用更加高级且兼具定制性。我们可在 Python 代码中定义表单布局，且无须担心每个字段在 HTML 的呈现方式。然而，如果需要添加特定的 HTML 属性或封装，其操作仍然十分简单。django-crispy-forms 所用的全部标记均位于针对特殊需求而覆写的模板中。

在当前示例中，我们将创建一个前端表单布局，并利用 Bootstrap 4 这一流行的前端框架编辑 idea，进而开发响应式的、移动优先的 Web 项目。

3.7.1　准备工作

在前述 ideas 示例应用程序的基础上，我们将逐一执行下列各项任务。

（1）确保创建了站点的 base.html 模板。关于 base.html 模板，读者可参考第 4 章的“安排 base.html 模板”示例。

（2）将 Bootstrap 4 前端框架 CSS 文件和 JS 文件（https://getbootstrap.com/docs/4.3/getting-started/introduction/）整合至 base.html 模板中。

（3）利用 pip 命令在虚拟环境中安装 django-crispy-forms（同时将其包含至 requirements/_base.txt 中）。

```
(env)$ pip install django-crispy-forms
```

（4）确保将"crispy_forms"添加至设置项的 INSTALLED_APPS 中，并且随后将"bootstrap4"设置为项目中所使用的模板包。

```
# myproject/settings/_base.py
INSTALLED_APPS = (
```

```
    # ...
    "crispy_forms",
    "ideas",
)
# ...
CRISPY_TEMPLATE_PACK = "bootstrap4"
```

3.7.2　实现方式

具体实现方式如下所示。

（1）调整 idea 的模型表单。

```
# myproject/apps/ideas/forms.py
from django import forms
from django.utils.translation import ugettext_lazy as _
from django.conf import settings
from django.db import models

from crispy_forms import bootstrap, helper, layout

from .models import Idea

class IdeaForm(forms.ModelForm):
    class Meta:
        model = Idea
        exclude = ["author"]

    def __init__(self, request, *args, **kwargs):
        self.request = request
        super().__init__(*args, **kwargs)

        self.fields["categories"].widget = forms.CheckboxSelectMultiple()

        title_field = layout.Field(
            "title", css_class="input-block-level"
        )
        content_field = layout.Field(
            "content", css_class="input-block-level", rows="3"
        )
        main_fieldset = layout.Fieldset(
            _("Main data"), title_field, content_field
        )
```

```
        picture_field = layout.Field(
            "picture", css_class="input-block-level"
        )
        format_html = layout.HTML(
            """{% include "ideas/includes
                /picture_guidelines.html" %}"""
        )

        picture_fieldset = layout.Fieldset(
            _("Picture"),
            picture_field,
            format_html,
            title=_("Image upload"),
            css_id="picture_fieldset",
        )

        categories_field = layout.Field(
            "categories", css_class="input-block-level"
        )
        categories_fieldset = layout.Fieldset(
            _("Categories"), categories_field,
            css_id="categories_fieldset"
        )

        submit_button = layout.Submit("save", _("Save"))
        actions = bootstrap.FormActions(submit_button)

        self.helper = helper.FormHelper()
        self.helper.form_action = self.request.path
        self.helper.form_method = "POST"
        self.helper.layout = layout.Layout(
            main_fieldset,
            picture_fieldset,
            categories_fieldset,
            actions,
        )

    def save(self, commit=True):
        instance = super().save(commit=False)
        instance.author = self.request.user
        if commit:
```

```
            instance.save()
            self.save_m2m()
        return instance
```

（2）创建包含以下内容的 picture_guidelines.html 模板。

```
{# ideas/includes/picture_guidelines.html #}
{% load i18n %}
<p class="form-text text-muted">
    {% trans "Available formats are JPG, GIF, and PNG." %}
    {% trans "Minimal size is 800 × 800 px." %}
</p>
```

（3）更新 idea 表单的模板。

```
{# ideas/idea_form.html #}
{% extends "base.html" %}
{% load i18n crispy_forms_tags static %}

{% block content %}
    <a href="{% url "ideas:idea_list" %}">{% trans "List of
     ideas" %}</a>
    <h1>
        {% if idea %}
            {% blocktrans trimmed with
             title=idea.translated_title %}
                Change Idea "{{ title }}
            {% endblocktrans %}
        {% else %}
            {% trans "Add Idea" %}
        {% endif %}
    </h1>
    {% crispy form %}
{% endblock %}
```

3.7.3　工作方式

在 idea 的模型表单中，我们创建了一个表单帮助方法，其布局包括一个主字段集、图片字段集、类别字段集和提交按钮。每个字段集由多个字段构成。任何字段集、字段或按钮都可包含附加的参数，这些参数最终变为该字段的属性，如 rows="3"或 placeholder=_("Please enter a title")。对于 HTML class 和 id 属性，则存在特殊的参数 css_class 和 css_id。

当前基于 idea 表单的页面如图 3.3 所示。

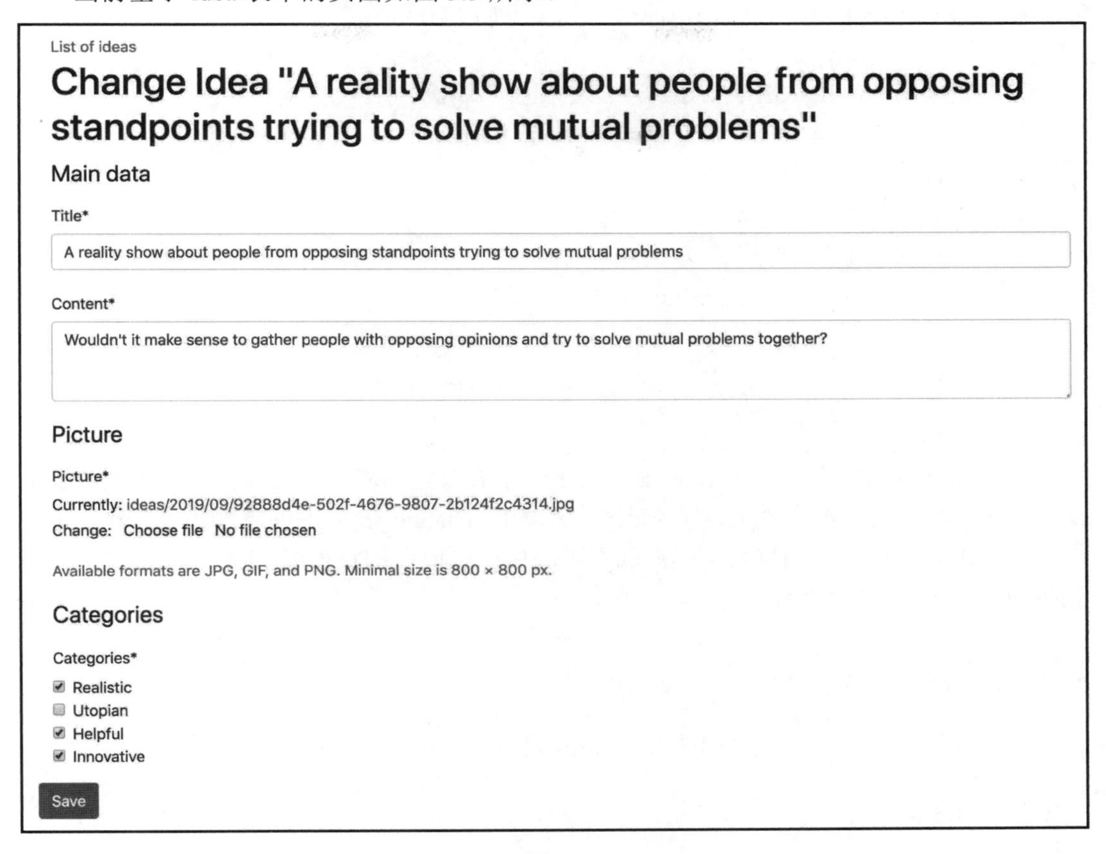

图 3.3

与前述示例类似，我们为类别字段调整了微件，并对图像字段加入了附加的帮助文本。

3.7.4　更多内容

对于基础应用，当前示例已然足够。然而，如果在项目中需要特定的表单标记，我们仍可覆写和调整 django-crispy-forms 应用程序的模板，因为 Python 文件中不存在标记硬编码，全部生成的标记均通过模板渲染。相应地，仅将 django-crispy-forms 应用程序中的模板复制至项目的模板目录中，并根据需要对其进行修改即可。

3.7.5 延伸阅读

- “利用 CRUDL 函数创建一个应用程序”示例。
- “利用自定义模板创建一个表单布局”示例。
- “过滤对象列表”示例。
- “管理分页列表”示例。
- “合成基于类的视图”示例。
- 第 4 章中的“安排 base.html 模板”示例。

3.8 与表单集协同工作

除了标准或模型表单，Django 还定义了一个表单集的概念，并存在相同类型的表单集合，允许我们可一次性地创建或修改多个实例。Django 表单集可通过 JavaScript 进一步充实，进而以动态方式将表单集添加至页面中。当前示例将扩展 idea 表单，并向同一页面的不同语言添加翻译结果。

3.8.1 准备工作

下面继续讨论上一个示例中的 IdeaForm 表单。

3.8.2 实现方式

具体实现方式包含下列步骤。

（1）调整 IdeaForm 的表单布局。

```
# myproject/apps/ideas/forms.py
from django import forms
from django.utils.translation import ugettext_lazy as _
from django.conf import settings
from django.db import models

from crispy_forms import bootstrap, helper, layout

from .models import Idea, IdeaTranslations
```

```
class IdeaForm(forms.ModelForm):
    class Meta:
        model = Idea
        exclude = ["author"]

    def __init__(self, request, *args, **kwargs):
        self.request = request
        super().__init__(*args, **kwargs)

        self.fields["categories"].widget = forms.CheckboxSelectMultiple()

        title_field = layout.Field(
            "title", css_class="input-block-level"
        )
        content_field = layout.Field(
            "content", css_class="input-block-level", rows="3"
        )
        main_fieldset = layout.Fieldset(
            _("Main data"), title_field, content_field
        )

        picture_field = layout.Field(
            "picture", css_class="input-block-level"
        )
        format_html = layout.HTML(
            """{% include "ideas/includes
                /picture_guidelines.html" %}"""
        )

        picture_fieldset = layout.Fieldset(
            _("Picture"),
            picture_field,
            format_html,
            title=_("Image upload"),
            css_id="picture_fieldset",
        )

        categories_field = layout.Field(
            "categories", css_class="input-block-level"
        )
        categories_fieldset = layout.Fieldset(
            _("Categories"), categories_field,
```

```
            css_id="categories_fieldset"
        )

        inline_translations = layout.HTML(
            """{% include "ideas/forms/translations.html" %}"""
        )

        submit_button = layout.Submit("save", _("Save"))
        actions = bootstrap.FormActions(submit_button)

        self.helper = helper.FormHelper()
        self.helper.form_action = self.request.path
        self.helper.form_method = "POST"
        self.helper.layout = layout.Layout(
            main_fieldset,
            inline_translations,
            picture_fieldset,
            categories_fieldset,
            actions,
        )

    def save(self, commit=True):
        instance = super().save(commit=False)
        instance.author = self.request.user
        if commit:
            instance.save()
            self.save_m2m()
        return instance
```

（2）在同一文件的尾部添加 IdeaTranslationsForm。

```
class IdeaTranslationsForm(forms.ModelForm):
    language = forms.ChoiceField(
        label=_("Language"),
        choices=settings.LANGUAGES_EXCEPT_THE_DEFAULT,
        required=True,
    )

    class Meta:
        model = IdeaTranslations
        exclude = ["idea"]

    def __init__(self, request, *args, **kwargs):
```

```
        self.request = request
        super().__init__(*args, **kwargs)

        id_field = layout.Field("id")
        language_field = layout.Field(
            "language", css_class="input-block-level"
        )
        title_field = layout.Field(
            "title", css_class="input-block-level"
        )
        content_field = layout.Field(
            "content", css_class="input-block-level", rows="3"
        )
        delete_field = layout.Field("DELETE")
        main_fieldset = layout.Fieldset(
            _("Main data"),
            id_field,
            language_field,
            title_field,
            content_field,
            delete_field,
        )

        self.helper = helper.FormHelper()
        self.helper.form_tag = False
        self.helper.disable_csrf = True
        self.helper.layout = layout.Layout(main_fieldset)
```

（3）调整视图，以添加或修改 idea，如下所示。

```
# myproject/apps/ideas/views.py
from django.contrib.auth.decorators import login_required
from django.shortcuts import render, redirect, get_object_or_404
from django.forms import modelformset_factory
from django.conf import settings

from .forms import IdeaForm, IdeaTranslationsForm
from .models import Idea, IdeaTranslations

@login_required
def add_or_change_idea(request, pk=None):
    idea = None
    if pk:
```

```
    idea = get_object_or_404(Idea, pk=pk)
IdeaTranslationsFormSet = modelformset_factory(
    IdeaTranslations, form=IdeaTranslationsForm,
    extra=0, can_delete=True
)
if request.method == "POST":
    form = IdeaForm(request, data=request.POST,
     files=request.FILES, instance=idea)
    translations_formset = IdeaTranslationsFormSet(
        queryset=IdeaTranslations.objects.filter(idea=idea),
        data=request.POST,
        files=request.FILES,
        prefix="translations",
        form_kwargs={"request": request},
    )
    if form.is_valid() and translations_formset.is_valid():
        idea = form.save()
        translations = translations_formset.save(
            commit=False
        )
        for translation in translations:
            translation.idea = idea
            translation.save()
        translations_formset.save_m2m()
        for translation in
         translations_formset.deleted_objects:
            translation.delete()
        return redirect("ideas:idea_detail", pk=idea.pk)
else:
    form = IdeaForm(request, instance=idea)
    translations_formset = IdeaTranslationsFormSet(
        queryset=IdeaTranslations.objects.filter(idea=idea),
        prefix="translations",
        form_kwargs={"request": request},
    )

context = {
    "idea": idea,
    "form": form,
    "translations_formset": translations_formset
}
return render(request, "ideas/idea_form.html", context)
```

（4）编辑 idea_form.html 模板，并在尾部添加指向 inlines.js 脚本文件的引用，如下所示。

```
{# ideas/idea_form.html #}
{% extends "base.html" %}
{% load i18n crispy_forms_tags static %}

{% block content %}
    <a href="{% url "ideas:idea_list" %}">{% trans "List of ideas" %}</a>
    <h1>
        {% if idea %}
            {% blocktrans trimmed with
              title=idea.translated_title %}
                Change Idea "{{ title }}"
            {% endblocktrans %}
        {% else %}
            {% trans "Add Idea" %}
        {% endif %}
    </h1>
    {% crispy form %}
{% endblock %}

{% block js %}
    <script src="{% static 'site/js/inlines.js' %}"></script>
{% endblock %}
```

（5）创建翻译表单集的模板。

```
{# ideas/forms/translations.html #}
{% load i18n crispy_forms_tags %}
<section id="translations_section" class="formset my-3">
    {{ translations_formset.management_form }}
    <h3>{% trans "Translations" %}</h3>
    <div class="formset-forms">
        {% for formset_form in translations_formset %}
            <div class="formset-form">
                {% crispy formset_form %}
            </div>
        {% endfor %}
    </div>
    <button type="button" class="btn btn-primary btn-sm
      add-inline-form">{% trans "Add translations to another
      language" %}</button>
```

```
    <div class="empty-form d-none">
        {% crispy translations_formset.empty_form %}
    </div>
</section>
```

（6）添加 JavaScript 以操控表单集。

```
/* site/js/inlines.js */
window.WIDGET_INIT_REGISTER = window.WIDGET_INIT_REGISTER || [];

$(function () {
    function reinit_widgets($formset_form) {
        $(window.WIDGET_INIT_REGISTER).each(function (index, func)
        {
            func($formset_form);
        });
    }

    function set_index_for_fields($formset_form, index) {
        $formset_form.find(':input').each(function () {
            var $field = $(this);
            if ($field.attr("id")) {
                $field.attr(
                    "id",
                    $field.attr("id").replace(/-__prefix__-/,
                    "-" + index + "-")
                );
            }
            if ($field.attr("name")) {
                $field.attr(
                    "name",
                    $field.attr("name").replace(
                        /-__prefix__-/, "-" + index + "-"
                    )
                );
            }
        });
        $formset_form.find('label').each(function () {
            var $field = $(this);
            if ($field.attr("for")) {
                $field.attr(
                    "for",
                    $field.attr("for").replace(
                        /-__prefix__-/, "-" + index + "-"
```

```
                )
            );
        }
    });
    $formset_form.find('div').each(function () {
        var $field = $(this);
        if ($field.attr("id")) {
            $field.attr(
                "id",
                $field.attr("id").replace(
                    /-__prefix__-/, "-" + index + "-"
                )
            );
        }
    });
}

function add_delete_button($formset_form) {
    $formset_form.find('input:checkbox[id$=DELETE]')
      .each(function () {
        var $checkbox = $(this);
        var $deleteLink = $(
            '<button class="delete btn btn-sm
              btn-danger mb-3">Remove</button>'
        );
        $formset_form.append($deleteLink);
        $checkbox.closest('.form-group').hide();
    });

}

$('.add-inline-form').click(function (e) {
    e.preventDefault();
    var $formset = $(this).closest('.formset');
    var $total_forms = $formset.find('[id$="TOTAL_FORMS"]');
    var $new_form = $formset.find('.empty-form')
    .clone(true).attr("id", null);
    $new_form.removeClass('empty-form d-none')
    .addClass('formset-form');
    set_index_for_fields($new_form,parseInt($total_forms.val(),10));
    $formset.find('.formset-forms').append($new_form);
    add_delete_button($new_form);
    $total_forms.val(parseInt($total_forms.val(), 10) + 1);
```

```
        reinit_widgets($new_form);
    });
    $('.formset-form').each(function () {
        $formset_form = $(this);
        add_delete_button($formset_form);
        reinit_widgets($formset_form);
    });
    $(document).on('click', '.delete', function (e) {
        e.preventDefault();
        var $formset = $(this).closest('.formset-form');
        var $checkbox =
        $formset.find('input:checkbox[id$=DELETE]');
        $checkbox.attr("checked", "checked");
        $formset.hide();
    });
});
```

3.8.3　工作方式

相信读者已对 Django 模型管理中的表单集有所了解。对于包含父模型外键的子模型，表单集以内联机制方式被使用。

在当前示例中，我们通过 django-crispy-forms 向 idea 表单中添加了表单集，对应结果如图 3.4 所示。

可以看到，不必将表单集插入至表单的结尾。相反，表单集在中间的任何地方可插入至任何有意义的地方。在当前示例中，在可翻译字段之后列出翻译结果是有意义的。

类似于 IdeaForm 布局，翻译表单的表单布局也包含主字段集。但除此之外，翻译表单还包含 id 和 DELETE 字段，这是识别每个模型实例以及从列表中删除它们的必要字段。DELETE 字段实际上是一个复选框；如果被选中，则从数据库中删除相应的条目。另外，翻译的表单帮助方法定义了 form_tag=False（不生成<form>标签）和 disable_csrf=True（不包含 CSRF 令牌），因为我们已经在父表单 IdeaForm 中定义了它们。

在视图中，如果请求由 POST 方法发出，且表单和表单集均为有效，然后保存表单并创建各自的翻译实例，而不是首先保存它们。这是通过 commit=False 属性实现的。对于每一个翻译实例，我们赋予 idea 并于随后将翻译结果保存至数据库中。最后，我们检查表单集中的表单是否标记为删除，并从数据库中删除表单。

在 translations.html 模板中，我们渲染表单集中的每一个表单，并于随后添加了一个额外的隐藏空表单。该表单由 JavaScript 使用，以生成新的、自动添加的表单集中的新表单。

List of ideas

Change Idea "A reality show about people from opposing standpoints trying to solve mutual problems"

Main data

Title*

A reality show about people from opposing standpoints trying to solve mutual problems

Content*

Wouldn't it make sense to gather people with opposing opinions and try to solve mutual problems together?

Translations

Main data

Language*

Lithuanian

Title*

Realybės šou apie žmonių su oponuojančiomis pažiūromis pastangas išspręsti bendras problemas

Content*

Argi nebūtų prasminga suburti žmones su skirtingomis nuomonėmis ir spręsti bendras problemas kartu?

Remove

Add translations to another language

Picture

Picture*

Currently: ideas/2019/09/92888d4e-502f-4676-9807-2b124f2c4314.jpg

Change: Choose file No file chosen

Available formats are JPG, GIF, and PNG. Minimal size is 800 × 800 px.

Categories

Categories*

- [x] Realistic
- [ ] Utopian
- [x] Helpful
- [x] Innovative

Save

图 3.4

每个表单集中的表单包含全部字段的前缀。例如，第 1 个表单集表单的 title 字段将包含一个 HTML 字段名"translations-0-title"; 同一表单集表单的 DELETE 字段将包含 HTML 字段名"translations-0-DELETE"; 空表单则包含一个单词"__prefix__", 如"translations-__prefix__-title", 而非索引。这是在 Django 层面上抽象出来的，但要使用 JavaScript 操控表单集，则需要对此有所了解。

inlines.js JavaScript 执行下列几项任务。

- 对于每个已有的表单集表单，初始化 JavaScript 微件（可使用工具提示、日期或颜色拾取器、地图等），并生成一个删除按钮，而非 DELETE 复选框。
- 当单击删除按钮时，检查 DELETE 复选框，并隐藏来自用户的表单集表单。
- 当单击添加按钮时，复制空表单并利用下一个有效索引替换"__prefix__", 向列表中添加新的表单，并初始化 JavaScript 微件。

3.8.4 更多内容

JavaScript 使用了一个数组 window.WIDGET_INIT_REGISTER，该数组包含了初始化微件（利用给定的表单集表单）时调用的函数。当在另一个 JavaScript 文件中注册一个新函数时，可执行下列操作。

```
/* site/js/main.js */
function apply_tooltips($formset_form) {
    $formset_form.find('[data-toggle="tooltip"]').tooltip();
}

/* register widget initialization for a formset form */
window.WIDGET_INIT_REGISTER = window.WIDGET_INIT_REGISTER || [];
window.WIDGET_INIT_REGISTER.push(apply_tooltips);
```

这将对所有在表单集表单中出现的内容应用工具提示功能，其中，标记中的标签具有 data-toggle="tooltip"和 title 属性，如下所示。

```
<button data-toggle="tooltip" title="{% trans 'Remove this
translation'%}">{% trans "Remove" %}</button>
```

3.8.5 延伸阅读

- “利用 django-crispy-forms 创建一个表单布局”示例。
- 第 4 章中的“安排 base.html 模板”示例。

3.9　过滤对象列表

在 Web 开发中，除了带有表单的视图之外，对象-列表视图和详细视图也是较为常见的视图。列表视图可简单地列出排序后的对象，如按照字母顺序或创建日期。但对于大量的数据，列表视图则缺乏友好性。出于可访问性和方便考虑，应能够按照各种可能的分类过滤内容。在当前示例中，我们所探讨的模式将能够按照任意数量的分类过滤列表视图。

具体来说，我们将创建一个 idea 列表视图，并能够按照作者、分类或评分进行过滤，如图 3.5 所示。

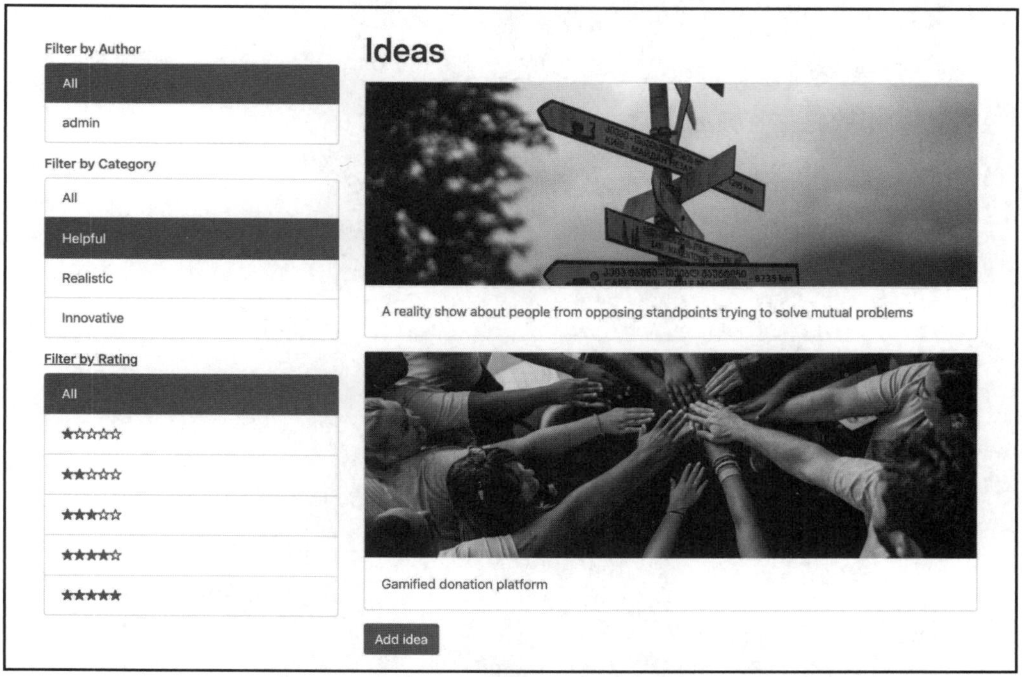

图 3.5

3.9.1　准备工作

针对过滤示例，我们将使用与作者和分类过滤条件相关的 Idea 模型。除此之外，我们还将按照评分机制执行过滤操作，即可选的 PositiveIntegerField。接下来将使用基于前述模型基础上的 idea 应用程序。

3.9.2　实现方式

具体实现方式包含下列步骤。

（1）创建包含所有可能过滤分类的 IdeaFilterForm。

```
# myproject/apps/ideas/forms.py
from django import forms
from django.utils.translation import ugettext_lazy as _
from django.db import models
from django.contrib.auth import get_user_model

from myproject.apps.categories.models import Category

from .models import RATING_CHOICES

User = get_user_model()

class IdeaFilterForm(forms.Form):
    author = forms.ModelChoiceField(
        label=_("Author"),
        required=False,
        queryset=User.objects.annotate(
            idea_count=models.Count("authored_ideas")
        ).filter(idea_count__gt=0),
    )
    category = forms.ModelChoiceField(
        label=_("Category"),
        required=False,
        queryset=Category.objects.annotate(
            idea_count=models.Count("category_ideas")
        ).filter(idea_count__gt=0),
    )
    rating = forms.ChoiceField(
        label=_("Rating"), required=False, choices=RATING_CHOICES
    )
```

（2）创建 idea_list 视图并列出过滤后的 idea。

```
# myproject/apps/ideas/views.py
from django.shortcuts import render, redirect, get_object_or_404
from django.conf import settings
```

```
from .forms import IdeaFilterForm
from .models import Idea, RATING_CHOICES

PAGE_SIZE = getattr(settings, "PAGE_SIZE", 24)

def idea_list(request):
    qs = Idea.objects.order_by("title")
    form = IdeaFilterForm(data=request.GET)

    facets = {
        "selected": {},
        "categories": {
            "authors": form.fields["author"].queryset,
            "categories": form.fields["category"].queryset,
            "ratings": RATING_CHOICES,
        },
    }

    if form.is_valid():
        filters = (
            # query parameter, filter parameter
            ("author", "author"),
            ("category", "categories"),
            ("rating", "rating"),
        )
        qs = filter_facets(facets, qs, form, filters)

    context = {"form": form, "facets": facets, "object_list": qs}
    return render(request, "ideas/idea_list.html", context)
```

（3）在同一个文件中，添加一个帮助函数 filter_facets()。

```
def filter_facets(facets, qs, form, filters):
    for query_param, filter_param in filters:
        value = form.cleaned_data[query_param]
        if value:
            selected_value = value
            if query_param == "rating":
                rating = int(value)
                selected_value = (rating, dict(RATING_CHOICES)[rating])
            facets["selected"][query_param] = selected_value
            filter_args = {filter_param: value}
            qs = qs.filter(**filter_args).distinct()
    return qs
```

（4）创建 base.html 模板（如果不存在，参见第 4 章）。

（5）利用下列内容创建 idea_list.html 模板。

```
{# ideas/idea_list.html #}
{% extends "base.html" %}
{% load i18n utility_tags %}

{% block sidebar %}
    {% include "ideas/includes/filters.html" %}
{% endblock %}

{% block main %}
    <h1>{% trans "Ideas" %}</h1>
    {% if object_list %}
        {% for idea in object_list %}
            <a href="{{ idea.get_url_path }}" class="d-block my-3">
                <div class="card">
                    <img src="{{ idea.picture_thumbnail.url }}" alt="" />
                    <div class="card-body">
                       <p class="card-text">{{ idea.translated_title }}</p>
                    </div>
                </div>
            </a>
        {% endfor %}
    {% else %}
        <p>{% trans "There are no ideas yet." %}</p>
    {% endif %}
    <a href="{% url 'ideas:add_idea' %}" class="btn btn-primary">
      {% trans "Add idea" %}</a>
{% endblock %}
```

（6）创建过滤器模板。该模板采用了第 5 章描述的{% modify_query %}模板标签生成过滤器的 URL。

```
{# ideas/includes/filters.html #}
{% load i18n utility_tags %}
<div class="filters panel-group" id="accordion">
    {% with title=_('Author') selected=facets.selected.author %}
        <div class="panel panel-default my-3">
            {% include "misc/includes/filter_heading.html" with
             title=title %}
            <div id="collapse-{{ title|slugify }}"
```

```
            class="panel-collapse{% if not selected %}
             collapse{% endif %}">
            <div class="panel-body"><div class="list-group">
                {% include "misc/includes/filter_all.html" with
                 param="author" %}
                {% for cat in facets.categories.authors %}
                    <a class="list-group-item
                        {% if selected == cat %}
                        active{% endif %}"
                         href="{% modify_query "page"
                          author=cat.pk %}">
                          {{ cat }}</a>
                {% endfor %}
            </div></div>
        </div>
    </div>
{% endwith %}
{% with title=_('Category') selected=facets.selected.category %}
    <div class="panel panel-default my-3">
        {% include "misc/includes/filter_heading.html" with
            title=title %}
        <div id="collapse-{{ title|slugify }}"
              class="panel-collapse{% if not selected %}
                collapse{% endif %}">
            <div class="panel-body"><div class="list-group">
                {% include "misc/includes/filter_all.html" with
                    param="category" %}
                {% for cat in facets.categories.categories %}
                    <a class="list-group-item
                        {% if selected == cat %}
                        active{% endif %}"
                         href="{% modify_query "page"
                          category=cat.pk %}">
                          {{ cat }}</a>
                {% endfor %}
            </div></div>
        </div>
    </div>
{% endwith %}
{% with title=_('Rating') selected=facets.selected.rating %}
    <div class="panel panel-default my-3">
        {% include "misc/includes/filter_heading.html" with
```

```
            title=title %}
        <div id="collapse-{{ title|slugify }}"
            class="panel-collapse{% if not selected %}
             collapse{% endif %}">
            <div class="panel-body"><div class="list-group">
                {% include "misc/includes/filter_all.html" with
                  param="rating" %}
                {% for r_val, r_display in facets.categories.ratings %}
                    <a class="list-group-item
                        {% if selected.0 == r_val %}
                        active{% endif %}"
                          href="{% modify_query "page"
                            rating=r_val %}">
                            {{ r_display }}</a>
                {% endfor %}
            </div></div>
        </div>
    </div>
    {% endwith %}
</div>
```

（7）每个分类都将遵循过滤器侧栏中的一个公共模式，因而可创建并包含基于公共部分的模板。首先，我们持有过滤标题（对应于 misc/includes/filter_heading.html），如下所示。

```
{# misc/includes/filter_heading.html #}
{% load i18n %}
<div class="panel-heading">
    <h6 class="panel-title">
        <a data-toggle="collapse" data-parent="#accordion"
            href="#collapse-{{ title|slugify }}">
              {% blocktrans trimmed %}
                    Filter by {{ title }}
              {% endblocktrans %}
        </a>
    </h6>
</div>
```

（8）每个过滤器包含一个链接，以重置对应分类的过滤行为，此处表示为 misc/includes/filter_all.html。该模板还使用了第 5 章描述的{% modify_query %}模板标签。

```
{# misc/includes/filter_all.html #}
```

```
{% load i18n utility_tags %}
<a class="list-group-item {% if not selected %}active{% endif %}"
   href="{% modify_query "page" param %}">
     {% trans "All" %}
</a>
```

（9）idea 列表需要被添加至 ideas 应用程序的 URL 中。

```
# myproject/apps/ideas/urls.py
from django.urls import path

from .views import idea_list

urlpatterns = [
   path("", idea_list, name="idea_list"),
   # other paths…
]
```

3.9.3 工作方式

这里，我们使用了 facets 字典，该字典被传递至模板上下文，以了解我们持有的过滤器和所选取的过滤器。进一步讲，facets 字典由两部分组成，即 categories 字典和 selected 字典。其中，categories 字典包含了 QuerySet 或所有可过滤分类的选择结果。selected 字典则包含了当前所选的每个分类值。在 IdeaFilterForm 中，我们确保仅列出至少包含一个 idea 的分类和作者。

在视图中，我们检查了表单中的查询参数是否有效，随后根据所选分类过滤对象的 QuerySet。除此之外，我们还将所选值设置为 facets 字典并传递至模板中。

在当前模板中，针对 facets 字典中的每个分类，我们列出了所有的分类，并将当前所选分类标记为活动分类。如果对于给定分类所选结果为空，那么可将默认的"All"链接标记为活动分类。

3.9.4 延伸阅读

- ❑ “管理分页列表”示例。
- ❑ “合成基于类的视图”示例。
- ❑ 第 4 章中的“安排 base.html 模板”示例。
- ❑ 第 5 章中的“创建模板标签以调整请求查询参数”示例。

3.10　管理分页列表

当动态修改对象列表，或者其数量大于 24 时，那么很可能需要分页以提供较好的用户体验。分页（而非全部 QuerySet）提供了数据集中特定数量的条目，且对应于页面的相应尺寸。此外，我们还显示了链接，并允许用户访问构成完整数据集的其他页面。Django 定义了相关类以管理分页后的数据，当前示例将对具体应用予以探讨。

3.10.1　准备工作

接下来启用 ideas 应用程序的模型、表单和视图。

3.10.2　实现方式

在向 ideas 应用程序的列表视图中添加分页时，可遵循下列各项步骤。

（1）将 Django 中所需的分页类导入至 views.py 文件中，并在过滤机制后向 idea_list 视图中加入分页管理；另外，通过将 page 赋予 object_list 键，还将对上下文字典稍作修改。

```
# myproject/apps/ideas/views.py
from django.shortcuts import render, redirect, get_object_or_404
from django.conf import settings
from django.core.paginator import (EmptyPage, PageNotAnInteger,Paginator)

from .forms import IdeaFilterForm
from .models import Idea, RATING_CHOICES

PAGE_SIZE = getattr(settings, "PAGE_SIZE", 24)

def idea_list(request):
    qs = Idea.objects.order_by("title")
    form = IdeaFilterForm(data=request.GET)

    facets = {
        "selected": {},
        "categories": {
            "authors": form.fields["author"].queryset,
            "categories": form.fields["category"].queryset,
            "ratings": RATING_CHOICES,
```

```
        },
    }

    if form.is_valid():
        filters = (
            # query parameter, filter parameter
            ("author", "author"),
            ("category", "categories"),
            ("rating", "rating"),
        )
        qs = filter_facets(facets, qs, form, filters)

    paginator = Paginator(qs, PAGE_SIZE)
    page_number = request.GET.get("page")
    try:
        page = paginator.page(page_number)
    except PageNotAnInteger:
        # If page is not an integer, show first page.
        page = paginator.page(1)
    except EmptyPage:
        # If page is out of range, show last existing page.
        page = paginator.page(paginator.num_pages)

    context = {
        "form": form,
        "facets": facets,
        "object_list": page,
    }
    return render(request, "ideas/idea_list.html", context)
```

（2）调整 idea_list.html 模板，如下所示。

```
{# ideas/idea_list.html #}
{% extends "base.html" %}
{% load i18n utility_tags %}

{% block sidebar %}
    {% include "ideas/includes/filters.html" %}
{% endblock %}

{% block main %}
    <h1>{% trans "Ideas" %}</h1>
    {% if object_list %}
```

```
        {% for idea in object_list %}
            <a href="{{ idea.get_url_path }}" class="d-block my-3">
                <div class="card">
                    <img src="{{ idea.picture_thumbnail.url }}" alt="" />
                    <div class="card-body">
                        <p class="card-text">{{ idea.translated_title
                         }}</p>
                    </div>
                </div>
            </a>
        {% endfor %}
        {% include "misc/includes/pagination.html" %}
    {% else %}
        <p>{% trans "There are no ideas yet." %}</p>
    {% endif %}
    <a href="{% url 'ideas:add_idea' %}" class="btn btn-primary">
     {% trans "Add idea" %}</a>
{% endblock %}
```

（3）创建分页微件模板。

```
{# misc/includes/pagination.html #}
{% load i18n utility_tags %}
{% if object_list.has_other_pages %}
    <nav aria-label="{% trans 'Page navigation' %}">

        <ul class="pagination">
            {% if object_list.has_previous %}
                <li class="page-item"><a class="page-link" href="{%
         modify_query page=object_list.previous_page_number %}">
                    {% trans "Previous" %}</a></li>
            {% else %}
                <li class="page-item disabled"><span class="page-
                 link">{%trans "Previous" %}</span></li>
            {% endif %}

            {% for page_number in object_list.paginator
             .page_range %}
                {% if page_number == object_list.number %}
                    <li class="page-item active">
                        <span class="page-link">{{ page_number }}
                            <span class="sr-only">{% trans
                             "(current)" %}</span>
```

```
                    </span>
                </li>
            {% else %}
                <li class="page-item">
                    <a class="page-link" href="{% modify_query
                     page=page_number %}">
                        {{ page_number }}</a>
                </li>
            {% endif %}
        {% endfor %}

        {% if object_list.has_next %}
            <li class="page-item"><a class="page-link" href="{%
        modify_query page=object_list.next_page_number %}">
                {% trans "Next" %}</a></li>
        {% else %}
            <li class="page-item disabled"><span class="pagelink">{%
             trans "Next" %}</span></li>
        {% endif %}
    </ul>
</nav>
{% endif %}
```

3.10.3　工作方式

当在浏览器中查看结果时，可以看到图 3.6 所示的分页控制。

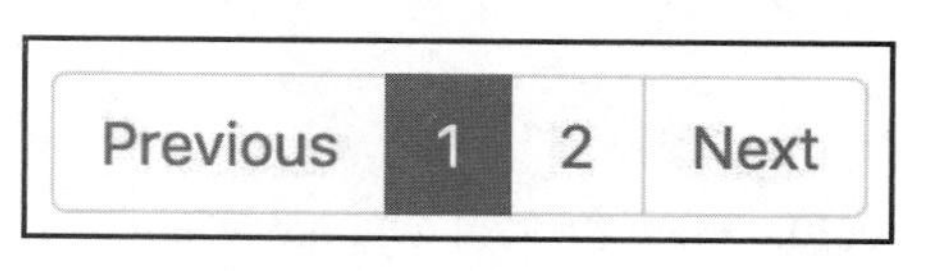

图 3.6

当 QuerySet 被过滤后，我们将创建一个分页器对象，并传递 QuerySet 和每页希望显示的最大条目数量（此处为 24），随后从查询参数中读取当前页码 page。接下来从分页器中检索当前页面对象。如果页码不是一个整数，我们将获取首页；如果页码超出了页面数量，则检索最后一页。针对图 3.6 所示的分页微件，页面对象包含了所需的方法和属性。另外，页面对象的行为类似于 QuerySet，以便可对其进行遍历，进而从部分页面中获取相关条目。

在模板中标记的代码片段采用 Bootstrap 4 前端框架的标记创建了一个分页微件。仅当页面数量超出当前页面，才显示分页控制，有到前一页或下一页的链接，以及微件中全部页面数量列表。另外，当前页码被标记为活动状态。当针对链接生成 URL 时，我们使用了{% modify_query %}模板标签，稍后将对此加以讨论。

3.10.4　延伸阅读

- “过滤对象列表”示例。
- “合成基于类的视图”示例。
- 第 5 章中的“创建模板标签以调整请求查询参数”示例。

3.11　合成基于类的视图

Django 视图是可调用的，该视图接收请求并返回响应结果。除了基于函数的视图，Django 还提供了一种替代方法将视图定义为类。该方法在创建可复用的模块化视图，或合成通用混入视图时十分有用。在当前示例中，我们将把之前基于函数的 idea_list 转换为基于类的 IdeaListView 视图。

3.11.1　准备工作

根据之前的示例创建模型、表单和模板。

3.11.2　实现方式

具体实现方式如下列步骤所示。

（1）基于类的视图 IdeaListView 继承自 Django View 类，并重载了 get()方法。

```
# myproject/apps/ideas/views.py
from django.shortcuts import render, redirect, get_object_or_404
from django.conf import settings
from django.core.paginator import (EmptyPage, PageNotAnInteger,Paginator)
from django.views.generic import View

from .forms import IdeaFilterForm
from .models import Idea, RATING_CHOICES

PAGE_SIZE = getattr(settings, "PAGE_SIZE", 24)
```

```
class IdeaListView(View):
    form_class = IdeaFilterForm
    template_name = "ideas/idea_list.html"

    def get(self, request, *args, **kwargs):
        form = self.form_class(data=request.GET)
        qs, facets = self.get_queryset_and_facets(form)
        page = self.get_page(request, qs)
        context = {"form": form, "facets": facets, "object_list": page}
        return render(request, self.template_name, context)

    def get_queryset_and_facets(self, form):
        qs = Idea.objects.order_by("title")
        facets = {
            "selected": {},
            "categories": {
                "authors": form.fields["author"].queryset,
                "categories": form.fields["category"].queryset,
                "ratings": RATING_CHOICES,
            },
        }
        if form.is_valid():
            filters = (
                # query parameter, filter parameter
                ("author", "author"),
                ("category", "categories"),
                ("rating", "rating"),
            )
            qs = self.filter_facets(facets, qs, form, filters)
        return qs, facets

    @staticmethod
    def filter_facets(facets, qs, form, filters):
        for query_param, filter_param in filters:
            value = form.cleaned_data[query_param]
            if value:
                selected_value = value
                if query_param == "rating":
                    rating = int(value)
                    selected_value = (rating, dict(RATING_CHOICES)[rating])
                facets["selected"][query_param] = selected_value
                filter_args = {filter_param: value}
                qs = qs.filter(**filter_args).distinct()
```

```
        return qs

    def get_page(self, request, qs):
        paginator = Paginator(qs, PAGE_SIZE)
        page_number = request.GET.get("page")
        try:
            page = paginator.page(page_number)
        except PageNotAnInteger:
            page = paginator.page(1)
        except EmptyPage:
            page = paginator.page(paginator.num_pages)
        return page
```

（2）我们需要利用基于类的视图在 URL 配置中创建一个 URL 规则。之前，我们可能针对基于函数的 idea_list 视图添加了一项类似的规则。当在 URL 规则中包含一个基于类的视图时，可采用 as_view()方法，如下所示。

```
# myproject/apps/ideas/urls.py
from django.urls import path

from .views import IdeaListView

urlpatterns = [
    path("", IdeaListView.as_view(), name="idea_list"),
    # other paths…
]
```

3.11.3　工作方式

下列内容解释了针对 HTTP GET 请求调用的 get()方法。

- 首先创建一个 form 对象，并向其中传递 request.GET 类似字典对象。request.GET 对象包含了利用 GET 方法传递的全部查询变量。
- form 对象被传递至 get_queryset_and_facets()方法中，并通过一个元组（两个 QuerySet 元素和 facets 字典）返回关联值。
- 当前请求对象和检索后的对象被传递至 get_page()方法中，该方法返回当前页面对象。
- 最后，创建一个 context 字典并渲染响应结果。

必要的话，还可提供一个 post()方法，该方法针对 HTTP POST 请求而调用。

3.11.4　更多内容

可以看到，get()和 get_page()方法很大程序上是通用的，因而可通过 core 应用程序中的这些方法创建一个泛型 FilterableListView 类。随后，在任何需要一个可过滤列表的应用程序中，我们可以创建一个扩展了 FilterableListView 的基于类的视图以处理此类情形。该扩展类仅定义 form_class 和 template_name 属性以及 get_queryset_and_facets()方法。这种模块化和可扩展性体现了基于类的视图的工作方式的两个优点。

3.11.5　延伸阅读

- “过滤对象列表”示例。
- “管理分页列表”示例。

3.12　提供 Open Graph 和 Twitter Card 数据

如果需要站点内容在社交网络上共享，那么至少应实现 Open Graph 和 Twitter Card 元标签。这些元标签定义了 Web 页面在 Facebook、Twitter Card 中的显示方式，包括显示的标题和描述、设置的图像以及 URL。在当前示例中，我们将准备用于社交共享的 idea_detail.html 模板。

3.12.1　准备工作

下面将继续完成之前的 ideas 应用程序。

3.12.2　实现方式

具体实现方式如下列步骤所示。

（1）确保持有图像字段和图像版本规范创建的 Idea 模型。更多信息可参见“利用 CRUDL 函数创建一个应用程序”和“上传图像”示例。

（2）确保持有 ideas 的详细视图。更多信息可参见“利用 CRUDL 函数创建一个应用程序”示例。

（3）将详细视图插入至 URL 配置中。具体内容可参见“利用 CRUDL 函数创建一个应用程序”示例。

（4）在特定环境设置中，将 WEBSITE_URL 和 MEDIA_URL 定义为多媒体文件的完整 URL。

```
# myproject/settings/dev.py
from ._base import *

DEBUG = True
WEBSITE_URL = "http://127.0.0.1:8000" # without trailing slash
MEDIA_URL = f"{WEBSITE_URL}/media/"
```

（5）在 core 应用程序中，创建一个上下文预处理器，并返回设置项中的 WEBSITE_URL 变量。

```
# myproject/apps/core/context_processors.py
from django.conf import settings

def website_url(request):
    return {
        "WEBSITE_URL": settings.WEBSITE_URL,
    }
```

（6）在设置项中插入上下文预处理器。

```
# myproject/settings/_base.py
TEMPLATES = [
    {
        "BACKEND":
        "django.template.backends.django.DjangoTemplates",
        "DIRS": [os.path.join(BASE_DIR, "myproject", "templates")],
        "APP_DIRS": True,
        "OPTIONS": {
            "context_processors": [
                "django.template.context_processors.debug",
                "django.template.context_processors.request",
                "django.contrib.auth.context_processors.auth",
                "django.contrib.messages.context_processors.messages",
                "django.template.context_processors.media",
                "django.template.context_processors.static",
                "myproject.apps.core.context_processors.website_url",
            ]
        },
    }
]
```

（7）利用下列内容创建 idea_detail.html 模板。

```
{# ideas/idea_detail.html #}
{% extends "base.html" %}
{% load i18n %}

{% block meta_tags %}
    <meta property="og:type" content="website" />
    <meta property="og:url" content="{{ WEBSITE_URL }}
     {{ request.path }}" />
    <meta property="og:title" content="{{ idea.translated_title }}"
     />
    {% if idea.picture_social %}
        <meta property="og:image" content=
         "{{ idea.picture_social.url }}" />
        <!-- Next tags are optional but recommended -->
        <meta property="og:image:width" content=
         "{{ idea.picture_social.width }}" />
        <meta property="og:image:height" content=
         "{{ idea.picture_social.height }}" />
    {% endif %}
    <meta property="og:description" content=
     "{{ idea.translated_content }}" />
    <meta property="og:site_name" content="MyProject" />
    <meta property="og:locale" content="{{ LANGUAGE_CODE }}" />

    <meta name="twitter:card" content="summary_large_image">
    <meta name="twitter:site" content="@DjangoTricks">
    <meta name="twitter:creator" content="@archatas">
    <meta name="twitter:url" content="{{ WEBSITE_URL }}
     {{ request.path }}">
    <meta name="twitter:title" content=
     "{{ idea.translated_title }}">
    <meta name="twitter:description" content=
     "{{ idea.translated_content }}">
    {% if idea.picture_social %}
        <meta name="twitter:image" content=
         "{{ idea.picture_social.url }}">
    {% endif %}
{% endblock %}

{% block content %}
```

```
    <a href="{% url "ideas:idea_list" %}">
     {% trans "List of ideas" %}</a>
    <h1>
        {% blocktrans trimmed with title=idea.translated_title %}
            Idea "{{ title }}"
        {% endblocktrans %}
    </h1>
    <img src="{{ idea.picture_large.url }}" alt="" />
    {{ idea.translated_content|linebreaks|urlize }}
    <p>
        {% for category in idea.categories.all %}
            <span class="badge badge-pill badge-info">
             {{ category.translated_title }}</span>
        {% endfor %}
    </p>
    <a href="{% url 'ideas:change_idea' pk=idea.pk %}"
     class="btn btn-primary">{% trans "Change this idea" %}</a>
    <a href="{% url 'ideas:delete_idea' pk=idea.pk %}"
     class="btn btn-danger">{% trans "Delete this idea" %}</a>
{% endblock %}
```

3.12.3　工作方式

Open Graph 标签是一类包含特殊名称（以 og:开始）的元标签；而 Twitter Card 标签也是一类包含特殊名称（以 twitter:开始）的元标签。这些元标签定义了 URL、标题、描述、当前页面的图像、站点名称、作者和区域。这里，提供完整的 URL 是十分重要的，而路径自身是远远不够的。

此处，我们使用了 picture_social 图像版本，这也是社交网络最优的尺寸，即 1024×512 px。

读者可访问 https://developers.facebook.com/tools/debug/sharing/验证 Open Graph 实现。另外，读者还可访问 https://cards-dev.twitter.com/validator 验证 Twitter Card 实现。

3.12.4　延伸阅读

- “利用 CRUDL 函数创建一个应用程序”示例。
- “上传图像”示例。
- “提供 schema.org 词汇表”示例。

3.13　提供 schema.org 词汇表

搜索引擎优化（Search Engine Optimization，SEO）包含语义标记是十分重要的。为了进一步提高搜索引擎排名，较好的做法是根据 schema.org 词汇表提供结构化数据。许多来自 Google、Microsoft、Pinterest、Yandex 等的应用程序均采用了 schema.org 结构以创建丰富可扩展的体验，如针对事件、电影、作者等的具有一致外观的特定卡片。

相应地，存在多种编码机制（包括 RDFa、Microdata 和 JSON-LD）可用于创建 schema.org 词汇表。在当前示例中，我们将采用 JSON-LD 格式准备 Idea 模型的结构化数据，这也是 Google 所推荐的首选方案。

3.13.1　准备工作

下面将 django-json-ld 包安装至项目的虚拟环境中（并将其包含在 requirements/_base.txt 中）。

```
(env)$ pip install django-json-ld==0.0.4
```

接下来将"django_json_ld"置于设置项的 INSTALLED_APPS 下。

```
# myproject/settings/_base.py
INSTALLED_APPS = [
    # other apps…
    "django_json_ld",
]
```

3.13.2　实现方式

具体实现方式如下列步骤所示。

（1）利用下列内容将 structured_data 属性添加至 Idea 模型中。

```
# myproject/apps/ideas/models.py
from django.db import models
from django.utils.translation import gettext_lazy as _

from myproject.apps.core.models import (
CreationModificationDateBase, UrlBase )
```

```
class Idea(CreationModificationDateBase, UrlBase):

# attributes, fields, properties, and methods…

@property
def structured_data(self):
    from django.utils.translation import get_language

    lang_code = get_language()
    data = {
        "@type": "CreativeWork",
        "name": self.translated_title,
        "description": self.translated_content,
        "inLanguage": lang_code,
    }
    if self.author:
        data["author"] = {
            "@type": "Person",
            "name": self.author.get_full_name() or self.author.username,
        }
    if self.picture:
        data["image"] = self.picture_social.url
    return data
```

（2）调整 idea_detail.html 模板。

```
{# ideas/idea_detail.html #}
{% extends "base.html" %}
{% load i18n json_ld %}

{% block meta_tags %}
    {# Open Graph and Twitter Card meta tags here… #}

    {% render_json_ld idea.structured_data %}
{% endblock %}

{% block content %}
    <a href="{% url "ideas:idea_list" %}">
     {% trans "List of ideas" %}</a>
    <h1>
        {% blocktrans trimmed with title=idea.translated_title %}
            Idea "{{ title }}"
        {% endblocktrans %}
```

```
    </h1>
    <img src="{{ idea.picture_large.url }}" alt="" />
    {{ idea.translated_content|linebreaks|urlize }}
    <p>
        {% for category in idea.categories.all %}
            <span class="badge badge-pill badge-info">
                {{ category.translated_title }}</span>
        {% endfor %}
    </p>
    <a href="{% url 'ideas:change_idea' pk=idea.pk %}"
     class="btn btn-primary">{% trans "Change this idea" %}</a>
    <a href="{% url 'ideas:delete_idea' pk=idea.pk %}"
     class="btn btn-danger">{% trans "Delete this idea" %}</a>
{% endblock %}
```

3.13.3　工作方式

{% render_json_ld %}模板标签将渲染脚本标签，如下所示。

```
<script type=application/ld+json>{"@type": "CreativeWork", "author":
{"@type": "Person", "name": "admin"}, "description": "Lots of African
countries have not enough water. Dig a water channel throughout Africa to
provide water to people who have no access to it.", "image":
"http://127.0.0.1:8000/media/CACHE/images/ideas/2019/09/b919eec5-c077-
41f0-afb4-35f221ab550c_bOFBDgv/9caa5e61fc832f65ff6382f3d482807a.jpg",
"inLanguage": "en", "name": "Dig a water channel throughout
Africa"}</script>
```

structured_data 属性根据 schema.org 词汇表返回一个嵌套的字典，大多数常见的搜索引擎均可较好地理解这些词汇表。

通过检查 https://schema.org/docs/schemas.html 上的官方文档来决定将哪些词汇表应用于模型中。

3.13.4　延伸阅读

- 第 2 章中的“创建一个模型混入以关注元标签”示例。
- “利用 CRUDL 函数创建一个应用程序”示例。
- “上传图像”示例。
- “提供 Open Graph 和 Twitter Card 数据”示例。

3.14　生成 PDF 文档

不仅仅是 HTML 页面，Django 视图还可生成更为丰富的内容，包括任意类型的文件。如在第 4 章中，视图可提供 JavaScript 文件输出，而非 HTML。另外，还可针对发票、票据、收据、预订确认等创建 PDF 文档。在当前示例中，我们将针对数据库中的每一个 idea 生成要打印的讲义，并采用 WeasyPrint 库从 HTML 模板中生成 PDF 文档。

3.14.1　准备工作

WeasyPrint 依赖于计算机上安装的多个库。在 macOS 上，可利用 Homebrew 进行安装，如下所示。

```
$ brew install python3 cairo pango gdk-pixbuf libffi
```

随后可在项目的虚拟环境中安装 WeasyPrint 自身。另外，还需要将 WeasyPrint 包含在 requirements/_base.txt 中。

```
(env)$ pip install WeasyPrint==48
```

对于其他操作系统，读者可访问 https://weasyprint.readthedocs.io/en/latest/nstall.html 查看安装说明。

另外，我们还将使用 django-qr-code 生成 QR 码，并链接回当前网站以供快速访问。同时，我们还将在虚拟环境中安装 django-qr-code（并将其包含在 requirements/_base.txt 中）。

```
(env)$ pip install django-qr-code==1.0.0
```

接下来，将"qr_code"添加至设置项的 INSTALLED_APPS 中。

```
# myproject/settings/_base.py
INSTALLED_APPS = [
    # Django apps…
    "qr_code",
]
```

3.14.2　实现方式

具体实现方式如下列步骤所示。

（1）创建生成 PDF 文档的视图。

```
# myproject/apps/ideas/views.py
from django.shortcuts import get_object_or_404
from .models import Idea

def idea_handout_pdf(request, pk):
    from django.template.loader import render_to_string
    from django.utils.timezone import now as timezone_now
    from django.utils.text import slugify
    from django.http import HttpResponse

    from weasyprint import HTML
    from weasyprint.fonts import FontConfiguration

    idea = get_object_or_404(Idea, pk=pk)
    context = {"idea": idea}
    html = render_to_string(
        "ideas/idea_handout_pdf.html", context
    )

    response = HttpResponse(content_type="application/pdf")
    response[
        "Content-Disposition"
    ] = "inline; filename={date}-{name}-handout.pdf".format(
        date=timezone_now().strftime("%Y-%m-%d"),
        name=slugify(idea.translated_title),
    )

    font_config = FontConfiguration()
    HTML(string=html).write_pdf(
        response, font_config=font_config
    )

    return response
```

（2）将视图插入至 URL 配置中。

```
# myproject/apps/ideas/urls.py
from django.urls import path

from .views import idea_handout_pdf

urlpatterns = [
    # URL configurations…
```

```
    path(
        "<uuid:pk>/handout/",
        idea_handout_pdf,
        name="idea_handout",
    ),
]
```

（3）创建 PDF 文档的模板。

```
{# ideas/idea_handout_pdf.html #}
{% extends "base_pdf.html" %}
{% load i18n qr_code %}

{% block content %}
    <h1 class="h3">{% trans "Handout" %}</h1>
    <h2 class="h1">{{ idea.translated_title }}</h2>
    <img src="{{ idea.picture_large.url }}" alt=""
     class="img-responsive w-100" />
    <div class="my-3">{{ idea.translated_content|linebreaks|
     urlize }}</div>
    <p>
        {% for category in idea.categories.all %}
            <span class="badge badge-pill badge-info">
             {{ category.translated_title }}</span>
        {% endfor %}
    </p>
    <h4>{% trans "See more information online:" %}</h4>
    {% qr_from_text idea.get_url size=20 border=0 as svg_code %}
    <img alt="" src="data:image/svg+xml,
     {{ svg_code|urlencode }}" />
    <p class="mt-3 text-break">{{ idea.get_url }}</p>
{% endblock %}
```

（4）生成 base_pdf.html 模板。

```
{# base_pdf.html #}
<!doctype html>
{% load i18n static %}
<html lang="en">
<head>
    <!-- Required meta tags -->
    <meta charset="utf-8">
    <meta name="viewport" content="width=device-width,
     initial-scale=1, shrink-to-fit=no">
```

```
    <!-- Bootstrap CSS -->
    <link rel="stylesheet"

    href="https://stackpath.bootstrapcdn.com
     /bootstrap/4.3.1/css/bootstrap.min.css"
        integrity="sha384-
         ggOyR0iXCbMQv3Xipma34MD+dH/1fQ784/j6cY
         /iJTQUOhcWr7x9JvoRxT2MZw1T" crossorigin="anonymous">

    <title>{% trans "Hello, World!" %}</title>

    <style>
    @page {
        size: "A4";
        margin: 2.5cm 1.5cm 3.5cm 1.5cm;
    }
    footer {
        position: fixed;
        bottom: -2.5cm;
        width: 100%;
        text-align: center;
        font-size: 10pt;
    }
    footer img {
        height: 1.5cm;
    }
    </style>

    {% block meta_tags %}{% endblock %}
</head>
<body>
    <main class="container">
        {% block content %}
        {% endblock %}
    </main>
    <footer>
        <img alt="" src="data:image/svg+xml,
         {# url-encoded SVG logo goes here #}" />
        <br />
         {% trans "Printed from MyProject" %}
    </footer>
```

```
</body>
</html>
```

3.14.3　工作方式

WeasyPrint 生成可随时打印且像素完美的文档。我们在演讲中向观众分发讲义的实例与此类似，如图 3.7 所示。

图 3.7

文档的布局定义于标记和 CSS 中。WeasyPrint 包含自身的渲染引擎。关于

WeasyPrint 的更多特性，读者可访问官方文档：https://weasyprint.readthedocs.io/en/latest/features.html。

我们可以使用 SVG 图像，它将被保存为矢量图形，而不是位图，因此在打印输出中更清晰。目前尚不支持 Inline SVG，但我们可结合数据源或外部 URL 使用<img>标签。在当前示例中，我们使用了 QR 码和页脚 Logo 的 SVG 图像。

接下来查看视图的代码。其中，作为一个 html 字符串，我们利用所选的 idea 渲染了 idea_handout_pdf.html 模板。随后，我们利用文件名（由当前日期和 slugified idea 标题构成）创建了一个 PDF 内容类型的 HttpResponse 对象。接下来创建了一个基于 HTML 内容的 WeasyPrint 的 HTML 对象，并将其写入至响应结果中（就像是写入一个文件中）。除此之外，我们还使用了 FontConfiguration 对象，该对象允许我们从布局中的 CSS 配置中绑定和使用 Web 字体。最后，我们返回了对应的响应对象。

3.14.4　延伸阅读

- “利用 CRUDL 函数创建一个应用程序”示例。
- “上传图像”示例。
- 第 4 章中的“公开 JavaScript 中的设置项”示例。

3.15　利用 Haystack 和 Whoosh 实现多语言搜索

内容驱动站点的主要功能之一是全文本搜索。Haystack 是一个模块化的搜索 API，并支持 Solr、Elasticsearch、Whoosh 和 Xapian 搜索引擎。对于项目中可查找到的每个模型，我们需要定义一个索引，进而从模型中读取文本信息并将其置于后台中。在当前示例中，我们将学习如何针对多语言站点利用 Haystack 和基于 Python 的搜索引擎完成一项搜索。

3.15.1　准备工作

当前示例将使用之前的 categories 和 ideas 应用程序。

另外，确保已经在虚拟环境中安装了 django-haystack 和 Whoosh（并将其包含至 requirements/_base.txt 中）。

```
(env)$ pip install django-haystack==2.8.1
(env)$ pip install Whoosh==2.7.4
```

3.15.2　实现方式

下列步骤将利用 Haystack and Whoosh 构建多语言搜索。

（1）创建一个 search 应用程序，其中包含 MultilingualWhooshEngine 和 ideas 的搜索索引。搜索引擎则位于 multilingual_whoosh_backend.py 文件中。

```
# myproject/apps/search/multilingual_whoosh_backend.py
from django.conf import settings
from django.utils import translation
from haystack.backends.whoosh_backend import (
    WhooshSearchBackend,
    WhooshSearchQuery,
    WhooshEngine,
)
from haystack import connections
from haystack.constants import DEFAULT_ALIAS

class MultilingualWhooshSearchBackend(WhooshSearchBackend):
    def update(self, index, iterable, commit=True,
     language_specific=False):
        if not language_specific and self.connection_alias ==
         "default":
            current_language = (translation.get_language() or
             settings.LANGUAGE_CODE)[
                :2
            ]
            for lang_code, lang_name in settings.LANGUAGES:
                lang_code_underscored = lang_code.replace("-", "_")
                using = f"default_{lang_code_underscored}"
                translation.activate(lang_code)
                backend = connections[using].get_backend()
                backend.update(index, iterable, commit,
                    language_specific=True)
            translation.activate(current_language)
        elif language_specific:
            super().update(index, iterable, commit)

class MultilingualWhooshSearchQuery(WhooshSearchQuery):
    def __init__(self, using=DEFAULT_ALIAS):
        lang_code_underscored =
        translation.get_language().replace("-", "_")
```

```
        using = f"default_{lang_code_underscored}"
        super().__init__(using=using)

class MultilingualWhooshEngine(WhooshEngine):
    backend = MultilingualWhooshSearchBackend
    query = MultilingualWhooshSearchQuery
```

（2）创建搜索索引，如下所示。

```
# myproject/apps/search/search_indexes.py
from haystack import indexes

from myproject.apps.ideas.models import Idea

class IdeaIndex(indexes.SearchIndex, indexes.Indexable):
    text = indexes.CharField(document=True)

    def get_model(self):
        return Idea

    def index_queryset(self, using=None):
        """
        Used when the entire index for model is updated.
        """
        return self.get_model().objects.all()

    def prepare_text(self, idea):
        """
        Called for each language / backend
        """
        fields = [
            idea.translated_title, idea.translated_content
        ]
        fields += [
            category.translated_title
            for category in idea.categories.all()
        ]
        return "\n".join(fields)
```

（3）配置设置项并使用 MultilingualWhooshEngine。

```
# myproject/settings/_base.py
import os
BASE_DIR = os.path.dirname(os.path.dirname(os.path.dirname(
```

```
    os.path.abspath(__file__)
)))

#…

INSTALLED_APPS = [
    # contributed
    # …
    # third-party
    # …
    "haystack",
    # local
    "myproject.apps.core",
    "myproject.apps.categories",
    "myproject.apps.ideas",
    "myproject.apps.search",
]

LANGUAGE_CODE = "en"

# All official languages of European Union
LANGUAGES = [
     ("bg", "Bulgarian"),
     ("hr", "Croatian"),
     ("cs", "Czech"),
     ("da", "Danish"),
     ("nl", "Dutch"),
     ("en", "English"),
     ("et", "Estonian"),
     ("fi", "Finnish"),
     ("fr", "French"),
     ("de", "German"),
     ("el", "Greek"),
     ("hu", "Hungarian"),
     ("ga", "Irish"),
     ("it", "Italian"),
     ("lv", "Latvian"),
     ("lt", "Lithuanian"),
     ("mt", "Maltese"),
     ("pl", "Polish"),
     ("pt", "Portuguese"),
     ("ro", "Romanian"),
     ("sk", "Slovak"),
```

```
    ("sl", "Slovene"),
    ("es", "Spanish"),
    ("sv", "Swedish"),
]

HAYSTACK_CONNECTIONS = {}
for lang_code, lang_name in LANGUAGES:
   lang_code_underscored = lang_code.replace("-", "_")
   HAYSTACK_CONNECTIONS[f"default_{lang_code_underscored}"] = {
   "ENGINE":
   "myproject.apps.search.multilingual_whoosh_backend
    .MultilingualWhooshEngine",
   "PATH": os.path.join(BASE_DIR, "tmp",
    f"whoosh_index_{lang_code_underscored}"),
  }
  lang_code_underscored = LANGUAGE_CODE.replace("-", "_")
  HAYSTACK_CONNECTIONS["default"] = HAYSTACK_CONNECTIONS[
  f"default_{lang_code_underscored}"
]
```

（4）向 URL 规则中添加路径。

```
# myproject/urls.py
from django.contrib import admin
from django.conf.urls.i18n import i18n_patterns
from django.urls import include, path
from django.conf import settings
from django.conf.urls.static import static
from django.shortcuts import redirect

urlpatterns = i18n_patterns(
    path("", lambda request: redirect("ideas:idea_list")),
    path("admin/", admin.site.urls),
    path("accounts/", include("django.contrib.auth.urls")),
    path("ideas/", include(("myproject.apps.ideas.urls", "ideas"),
    namespace="ideas")),
    path("search/", include("haystack.urls")),
)
urlpatterns += static(settings.STATIC_URL,
document_root=settings.STATIC_ROOT)
urlpatterns += static("/media/", document_root=settings.MEDIA_ROOT)
```

（5）搜索表单和搜索结果的模板如下所示。

```
{# search/search.html #}
{% extends "base.html" %}
{% load i18n %}

{% block sidebar %}
    <form method="get" action="{{ request.path }}">
        <div class="well clearfix">
            {{ form.as_p }}
            <p class="pull-right">
                <button type="submit" class="btn btn-primary">
                 {% trans "Search" %}</button>
            </p>
        </div>
    </form>
{% endblock %}

{% block main %}
    {% if query %}
        <h1>{% trans "Search Results" %}</h1>

        {% for result in page.object_list %}
            {% with idea=result.object %}
                <a href="{{ idea.get_url_path }}"
                 class="d-block my-3">
                    <div class="card">
                        <img src="{{ idea.picture_thumbnail.url }}"
                         alt="" />
                        <div class="card-body">
                            <p class="card-text">
                             {{ idea.translated_title }}</p>
                        </div>
                    </div>
                </a>
            {% endwith %}
        {% empty %}
            <p>{% trans "No results found." %}</p>
        {% endfor %}

        {% include "misc/includes/pagination.html" with
         object_list=page %}
    {% endif %}
{% endblock %}
```

（6）类似于“管理分页列表”示例，在 misc/includes/pagination.html 处添加一个分

页模板。

（7）调用 rebuild_index 管理命令，索引数据库数据并准备所用的全文本搜索。

```
(env)$ python manage.py rebuild_index --noinput
```

3.15.3　工作方式

MultilingualWhooshEngine 指定了两个自定义属性。

（1）backend 指向 MultilingualWhooshSearchBackend，针对 LANGUAGES 设置项中给定的每门语言，确保条目将被索引，并置于 HAYSTACK_CONNECTIONS 定义的关联 Haystack 索引位置之下。

（2）query 引用了 MultilingualWhooshSearchQuery，其职责是确保当搜索关键字时，特定于当前语言的连接将被使用。

每一个索引均包含一个 text 字段，并于其中存储模型特定语言的全文本。该索引的模型由 get_model()方法确定，index_queryset()方法则定义要索引的 QuerySet 内容，而要搜索的内容在 prepare_text()方法中定义为一个换行分隔的字符串。

对于模板，我们在 Bootstrap 4 中加入了一些元素，使用了表单的可用的渲染功能。

最终的搜索页面在侧栏中包含了一个表单，并在主列中显示搜索结果，如图 3.8 所示。

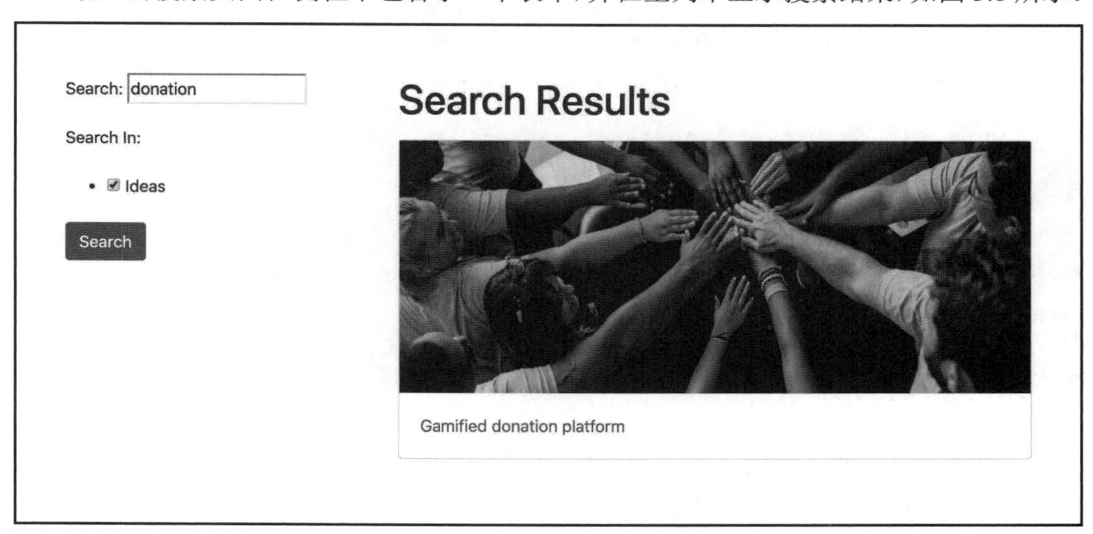

图 3.8

定期更新搜索索引的最简单的方法是调用 rebuild_index 管理命令，这可以通过每晚的 cron 作业予以实现。对此，读者可参考第 13 章中的“设置定期任务的 cron 作业”示例。

3.15.4　延伸阅读

- “利用 django-crispy-forms 创建一个表单布局”示例。
- “管理分页列表”示例。
- 第 13 章中的“设置常规作业的定时任务”示例。

3.16　利用 Elasticsearch DSL 实现多语言搜索

基于 Whoosh 的 Haystack 可视为一类稳定的搜索机制，且仅需要一些 Python 模块即可。对于较好的性能，我们建议使用 Elasticsearch。在当前示例中，我们将展示如何针对多语言搜索使用 Elasticsearch。

3.16.1　准备工作

下面首先安装Elasticsearch服务器。在macOS上，可通过Homebrew安装Elasticsearch。

```
$ brew install elasticsearch
```

在本书编写时，Homebrew 中最新稳定的 Elasticsearch 版本是 6.8.2。

在虚拟环境中安装 django-elasticsearch-dsl（并将其包含至 requirements/_base.txt 中）。

```
(env)$ pip install django-elasticsearch-dsl==6.4.1
```

注意：

安装匹配的 django-elasticsearch-dsl 版本十分重要。否则，在尝试连接至 Elasticsearch 服务器或构建索引时将会出现错误。读者可访问 https://github.com/sabricot/django-elasticsearch-dsl 查看一个版本兼容表。

3.16.2　实现方式

下列步骤将设置基于 Elasticsearch DSL 的多语言搜索。

（1）调整设置项文件并将"django_elasticsearch_dsl"添加至 INSTALLED_APPS 中，随后设置 ELASTICSEARCH_DSL 设置项，如下所示。

```
# myproject/settings/_base.py

INSTALLED_APPS = [
```

```
    # other apps…
    "django_elasticsearch_dsl",
]

ELASTICSEARCH_DSL={
    'default': {
        'hosts': 'localhost:9200'
    },
}
```

（2）在 ideas 应用程序中，利用 idea 搜索索引的 IdeaDocument 创建一个文件。

```
# myproject/apps/ideas/documents.py
from django.conf import settings
from django.utils.translation import get_language, activate
from django.db import models

from django_elasticsearch_dsl import fields
from django_elasticsearch_dsl.documents import (
    Document,
    model_field_class_to_field_class,
)
from django_elasticsearch_dsl.registries import registry

from myproject.apps.categories.models import Category
from .models import Idea

def _get_url_path(instance, language):
    current_language = get_language()
    activate(language)
    url_path = instance.get_url_path()
    activate(current_language)
    return url_path

@registry.register_document
class IdeaDocument(Document):
    author = fields.NestedField(
        properties={
            "first_name": fields.StringField(),
            "last_name": fields.StringField(),
            "username": fields.StringField(),
            "pk": fields.IntegerField(),
        },
```

```
        include_in_root=True,
    )
    title_bg = fields.StringField()
    title_hr = fields.StringField()
    # other title_* fields for each language in the LANGUAGES setting…
    content_bg = fields.StringField()
    content_hr = fields.StringField()
    # other content_* fields for each language in the LANGUAGES setting…

    picture_thumbnail_url = fields.StringField()

    categories = fields.NestedField(
        properties=dict(
            pk=fields.IntegerField(),
            title_bg=fields.StringField(),
            title_hr=fields.StringField(),
            # other title_* definitions for each language in the
              LANGUAGES setting…
        ),
        include_in_root=True,
    )

    url_path_bg = fields.StringField()
    url_path_hr = fields.StringField()
    # other url_path_* fields for each language in the LANGUAGES setting…

    class Index:
        name = "ideas"
        settings = {"number_of_shards": 1, "number_of_replicas": 0}

    class Django:
        model = Idea
        # The fields of the model you want to be indexed in Elasticsearch
        fields = ["uuid", "rating"]
        related_models = [Category]

    def get_instances_from_related(self, related_instance):
        if isinstance(related_instance, Category):
            category = related_instance
            return category.category_ideas.all()
```

（3）向 IdeaDocument 中添加 prepare_*方法，以准备索引的数据。

```
def prepare(self, instance):
    lang_code_underscored = settings.LANGUAGE_CODE.replace
     ("-", "_")
    setattr(instance, f"title_{lang_code_underscored}", instance.title)
    setattr(instance, f"content_{lang_code_underscored}",
     instance.content)
    setattr(
        instance,
        f"url_path_{lang_code_underscored}",
        _get_url_path(instance=instance,
         language=settings.LANGUAGE_CODE),
    )
    for lang_code, lang_name in
    settings.LANGUAGES_EXCEPT_THE_DEFAULT:
        lang_code_underscored = lang_code.replace("-", "_")
        setattr(instance, f"title_{lang_code_underscored}", "")
        setattr(instance, f"content_{lang_code_underscored}", "")
        translations = instance.translations.filter(language=
         lang_code).first()
        if translations:
            setattr(instance, f"title_{lang_code_underscored}",
             translations.title)
            setattr(
                instance, f"content_{lang_code_underscored}",
                 translations.content
            )
        setattr(
            instance,
            f"url_path_{lang_code_underscored}",
            _get_url_path(instance=instance,
              language=lang_code),
        )
    data = super().prepare(instance=instance)
    return data

def prepare_picture_thumbnail_url(self, instance):
    if not instance.picture:
        return ""
    return instance.picture_thumbnail.url

def prepare_author(self, instance):
    author = instance.author
```

```
    if not author:
        return []
    author_dict = {
        "pk": author.pk,
        "first_name": author.first_name,
        "last_name": author.last_name,
        "username": author.username,
    }
    return [author_dict]

def prepare_categories(self, instance):
    categories = []
    for category in instance.categories.all():
        category_dict = {"pk": category.pk}
        lang_code_underscored =
         settings.LANGUAGE_CODE.replace("-", "_")
        category_dict[f"title_{lang_code_underscored}"] =
         category.title
        for lang_code, lang_name in
         settings.LANGUAGES_EXCEPT_THE_DEFAULT:
            lang_code_underscored = lang_code.replace("-", "_")
            category_dict[f"title_{lang_code_underscored}"] =
              ""
            translations =
              category.translations.filter(language=
                lang_code).first()
            if translations:
                category_dict[f"title_{lang_code_underscored}"]
                = translations.title
        categories.append(category_dict)
    return categories
```

（4）向 IdeaDocument 中添加某些属性和方法，并从索引的文档中返回翻译后的内容。

```
@property
def translated_title(self):
    lang_code_underscored = get_language().replace("-", "_")
    return getattr(self, f"title_{lang_code_underscored}", "")

@property
def translated_content(self):
    lang_code_underscored = get_language().replace("-", "_")
    return getattr(self, f"content_{lang_code_underscored}", "")
```

```
def get_url_path(self):
  lang_code_underscored = get_language().replace("-", "_")
  return getattr(self, f"url_path_{lang_code_underscored}", "")

def get_categories(self):
  lang_code_underscored = get_language().replace("-", "_")
  return [
    dict(
      translated_title=category_dict[f"title_{lang_code_underscored}"],
      **category_dict,
    )
    for category_dict in self.categories
  ]
```

（5）在 documents.py 文件中，还要对 UUIDField 映射添加猴子补丁（monkey-patch）。因为默认状态下，Django Elasticsearch DSL 尚不支持 UUIDField 映射。对此，可在导入语句之后插入下列语句。

```
model_field_class_to_field_class[models.UUIDField] = fields.TextField
```

（6）在 ideas 应用程序中，创建 forms.py 下的 IdeaSearchForm。

```
# myproject/apps/ideas/forms.py
from django import forms
from django.utils.translation import ugettext_lazy as _

from crispy_forms import helper, layout

class IdeaSearchForm(forms.Form):
   q = forms.CharField(label=_("Search for"), required=False)

   def __init__(self, request, *args, **kwargs):
      self.request = request
      super().__init__(*args, **kwargs)

      self.helper = helper.FormHelper()
      self.helper.form_action = self.request.path
      self.helper.form_method = "GET"
      self.helper.layout = layout.Layout(
         layout.Field("q", css_class="input-block-level"),
         layout.Submit("search", _("Search")),
      )
```

（7）添加基于 Elasticsearch 搜索的视图。

```
# myproject/apps/ideas/views.py
from django.shortcuts import render
from django.conf import settings
from django.core.paginator import EmptyPage, PageNotAnInteger,
Paginator
from django.utils.functional import LazyObject

from .forms import IdeaSearchForm

PAGE_SIZE = getattr(settings, "PAGE_SIZE", 24)

class SearchResults(LazyObject):
    def __init__(self, search_object):
        self._wrapped = search_object

    def __len__(self):
        return self._wrapped.count()

    def __getitem__(self, index):
        search_results = self._wrapped[index]
        if isinstance(index, slice):
            search_results = list(search_results)
        return search_results

def search_with_elasticsearch(request):
    from .documents import IdeaDocument
    from elasticsearch_dsl.query import Q

    form = IdeaSearchForm(request, data=request.GET)

    search = IdeaDocument.search()

    if form.is_valid():
        value = form.cleaned_data["q"]
        lang_code_underscored = request.LANGUAGE_CODE.replace("-", "_")
        search = search.query(
            Q("match_phrase", **{f"title_{
             lang_code_underscored}":
             value})
            | Q("match_phrase", **{f"content_{
              lang_code_underscored}": value})
```

```
        | Q(
            "nested",
            path="categories",
            query=Q(
                "match_phrase",
                **{f"categories__title_{
                lang_code_underscored}": value},
            ),
        )
    )
search_results = SearchResults(search)

paginator = Paginator(search_results, PAGE_SIZE)
page_number = request.GET.get("page")
try:
    page = paginator.page(page_number)
except PageNotAnInteger:
    # If page is not an integer, show first page.
    page = paginator.page(1)
except EmptyPage:
    # If page is out of range, show last existing page.
    page = paginator.page(paginator.num_pages)

context = {"form": form, "object_list": page}
return render(request, "ideas/idea_search.html", context)
```

（8）创建搜索表单和搜索结果的 idea_search.html 模板

```
{# ideas/idea_search.html #}
{% extends "base.html" %}
{% load i18n crispy_forms_tags %}

{% block sidebar %}
    {% crispy form %}
{% endblock %}

{% block main %}
    <h1>{% trans "Search Results" %}</h1>
    {% if object_list %}
        {% for idea in object_list %}
            <a href="{{ idea.get_url_path }}" class="d-block my-3">
                <div class="card">
                    <img src="{{ idea.picture_thumbnail_url }}"
```

```
                    alt="" />
                <div class="card-body">
                    <p class="card-text">{{ idea.translated_title
                        }}</p>
                </div>
            </div>
        </a>
    {% endfor %}
    {% include "misc/includes/pagination.html" %}
  {% else %}
    <p>{% trans "No ideas found." %}</p>
  {% endif %}
{% endblock %}
```

（9）类似于“管理分页列表”示例，在 misc/includes/pagination.html 处添加分页模板。

（10）调用 search_index --rebuild 管理命令，索引数据库数据并准备所用的完整文本搜索。

```
(env)$ python manage.py search_index --rebuild
```

3.16.3　工作方式

Django Elasticsearch DSL 文档类似于模型表单。这里，可以定义模型的哪些字段要保存到索引中，这些索引稍后将用于搜索查询。在当前 IdeaDocument 示例中，我们将要保存 UUID、评分、作者、分类、标题、内容、全部语言中的 URL 路径，以及图像缩略图 URL。Index 类定义了这一类文档的 Elasticsearch 索引设置。Django 类则定义了索引字段的填充位置。related_models 设置负责通知在哪个模型更改之后也要更新这个索引，在当前示例中为 Category 模型。注意，当使用 django-elasticsearch-dsl 并保存模型时，索引将自动被更新。这是通过相关信号完成的。

get_instances_from_related()方法通知在 Category 实例变化时如何检索 Idea 模型实例。

IdeaDocument 的 prepare()和 prepare_*()方法通知获取数据的位置，以及如何保存特定字段的数据。例如，我们将读取 IdeaTranslations 模型的 title 字段中的 title_lt 的数据（其中，language 等于"lt"）。

IdeaDocument 的最后一个属性和方法用于检索当前活动语言的索引中的信息。

接下来是包含搜索表单的视图。该视图中有一个名为 q 的查询字段。提交后，将在当前语言的标题、内容或分类标题字段中搜索查询单词。随后，通过延迟评估的 SearchResults 类，我们将封装搜索结果，以便将其与默认的 Django 分页器结合使用。

视图模板将在侧栏中包含搜索表单，搜索结果则显示于主列中，如图 3.9 所示。

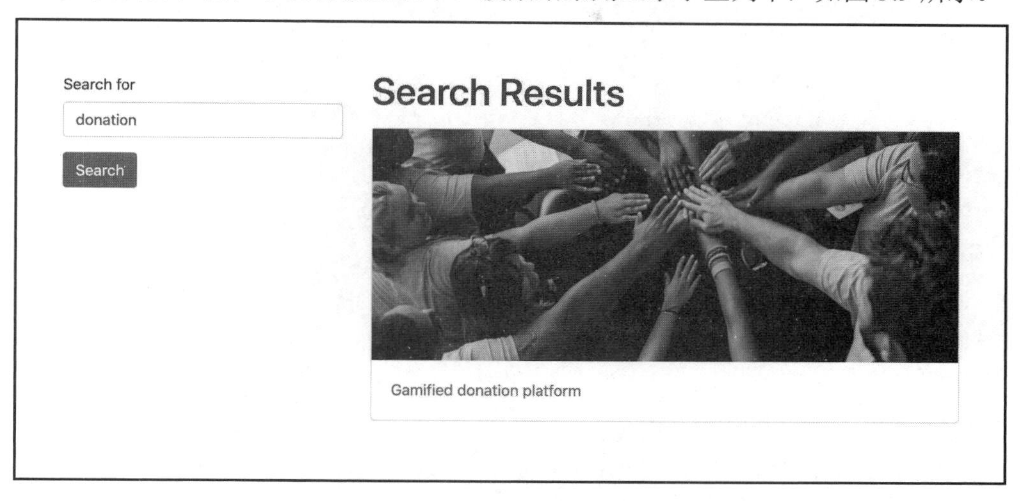

图 3.9

3.16.4　延伸阅读

- “利用 CRUDL 函数创建一个应用程序”示例。
- “利用 Haystack 和 Whoosh 实现多语言搜索”示例。
- “利用 django-crispy-forms 创建一个表单布局”示例。
- “管理分页列表”示例。

第 4 章　模板和 JavaScript

本章主要涉及下列主题。

- ❑ 安排 base.html 模板。
- ❑ 使用 Django Sekizai。
- ❑ 公开 JavaScript 中的设置项。
- ❑ 使用 HTML 5 数据属性。
- ❑ 提供响应式图像。
- ❑ 实现连续的滚动机制。
- ❑ 在模式对话框中打开对象详细信息。
- ❑ 实现 Like 微件。
- ❑ 通过 Ajax 上传图像。

4.1　简　　介

静态站点对于静态内容十分有用，如传统的文档、在线书籍和教程。然而，当今大多数交互式 Web 应用程序和平台均涵盖了动态内容，进而提升了用户的访问体验。在本章中，我们将学习如何在 Django 模板的基础上使用 JavaScript 和 CSS。其间，我们将使用 Bootstrap 4 前端框架用以响应式布局，并针对生产性脚本使用 jQuery JavaScript 框架。

4.2　技 术 需 求

如前所述，当与本章代码协同工作时，需要安装最新稳定版本的 Python、MySQL 或 PostgreSQL 数据库，以及基于虚拟环境的 Django 项目。另外，某些示例需要特定的 Python 依赖项和额外的 JavaScript 库。

读者可访问 GitHub 储存库的 ch04 目录查看本章代码，对应网址为 https://github.com/PacktPublishing/Django-3-Web-Development-Cookbook-Fourth-Edition。

4.3　安排 base.html 模板

当开始处理模板时，首先需要创建 base.html 样板代码，进而被项目中大多数页面模板所扩展。在当前示例中，我们将描述如何针对多语言 HTML5 站点创建此类模板，并考虑到响应性。

提示：

无论访问者使用桌面浏览器、平板电脑或手机，响应式站点将为所有设备提供相同的基本内容，以及样式适宜的视口。这与自适应站点不同。在自适应站点中，服务器试图根据用户代理确定设备类型，然后根据用户代理的分类方式提供完全不同的内容、标记，甚至功能。

4.3.1　准备工作

在项目中创建 templates 目录，并在设置项中设置模板目录以包含该目录，如下所示。

```
# myproject/settings/_base.py
TEMPLATES = [
    {
        "BACKEND": "django.template.backends.django.DjangoTemplates",
        "DIRS": [os.path.join(BASE_DIR, "myproject", "templates")],
        "APP_DIRS": True,
        "OPTIONS": {
            "context_processors": [
                "django.template.context_processors.debug",
                "django.template.context_processors.request",
                "django.contrib.auth.context_processors.auth",
                "django.contrib.messages.context_processors.messages",
                "django.template.context_processors.media",
                "django.template.context_processors.static",
            ]
        },
    }
]
```

4.3.2　实现方式

具体实现方式如下列步骤所示。

（1）在模板的根目录中，利用下列内容创建 base.html 文件。

```
{# base.html #}
<!doctype html>
{% load i18n static %}
<html lang="en">
<head>
    <meta charset="utf-8" />
    <meta name="viewport" content="width=device-width,
    Initial-scale=1, shrink-to-fit=no" />
    <title>{% block head_title %}{% endblock %}</title>
    {% include "misc/includes/favicons.html" %}
    {% block meta_tags %}{% endblock %}

    <link rel="stylesheet"
        href="https://stackpath.bootstrapcdn.com/bootstrap
         /4.3.1/css/bootstrap.min.css"
        integrity="sha384-ggOyR0iXCbMQv3Xipma34MD+dH/1fQ784
         /j6cY/iJTQUOhcWr7x9JvoRxT2MZw1T"
        crossorigin="anonymous" />
    <link rel="stylesheet"
        href="{% static 'site/css/style.css' %}"
        crossorigin="anonymous" />

    {% block css %}{% endblock %}
    {% block extra_head %}{% endblock %}
</head>
<body>
    {% include "misc/includes/header.html" %}
    <div class="container my-5">
        {% block content %}
            <div class="row">
                <div class="col-lg-4">{% block sidebar %}
                 {% endblock %}</div>
                <div class="col-lg-8">{% block main %}
                 {% endblock %}</div>
            </div>
        {% endblock %}
    </div>
    {% include "misc/includes/footer.html" %}
    <script src="https://code.jquery.com/jquery-3.4.1.min.js"
          crossorigin="anonymous"></script>
    <script src="https://cdnjs.cloudflare.com/ajax/libs/popper.js
```

```
    /1.14.7/umd/popper.min.js"
          integrity="sha384-UO2eT0CpHqdSJQ6hJty5KVphtPhzWj
           9WO1clHTMGa3JDZwrnQq4sF86dIHNDz0W1"
          crossorigin="anonymous"></script>
    <script src="https://stackpath.bootstrapcdn.com/bootstrap
     /4.3.1/js/bootstrap.min.js"
          integrity="sha384-JjSmVgyd0p3pXB1rRibZUAYoIIy6OrQ6Vrj
           IEaFf/nJGzIxFDsf4x0xIM+B07jRM"
          crossorigin="anonymous"></script>
    {% block js %}{% endblock %}
    {% block extra_body %}{% endblock %}
</body>
</html>
```

（2）在 misc/includes 下，创建一个包含全部版本收藏夹图标的模板。

```
{# misc/includes/favicon.html #}
{% load static %}
<link rel="icon" type="image/png" href="{% static
'site/img/favicon-32x32.png' %}" sizes="32x32"/>
<link rel="icon" type="image/png" href="{% static
'site/img/favicon-16x16.png' %}" sizes="16x16"/>
```

提示：

收藏夹图标是一幅较小的图像，经常可以在浏览器选项卡、最近访问过的站点图标和桌面快捷方式中看到。我们可采用在线生成器生成不同的收藏夹图标版本（包括不同用例、浏览器和平台的 Logo），如 https://favicomatic.com/和 https://realfavicongenerator.net/。

（3）创建基于站点页眉和页脚的模板 misc/includes/header.html 和 misc/includes/footer.html。当前，我们仅可创建空的文件。

4.3.3　工作方式

基础模板包含 HTML 文档的<head>和<body>部分，以及站点每个页面上重复使用的所有细节内容。根据网页的设计需求，我们可针对不同的布局持有额外的基础模板。例如，我们可添加 base_simple.html 文件，该文件包含相同的 HTML <head>部分，以及简约的<body>部分，这可用于登录屏幕、密码重置或其他简单的页面。另外，对于其他布局，还可持有独立的基础模板，如单列、两列和三列布局，其中每个布局都扩展了 base.html 并根据需要覆写相应的块。

接下来探讨之前定义的 base.html 模板中的细节内容。<head>部分的细节内容如下所示。

- 我们将 UTF-8 定义为默认的编码机制，以支持多语言内容。
- 随后定义在浏览器中可缩放站点的视口并使用全部宽度。这对于较小屏幕的设备是必要的，这一类设备利用 Bootstrap 前端框架生成了特定的屏幕布局。
- 存在一个可自定义的站点标题，并用于浏览器选项卡中以搜索结果。
- 随后是元标签块，可用于搜索引擎优化（SEO）、Open Graph 和 Twitter Card 中。
- 包含不同格式和尺寸的收藏夹图标。
- 包含了默认的 Bootstrap 和自定义站点样式。由于打算设置响应式布局，因而加载了 Bootstrap CSS。此外，对于浏览器间的一致性而言，这还将标准化基本样式，以实现所有元素的一致性。
- 最后，我们还对元标签、样式表以及<head>部分所需的其他内容扩展了块。

<body>部分的细节内容如下所示。

- 首先包含站点的标头。这也是放置 Logo、站点标题和主导航之处。
- 随后是包含内容块占位符的主容器，并由扩展的模板所填充。
- 在容器内，存在一个 content 块，其中包含了 sidebar 和 main 块。在子模板中，当需要一个包含侧栏的布局时，我们将覆写 sidebar 和 main 块。但是，当需要全宽度内容时，我们需要覆写 content 块。
- 随后包含站点的页脚。这里，我们可以放置版权信息，以及指向重要元页面的链接，如隐私策略、使用条款和联系表单等。
- 接下来加载 jQuery 和 Bootstrap 脚本。遵循页面加载性能的最佳实践，可扩展的 JavaScript 块被包含在<body>的末尾，这与<head>中包含的样式表非常相似。
- 最后是附加的 JavaScript 和 HTML 块，如 JavaScript 的 HTML 模板，或隐藏模式对话框，稍后将对此加以讨论。

我们创建的基本模板并不是一个静态的不可更改的模板。我们可以调整标记结构，或者向其中添加所需的元素。例如，body 属性的模板块、Google Analytics 代码片段、公共 JavaScript 文件、iPhone 书签的 Apple 触摸图标，Open Graph 元标签、Twitter Card 标签、schema.org 属性等。除此之外，根据项目的具体需求，还可以定义其他块，甚至还可封装全部 body 内容，以便可在子模板中覆写它。

4.3.4 延伸阅读

- “使用 Django Sekizai”示例。
- “公开 JavaScript 中的设置项”示例。

4.4　使用 Django Sekizai

在 Django 模板中，通常情况下，我们应采用模板继承覆写父模板中的块，进而将样式或脚本包含至 HTML 文档中。这意味着，每个视图的每个主模板都应该了解其中的所有内容。某些时候，令包含的模板决定加载何种样式和脚本将会方便得多。Django Sekizai 满足这一要求，当前示例将探讨其应用方式。

4.4.1　准备工作

在开始当前示例之前，首先需要执行下列步骤。

（1）将 django-classy-tags 和 django-sekizai 安装至虚拟环境中（并将其添加至 requirements/_base.txt 中）。

```
(env)$ pip install -e
git+https://github.com/divio/django-classy-tags.git@4c94d0354eca160
0ad2ead9c3c151ad57af398a4#egg=django-classy-tags
(env)$ pip install django-sekizai==1.0.0
```

（2）在设置项中，将 sekizai 添加至安装后的应用程序中。

```
# myproject/settings/_base.py
INSTALLED_APPS = [
    # …
    "sekizai",
    # …
]
```

（3）在设置项中，将 sekizai 上下文预处理器添加至模板配置中。

```
# myproject/settings/_base.py
TEMPLATES = [
    {
        "BACKEND":
        "django.template.backends.django.DjangoTemplates",
        "DIRS": [os.path.join(BASE_DIR, "myproject", "templates")],
        "APP_DIRS": True,
        "OPTIONS": {
            "context_processors": [
                "django.template.context_processors.debug",
```

```
                "django.template.context_processors.request",
                "django.contrib.auth.context_processors.auth",
                "django.contrib.messages.context_processors.messages",
                "django.template.context_processors.media",
                "django.template.context_processors.static",
                "sekizai.context_processors.sekizai",
            ]
        },
    }
]
```

4.4.2 实现方式

具体实现方式如下列步骤所示。

（1）在 base.html 模板开始处，加载 sekizai_tags 库。

```
{# base.html #}
<!doctype html>
{% load i18n static sekizai_tags %}
```

（2）在同一个文件中，在<head>部分的结尾处，添加模板标签{% render_block "css" %}，如下所示。

```
    {% block css %}{% endblock %}
    {% render_block "css" %}
    {% block extra_head %}{% endblock %}
</head>
```

（3）在<body>部分的结尾处，添加模板标签{% render_block "js" %}，如下所示。

```
    {% block js %}{% endblock %}
    {% render_block "js" %}
    {% block extra_body %}{% endblock %}
</body>
```

（4）在任何所包含的模板中，当打算添加一些样式或 JavaScript 时，可使用{% addtoblock %}模板标签，如下所示。

```
{% load static sekizai_tags %}

<div>Sample widget</div>

{% addtoblock "css" %}
```

```
<link rel="stylesheet" href="{% static 'site/css/sample-widget.css' %}"/>
{% endaddtoblock %}

{% addtoblock "js" %}
<script src="{% static 'site/js/sample-widget.js' %}"></script>
{% endaddtoblock %}
```

4.4.3　工作方式

Django Sekizai 与{% include %}包含的模板标签、利用模板渲染的自定义模板标签，或者表单微件的模板协同工作。{% addtoblock %}模板标签定义了打算添加 HTML 内容的 Sekizai 块。

当向 Sekizai 块中添加某些内容时，django-sekizai 仅于此处设置一次。这意味着，我们可持有同一类型的多个所包含的微件，但其 CSS 和 JavaScript 仅被加载和执行一次。

4.4.4　延伸阅读

- “实现 Like 微件”示例。
- “通过 Ajax 上传图像”示例。

4.5　公开 JavaScript 中的设置项

Django 项目在设置项文件中包含自身的配制集，如用于开发环境的 myproject/settings/dev.py（参见第 1 章）。其中，一些配置值对于浏览器中的功能十分有用，因而也需要在 JavaScript 中被设置。对此，我们需要一个单一位置定义项目设置项。因此，在当前示例中，我们将探讨如何在 Django 服务器和浏览器之间传递某些配置值。

4.5.1　准备工作

确保将 request 上下文预处理器包含至 TEMPLATES['OPTIONS']['context_processors']设置项中。

```
# myproject/settings/_base.py
TEMPLATES = [
    {
        "BACKEND": "django.template.backends.django.DjangoTemplates",
```

```
        "DIRS": [os.path.join(BASE_DIR, "myproject", "templates")],
        "APP_DIRS": True,
        "OPTIONS": {
            "context_processors": [
                "django.template.context_processors.debug",
                "django.template.context_processors.request",
                "django.contrib.auth.context_processors.auth",
                "django.contrib.messages.context_processors.messages",
                "django.template.context_processors.media",
                "django.template.context_processors.static",
                "sekizai.context_processors.sekizai",
            ]
        },
    }
]
```

除此之外，还应创建一个 core 应用程序（如果不存在），并将其置于设置项的 INSTALLED_APPS 中。

```
INSTALLED_APPS = [
    # …
    "myproject.apps.core",
    # …
]
```

4.5.2　实现方式

下列步骤将创建和包含 JavaScript 设置项。

（1）在 core 应用程序的 views.py 中，创建一个 js_settings()视图，并返回一个 JavaScript 内容类型的响应结果，如下所示。

```
# myproject/apps/core/views.py
import json
from django.http import HttpResponse
from django.template import Template, Context
from django.views.decorators.cache import cache_page
from django.conf import settings

JS_SETTINGS_TEMPLATE = """
window.settings = JSON.parse('{{ json_data|escapejs }}');
"""
```

```
@cache_page(60 * 15)
def js_settings(request):
    data = {
        "MEDIA_URL": settings.MEDIA_URL,
        "STATIC_URL": settings.STATIC_URL,
        "DEBUG": settings.DEBUG,
        "LANGUAGES": settings.LANGUAGES,
        "DEFAULT_LANGUAGE_CODE": settings.LANGUAGE_CODE,
        "CURRENT_LANGUAGE_CODE": request.LANGUAGE_CODE,
    }
    json_data = json.dumps(data)
    template = Template(JS_SETTINGS_TEMPLATE)
    context = Context({"json_data": json_data})
    response = HttpResponse(
        content=template.render(context),
        content_type="application/javascript; charset=UTF-8",
    )
    return response
```

（2）将当前视图插入至 URL 配置下。

```
# myproject/urls.py
from django.conf.urls.i18n import i18n_patterns
from django.urls import include, path
from django.conf import settings
from django.conf.urls.static import static

from myproject.apps.core import views as core_views

urlpatterns = i18n_patterns(
    # other URL configuration rules…
    path("js-settings/", core_views.js_settings, name="js_settings"),
)

urlpatterns += static(settings.STATIC_URL,
document_root=settings.STATIC_ROOT)
urlpatterns += static("/media/", document_root=settings.MEDIA_ROOT)
```

（3）将基于 JavaScript 的视图添加至 base.html 模板的结尾处，进而将该视图加载至前端。

```
{# base.html #}

    {# … #}
```

```
    <script src="{% url 'js_settings' %}"></script>
    {% block js %}{% endblock %}
    {% render_block "js" %}
    {% block extra_body %}{% endblock %}
</body>
</html>
```

（4）当前，我们可访问任何 JavaScript 文件中所指定的设置项。

```
if (window.settings.DEBUG) {
    console.warn('The website is running in DEBUG mode!');
}
```

4.5.3 工作方式

在 js_settings 视图中，我们构建了一个传递至浏览器的设置项字典、将字典转换为 JSON，并渲染 JavaScript 文件的模板（解析 JSON，并将结果赋予至 window.settings 变量中）。通过将字典转换为 JSON 字符串，并将其解析至 JavaScript 文件中，我们可以确定最后一个元素后面的逗号不会有任何问题——这在 Python 中是允许的，但在 JavaScript 中却是无效的。

渲染后的 JavaScript 文件如下所示。

```
# http://127.0.0.1:8000/en/js-settings/
window.settings = JSON.parse('{\u0022MEDIA_URL\u0022:
\u0022http://127.0.0.1:8000/media/\u0022, \u0022STATIC_URL\u0022:
\u0022/static/20191001004640/\u0022, \u0022DEBUG\u0022: true,
\u0022LANGUAGES\u0022: [[\u0022bg\u0022, \u0022Bulgarian\u0022],
[\u0022hr\u0022, \u0022Croatian\u0022], [\u0022cs\u0022,
\u0022Czech\u0022], [\u0022da\u0022, \u0022Danish\u0022], [\u0022nl\u0022,
\u0022Dutch\u0022], [\u0022en\u0022, \u0022English\u0022], [\u0022et\u0022,
\u0022Estonian\u0022], [\u0022fi\u0022, \u0022Finnish\u0022],
[\u0022fr\u0022, \u0022French\u0022], [\u0022de\u0022, \u0022German\u0022],
[\u0022el\u0022, \u0022Greek\u0022], [\u0022hu\u0022,
\u0022Hungarian\u0022], [\u0022ga\u0022, \u0022Irish\u0022],
[\u0022it\u0022, \u0022Italian\u0022], [\u0022lv\u0022,
\u0022Latvian\u0022], [\u0022lt\u0022, \u0022Lithuanian\u0022],
[\u0022mt\u0022, \u0022Maltese\u0022], [\u0022pl\u0022,
\u0022Polish\u0022], [\u0022pt\u0022, \u0022Portuguese\u0022],
[\u0022ro\u0022, \u0022Romanian\u0022], [\u0022sk\u0022,
\u0022Slovak\u0022], [\u0022sl\u0022, \u0022Slovene\u0022],
```

```
[\u0022es\u0022, \u0022Spanish\u0022], [\u0022sv\u0022,
\u0022Swedish\u0022]], \u0022DEFAULT_LANGUAGE_CODE\u0022: \u0022en\u0022,
\u0022CURRENT_LANGUAGE_CODE\u0022: \u0022en\u0022}');
```

4.5.4 延伸阅读

- 第 1 章中的“针对开发、测试、预发布和产品环境，配置设置项”示例。
- “安排 base.html 模板”示例。
- “使用 HTML 5 数据属性”示例。

4.6 使用 HTML 5 数据属性

HTML 5 引入了 data-*属性，并在 Web 服务器和 JavaScript 与 CSS 之间传递与特定 HTML 元素相关的数据。在当前示例中，我们将探讨一种方式将 Django 中的数据高效地与自定义 HTML 5 数据属性进行绑定，并于随后用一个实例描述如何从 JavaScript 中读取数据。我们将利用特定地理位置处的标记渲染 Google Map；当单击该标记时，将在信息窗口中显示对应的地址。

4.6.1 准备工作

相应的准备工作如下列步骤所示。

（1）使用带有 PostGIS 扩展名的 PostgreSQL 数据库。读者可访问官方文档查看 PostGIS 扩展名的安装方式，对应网址为 https://docs.djangoproject.com/en/2.2/ref/contrib/gis/install/postgis/。

（2）确保针对 Django 项目使用 postgis 数据库后端。

```
# myproject/settings/_base.py
DATABASES = {
    "default": {
        "ENGINE": "django.contrib.gis.db.backends.postgis",
        "NAME": get_secret("DATABASE_NAME"),
        "USER": get_secret("DATABASE_USER"),
        "PASSWORD": get_secret("DATABASE_PASSWORD"),
        "HOST": "localhost",
        "PORT": "5432",
    }
}
```

（3）利用 Location 模型创建一个 locations 应用程序，这将包含一个 UUID 主键、名称的字符字段、街道地址、城市、国家、邮政编码、一个与 PostGIS 相关的 Geoposition 字段，以及 Description 文本字段。

```
# myproject/apps/locations/models.py
import uuid
from collections import namedtuple
from django.contrib.gis.db import models
from django.urls import reverse
from django.conf import settings
from django.utils.translation import gettext_lazy as _
from myproject.apps.core.models import (
    CreationModificationDateBase, UrlBase
)

COUNTRY_CHOICES = getattr(settings, "COUNTRY_CHOICES", [])

Geoposition = namedtuple("Geoposition", ["longitude", "latitude"])

class Location(CreationModificationDateBase, UrlBase):
    uuid = models.UUIDField(primary_key=True, default=None,
     editable=False)
    name = models.CharField(_("Name"), max_length=200)
    description = models.TextField(_("Description"))
    street_address = models.CharField(_("Street address"),
     max_length=255, blank=True)
    street_address2 = models.CharField(
        _("Street address (2nd line)"), max_length=255, blank=True
    )
    postal_code = models.CharField(_("Postal code"),
     max_length=255, blank=True)
    city = models.CharField(_("City"), max_length=255, blank=True)
    country = models.CharField(
        _("Country"), choices=COUNTRY_CHOICES, max_length=255, blank=True
    )
    geoposition = models.PointField(blank=True, null=True)

    class Meta:
        verbose_name = _("Location")
        verbose_name_plural = _("Locations")

    def __str__(self):
```

```
        return self.name

    def get_url_path(self):
        return reverse("locations:location_detail", kwargs={"pk":self.pk})
```

（4）覆写 save()方法，并在创建一个位置时生成唯一的 UUID 字段值。

```
def save(self, *args, **kwargs):
    if self.pk is None:
        self.pk = uuid.uuid4()
    super().save(*args, **kwargs)
```

（5）创建方法，以获取一个字符串中位置的完整地址。

```
def get_field_value(self, field_name):
    if isinstance(field_name, str):
        value = getattr(self, field_name)
        if callable(value):
            value = value()
        return value
    elif isinstance(field_name, (list, tuple)):
        field_names = field_name
        values = []
        for field_name in field_names:
            value = self.get_field_value(field_name)
            if value:
                values.append(value)
        return " ".join(values)
    return ""

def get_full_address(self):
    field_names = [
        "name",
        "street_address",
        "street_address",
        ("postal_code", "city"),
        "get_country_display",
    ]
    full_address = []
    for field_name in field_names:
        value = self.get_field_value(field_name)
        if value:
            full_address.append(value)
    return ", ".join(full_address)
```

（6）编写相关函数，并通过 latitude 和 longitude 获取或设置地理位置。在数据库中，geoposition 保存为一个 Point 字段。我们可在 Django shell、表单、管理命令、数据迁移等处使用这些函数。

```
def get_geoposition(self):
    if not self.geoposition:
        return None
    return Geoposition(
        self.geoposition.coords[0], self.geoposition.coords[1]
    )

def set_geoposition(self, longitude, latitude):
    from django.contrib.gis.geos import Point
    self.geoposition = Point(longitude, latitude, srid=4326)
```

（7）记住在更新模型后生成并运行迁移。

（8）创建一个模型管理以添加和修改位置。相应地，我们将采用 gis 应用程序中的 OSMGeoAdmin，而非标准的 ModelAdmin。这将利用 OpenStreetMap 渲染一幅地图并设置 geoposition，读者可访问 https://www.openstreetmap.org 予以查看。

```
# myproject/apps/locations/admin.py
from django.contrib.gis import admin
from .models import Location

@admin.register(Location)
class LocationAdmin(admin.OSMGeoAdmin):
    pass
```

（9）将一些位置添加至管理中以供后续使用。

在后续示例中，我们还将继续使用并进一步丰富 locations 应用程序。

4.6.2　实现方式

具体实现方式如下列步骤所示。

（1）注册 Google Maps API 密钥。对此，读者可查看 Google 开发者文档以了解相关信息，对应网址为 https://developers.google.com/maps/documentation/javascript/get-api-key。

（2）将 Google Maps API 密钥添加至密码中，随后在设置项中进行读取。

```
# myproject/settings/_base.py
# ...
GOOGLE_MAPS_API_KEY = get_secret("GOOGLE_MAPS_API_KEY")
```

（3）在 core 应用程序中，创建一个上下文预处理器，并向模板公开 GOOGLE_MAPS_API_KEY。

```
# myproject/apps/core/context_processors.py
from django.conf import settings

def google_maps(request):
    return {
        "GOOGLE_MAPS_API_KEY": settings.GOOGLE_MAPS_API_KEY,
    }
```

（4）在模板设置项中引用该上下文预处理器。

```
# myproject/settings/_base.py
TEMPLATES = [
    {
        "BACKEND":
        "django.template.backends.django.DjangoTemplates",
        "DIRS": [os.path.join(BASE_DIR, "myproject", "templates")],
        "APP_DIRS": True,
        "OPTIONS": {
            "context_processors": [
                "django.template.context_processors.debug",
                "django.template.context_processors.request",
                "django.contrib.auth.context_processors.auth",
                "django.contrib.messages.context_processors.messages",
                "django.template.context_processors.media",
                "django.template.context_processors.static",
                "sekizai.context_processors.sekizai",
                "myproject.apps.core.context_processors.google_maps",
            ]
        },
    }
]
```

（5）创建位置的列表和详细视图。

```
# myproject/apps/locations/views.py
from django.views.generic import ListView, DetailView
from .models import Location

class LocationList(ListView):
    model = Location
    paginate_by = 10
```

```
class LocationDetail(DetailView):
    model = Location
    context_object_name = "location"
```

（6）创建 locations 应用程序的 URL 配置。

```
# myproject/apps/locations/urls.py
from django.urls import path
from .views import LocationList, LocationDetail

urlpatterns = [
    path("", LocationList.as_view(), name="location_list"),
    path("<uuid:pk>/", LocationDetail.as_view(), name="location_detail"),
]
```

（7）将位置的 URL 包含至项目的 URL 配置中。

```
# myproject/urls.py
from django.contrib import admin
from django.conf.urls.i18n import i18n_patterns
from django.urls import include, path
from django.conf import settings
from django.conf.urls.static import static
from django.shortcuts import redirect

from myproject.apps.core import views as core_views

urlpatterns = i18n_patterns(
    path("", lambda request: redirect("locations:location_list")),
    path("admin/", admin.site.urls),
    path("accounts/", include("django.contrib.auth.urls")),
    path("locations/", include(("myproject.apps.locations.urls",
    "locations"), namespace="locations")),
    path("js-settings/", core_views.js_settings, name="js_settings"),
)
urlpatterns += static(settings.STATIC_URL,
document_root=settings.STATIC_ROOT)
urlpatterns += static("/media/", document_root=settings.MEDIA_ROOT)
```

（8）创建位置列表和位置详细视图的模板。目前，位置列表较为简单，仅需浏览相关位置并访问位置详细视图即可。

```
{# locations/location_list.html #}
{% extends "base.html" %}
{% load i18n %}
```

```
{% block content %}
    <h1>{% trans "Interesting Locations" %}</h1>
    {% if object_list %}
        <ul>
            {% for location in object_list %}
                <li><a href="{{ location.get_url_path }}">
                    {{ location.name }}
                </a></li>
            {% endfor %}
        </ul>
    {% else %}
        <p>{% trans "There are no locations yet." %}</p>
    {% endif %}
{% endblock %}
```

（9）通过扩展 base.html 并覆写 content 块，创建位置详细信息模板。

```
{# locations/location_detail.html #}
{% extends "base.html" %}
{% load i18n static %}

{% block content %}
    <a href="{% url "locations:location_list" %}">{% trans
     "Interesting Locations" %}</a>
    <h1 class="map-title">{{ location.name }}</h1>
    <div class="my-3">
        {{ location.description|linebreaks|urlize }}
    </div>
    {% with geoposition=location.get_geoposition %}
<div id="map" class="mb-3"
data-latitude="{{ geoposition.latitude|stringformat:"f" }}"
data-longitude="{{ geoposition.longitude|stringformat:"f" }}"
data-address="{{ location.get_full_address }}"></div>
{% endwith %}
{% endblock %}
```

（10）在同一个模板中，覆写 js 块。

```
{% block js %}
  <script src="{% static 'site/js/location_detail.js' %}"></script>
  <script async defer src="https://maps-api-ssl.
    google.com/maps/api/js?key={{ GOOGLE_MAPS_API_KEY
}}&callback=Location.init"></script>
{% endblock %}
```

（11）除模板外，我们还需要 JavaScript 文件读取 HTML 5 数据属性，并以此通过地图上的标记渲染一幅地图。

```
/* site_static/site/js/location_detail.js */
(function(window) {
    "use strict";

    function Location() {
        this.case = document.getElementById("map");
        if (this.case) {
            this.getCoordinates();
            this.getAddress();
            this.getMap();
            this.getMarker();
            this.getInfoWindow();
        }
    }

    Location.prototype.getCoordinates = function() {
        this.coords = {
            lat: parseFloat(this.case.getAttribute("data-latitude")),
            lng: parseFloat(this.case.getAttribute("data-longitude"))
        };
    };

    Location.prototype.getAddress = function() {
        this.address = this.case.getAttribute("data-address");
    };

    Location.prototype.getMap = function() {
        this.map = new google.maps.Map(this.case, {
            zoom: 15,
            center: this.coords
        });
    };

    Location.prototype.getMarker = function() {
        this.marker = new google.maps.Marker({
            position: this.coords,
            map: this.map
        });
    };
```

```
    Location.prototype.getInfoWindow = function() {
        var self = this;
        var wrap = this.case.parentNode;
        var title = wrap.querySelector(".map-title").textContent;

        this.infoWindow = new google.maps.InfoWindow({
            content: "<h3>"+title+"</h3><p>"+this.address+"</p>"
        });

        this.marker.addListener("click", function() {
            self.infoWindow.open(self.map, self.marker);
        });
    };

    var instance;
    Location.init = function() {
        // called by Google Maps service automatically once loaded
        // but is designed so that Location is a singleton
        if (!instance) {
            instance = new Location();
        }
    };

    // expose in the global namespace
    window.Location = Location;
}(window));
```

（12）为了更好地显示地图，还需要设置一些 CSS，如下所示。

```
/* site_static/site/css/style.css */
#map {
    box-sizing: padding-box;
    height: 0;
    padding-bottom: calc(9 / 16 * 100%); /* 16:9 aspect ratio */
    width: 100%;
}
@media screen and (max-width: 480px) {
    #map {
        display: none; /* hide on mobile devices (esp. portrait) */
    }
}
```

4.6.3　工作方式

当运行本地开发服务器并浏览位置的详细视图时，我们将导航至包含一幅地图和一个标记的页面。当单击该标记时，将会弹出包含地址信息的提示内容，如图 4.1 所示。

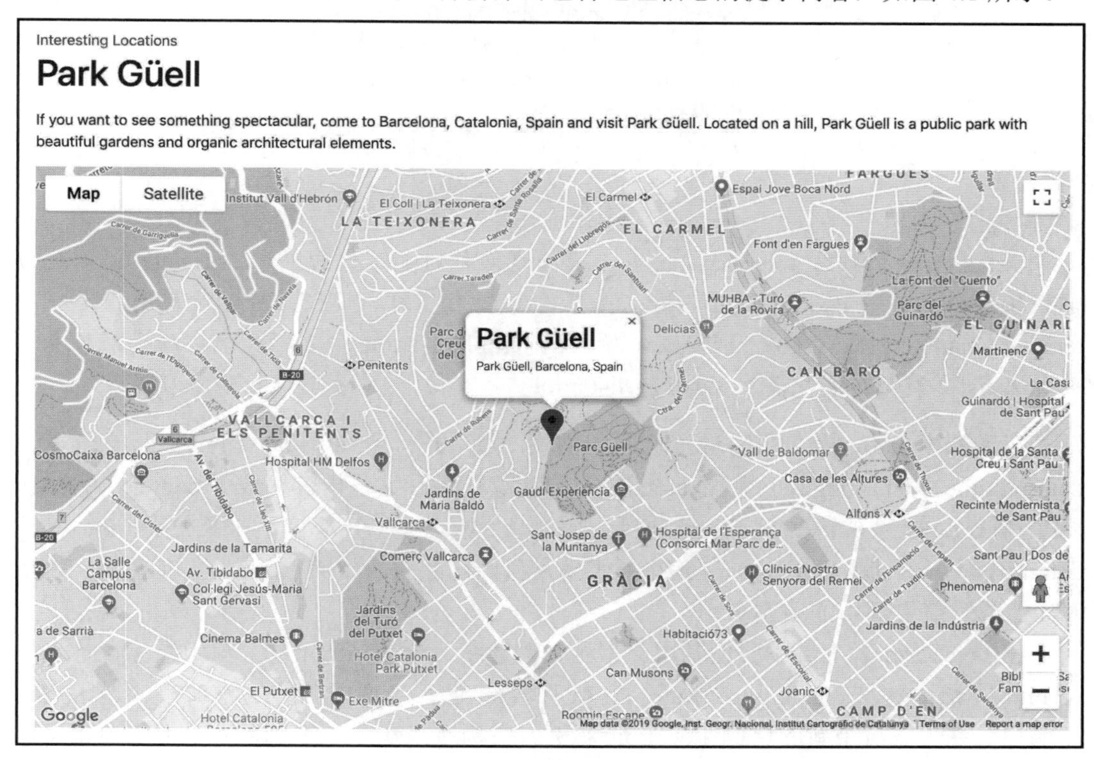

图 4.1

由于在移动设备上滚动地图可能会因为“滚动中滚动”（scroll-within-scroll）而出现问题，我们选择在小屏幕上隐藏地图（小于或等于 480 像素宽度），这样当我们调整屏幕大小时，地图最终会变得不可见，如图 4.2 所示。

接下来探讨代码内容。在前几个步骤中，我们添加了 Google Maps API 密钥，并将其向所有的模板公开。随后，我们创建了视图来浏览位置，并将其插入至 URL 配置中。接下来，我们创建了列表和详细模板。

Interesting Locations

Park Güell

If you want to see something spectacular, come to Barcelona, Catalonia, Spain and visit Park Güell. Located on a hill, Park Güell is a public park with beautiful gardens and organic architectural elements.

图 4.2

注意：

DetailView 的默认 template_name 源自模型名称的小写版本，并加上 detail，因此当前模板被命名为 location_detail.html。如果需要使用不同的模板，可针对视图指定一个 template_name 属性。同样，ListView 的默认 template_name 源自模型名称的小写版本，并加上 list，因而其被命名为 location_list.html。

在详细模板中，我们持有位置标题和描述，随后是一个 id="map"的<div>元素，以及 data-latitude、data-longitude 和 data-address 自定义属性。这些内容构成了 content 块元素。两个<script>标签被添加至<body>结尾处的 js 块——一个是接下来描述的 location_detail.js，另一个是 Google Maps API 的脚本，我们已经向其传递了 Maps API 密钥，以及 API 加载时需要调用的回调名称。

在 JavaScript 文件中，我们利用原型函数创建了一个 Location 类。该函数包含一个静态 init()方法，这是作为 Google Maps API 回调给出的。当调用 init()方法时，构造函数将被调用，并生成一个新的单例 Location 实例。在构造函数中，将执行一系列的步骤设置地图及其特性。

（1）地图案例（容器）是通过它的 ID 找到的，只有找到该元素才继续。

（2）利用 data-latitude 和 data-longitude 属性查找地理坐标，并将坐标作为位置的 coords 存储在字典中。该对象为 Google Maps API 所理解和后续使用的表单。

（3）读取 data-address 并作为位置的地址属性直接被存储。

（4）自地图起开始构建，为了确保位置可见，我们通过数据属性中的 coord 设置中心。

（5）标记使得位置在地图上可见，并通过相同的 coords 进行定位。

（6）构建信息窗口，该窗口是弹出式的，并可利用 API 直接在地图上显示。除了之前检索的地址之外，我们还将根据模板中给出的.map-title 类查找位置标题。这将作为<h1>

标题添加至窗口中，随后是作为<p>段落的地址。为了使窗口能够显示，我们向打开窗口的标记中添加一个单击事件监听器。

4.6.4　延伸阅读

- “公开 JavaScript 中的设置项”示例。
- “安排 base.html 模板”示例。
- “提供响应式图像”示例。
- “在模式对话框中打开对象的细节信息”示例。
- 第 6 章中的“将一幅地图插入至变化表单中”示例。

4.7　提供响应式图像

随着响应式站点越来越普及，当同时向移动设备和桌面计算机提供相同的内容时，会出现许多性能问题。在小型设备上，一种减少响应式站点加载时间的最简单的方法是提供较小的图像。对此，可采用响应式图像的关键组件 srcset 和 sizes。

4.7.1　准备工作

当前示例将在之前的 locations 示例的基础上完成。

4.7.2　实现方式

添加响应式图像的具体步骤如下所示。

（1）在虚拟环境中安装 django-imagekit，并将其添加至 requirements/_base.txt 中，进而以此将原始图像的尺寸重置为特定尺寸。

```
(env)$ pip install django-imagekit==4.0.2
```

（2）在设置项中，将"imagekit"置于 INSTALLED_APPS 中。

```
# myproject/settings/_base.py
INSTALLED_APPS = [
    # …
    "imagekit",
```

```
    # ...
]
```

（3）在 models.py 文件开始处，导入某些库供图像版本使用，并定义一个函数处理目录和图像文件的文件名。

```
# myproject/apps/locations/models.py
import contextlib
import os
# ...
from imagekit.models import ImageSpecField
from pilkit.processors import ResizeToFill
# ...

def upload_to(instance, filename):
    now = timezone_now()
    base, extension = os.path.splitext(filename)
    extension = extension.lower()
    return f"locations/{now:%Y/%m}/{instance.pk}{extension}"
```

（4）连同图像版本定义，在同一文件中将 picture 字段添加至 Location 模型中。

```
class Location(CreationModificationDateBase, UrlBase):
    # ...
    picture = models.ImageField(_("Picture"), upload_to=upload_to)
    picture_desktop = ImageSpecField(
        source="picture",
        processors=[ResizeToFill(1200, 600)],
        format="JPEG",
        options={"quality": 100},
    )
    picture_tablet = ImageSpecField(
        source="picture", processors=[ResizeToFill(768, 384)],
         format="PNG"
    )
    picture_mobile = ImageSpecField(
        source="picture", processors=[ResizeToFill(640, 320)],
         format="PNG"
    )
```

（5）覆写 Location 模型的 delete()方法，并在删除模型实例时删除生成的版本。

```
def delete(self, *args, **kwargs):
    from django.core.files.storage import default_storage
```

```
    if self.picture:
        with contextlib.suppress(FileNotFoundError):
            default_storage.delete(self.picture_desktop.path)
            default_storage.delete(self.picture_tablet.path)
            default_storage.delete(self.picture_mobile.path)
        self.picture.delete()

    super().delete(*args, **kwargs)
```

（6）生成并运行迁移，将新的 picture 字段添加至数据库模式中。

（7）更新位置细节模板并包含图像。

```
{# locations/location_detail.html #}
{% extends "base.html" %}
{% load i18n static %}

{% block content %}
    <a href="{% url "locations:location_list" %}">{% trans
    "Interesting Locations" %}</a>
    <h1 class="map-title">{{ location.name }}</h1>
    {% if location.picture %}
        <picture class="img-fluid">
            <source
                media="(max-width: 480px)"
                srcset="{{ location.picture_mobile.url }}" />
            <source
                media="(max-width: 768px)"
                srcset="{{ location.picture_tablet.url }}" />
            <img
                src="{{ location.picture_desktop.url }}"
                alt="{{ location.name }}"
                class="img-fluid"
            />
        </picture>
    {% endif %}
    {# … #}
{% endblock %}

{% block js %}
    {# … #}
{% endblock %}
```

（8）针对管理中的位置添加一些图像。

4.7.3 工作方式

响应式图像非常强大，其基本功能是根据媒体规则提供不同的图像。这里，媒体规则指出每幅图像所依据的显示特征。此处，首先需要添加 django-imagekit 应用程序，进而生成所需的不同图像。

显然，我们还需要原始的图像源。因此，在 Location 模型中，我们添加了一个名为 picture 的图像字段。在 upload_to()函数中，我们构建了上传路径和基于年份和月份的文件名、位置的 UUID，以及与上传文件相同的文件扩展名。此外还定义了图像版本规范，如下所示。

- ❑ picture_desktop 的尺寸为 1200×600，适用于桌面布局。
- ❑ picture_tablet 的尺寸为 768×324，适用于平板电脑。
- ❑ picture_mobile 的尺寸为 640×320，适用于智能手机。

在位置的 delete()方法中，我们检查 picture 字段是否有任何值，并在删除位置自身之前尝试删除该值和图像版本。如果文件未在磁盘上找到，此处将使用 contextlib.Suppress (FileNotFoundError)并暂时忽略任何错误。

若位置图像存在，随后将构建<picture>元素。从表面上看，这基本上是一个容器。实际上，除了出现在模板末尾的默认<img>标签，其内部可能不存在其他内容，因而没有太多用处。除了默认的图像，我们还生成了其他宽度的缩略图，如 480px 和 768px，随后用于构建附加的<source>元素。每个<source>元素都包含相应的媒体规则，其中包含了从 srcset 属性值中选择图像的条件。当前示例仅对每个<source>提供了一幅图像。当前，位置详细页面将包含地图上的图像，如图 4.3 所示。

当浏览器加载标记时，将通过一系列的步骤确定加载哪一幅图像。

- ❑ 每个<source>的媒体规则依次被检查，并查看任何一条规则是否与当前视口匹配。
- ❑ 当某条规则匹配，srcset 将被读取，并加载和显示相应的图像 URL。
- ❑ 如果规则不匹配，则加载最终的默认图像的 src。

最终，较小的图像将被加载至较小的视口中。例如，图 4.4 显示了加载的最小视口图像仅有 375 像素宽。

对于无法理解<picture>和<source>标签的浏览器，仍可加载默认的图像，因为这只是一个普通的<img>标签。

Interesting Locations
Park Güell
If you want to see something spectacular, come to Barcelona, Catalonia, Spain and visit Park Güell. Located on a hill, Park Güell is a public park with beautiful gardens and organic architectural elements.
Map
Satellite
Institut Vall d'Hebrón
El Coll | La Teixonera
LA TEIXONERA
El Carmel
EL CARMEL
Espai Jove Boca Nord
Font d'en Fargues
La Font del "Cuento"
MUHBA - Turó de la Rovira
Parc del Guinardó
EL GUINARD
Delicias
Parc de La Creueta

图 4.3

Interesting Locations
Park Güell
If you want to see something spectacular, come to Barcelona, Catalonia, Spain and visit Park Güell. Located on a hill, Park Güell is a public park with beautiful gardens and organic architectural elements.

图 4.4

4.7.4　更多内容

响应式图像不仅可提供目标图像尺寸，还可区分像素密度，并在任何给定的视口尺寸下提供显式制定的设计图像。这被称为艺术指导。感兴趣的读者还可查看 Mozilla Developer Network（MDN）提供的一篇关于这一话题的完整文章，对应网址为 https://developer.mozilla.org/en-US/docs/Learn/HTML/Multimedia_and_embedding/Responsive_images。

4.7.5　延伸阅读

- “安排 base.html 模板”示例。
- “使用 HTML 5 数据属性”示例。
- “在模式对话框中打开对象的细节信息”示例。
- 第 6 章中的“将一幅地图插入至修改表单中”示例。

4.8　实现连续的滚动

社交网站通常会包含连续滚动这一特性，也称作无限滚动，并可视为分页机制的一种替代方案。其中包含了较长的条目集，而非单独添加链接以查看额外的条目集。当滚动页面时，新条目将被加载并自动绑定至底部。在当前示例中，我们将探讨如何利用 Django 和 jScroll jQuery 插件实现此类效果。

注意：

读者可下载 jScroll 脚本，还可以访问 https://jscroll.com/查找关于该插件的大量文档。

4.8.1　准备工作

当前示例将在之前的 locations 应用程序的基础上完成。

为了在列表视图中显示更多的数据，可向 Location 模型中添加 ratings 字段，如下所示。

```
# myproject/apps/locations/models.py
# …
RATING_CHOICES = ((1, "★☆☆☆☆"), (2, "★★☆☆☆"), (3, "★★★☆☆"),
(4,"★★★★☆"), (5, "★★★★★"))

class Location(CreationModificationDateBase, UrlBase):
```

```
    # ...
    rating = models.PositiveIntegerField(
        _("Rating"), choices=RATING_CHOICES, blank=True, null=True
    )

    # ...
    def get_rating_percentage(self):
        return self.rating * 20 if self.rating is not None else None
```

get_rating_percentage()方法将以百分比的形式返回评级。

不要忘记生成和运行迁移，并于随后针对管理中的位置添加评级。

4.8.2　实现方式

创建连续的滚动页面如下列步骤所示。

（1）在管理中添加足够的位置。在“使用 HTML 5 数据属性”示例中可以看到，我们将以每个页面 10 个条目的方式对 LocationList 视图进行分页，因而至少需要 11 个位置查看连续滚动是否按照期望方式工作。

（2）调整位置列表视图模板，如下所示。

```
{# locations/location_list.html #}
{% extends "base.html" %}
{% load i18n static utility_tags %}

{% block content %}
    <div class="row">
        <div class="col-lg-8">
            <h1>{% trans "Interesting Locations" %}</h1>
            {% if object_list %}
                <div class="item-list">
                    {% for location in object_list %}
                        <a href="{{ location.get_url_path }}"
                          class="item d-block my-3">
                            <div class="card">
                                <div class="card-body">
                                    <div class="float-right">
                                        <div class="rating" aria-label="{%
                                          blocktrans with stars=location.rating %}
                                          {{ stars }} of 5 stars
                                           {% endblocktrans %}">
                                            <span style="width:{{
```

```
                                    location.get_rating
                                _percentage }}%"></span>
                                </div>
                            </div>
                            <p class="card-text">{{
                                location.name }}<br/>
                                <small>{{ location.city }},
                                 {{location.get_country
                                   _display }}</small>
                            </p>
                        </div>
                    </div>
                </a>
            {% endfor %}
            {% if page_obj.has_next %}
                <div class="text-center">
                    <div class="loading-indicator"></div>
                </div>
                <p class="pagination">
                    <a class="next-page"
                        href="{% modify_query
                        page=page_obj.next_page_number %}">
                          {% trans "More..." %}</a>
                </p>
            {% endif %}
        </div>
    {% else %}
        <p>{% trans "There are no locations yet." %}</p>
    {% endif %}
    </div>
    <div class="col-lg-4">
        {% include "locations/includes/navigation.html" %}
    </div>
  </div>
{% endblock %}
```

（3）在同一模板中，利用下列标记覆写 css 和 js 块。

```
{% block css %}
  <link rel="stylesheet" type="text/css"
        href="{% static 'site/css/rating.css' %}">
{% endblock %}
```

```
{% block js %}
    <script src="https://cdnjs.cloudflare.com/ajax
     /libs/jscroll/2.3.9/jquery.jscroll.min.js"></script>
    <script src="{% static 'site/js/list.js' %}"></script>
{% endblock %}
```

（4）利用加载指示器的 JavaScript 模板覆写 extra_body 块。

```
{% block extra_body %}
    <script type="text/template" class="loader">
        <div class="text-center">
            <div class="loading-indicator"></div>
        </div>
    </script>
{% endblock %}
```

（5）在 locations/includes/navigation.html 创建页面的导航。当前，仅需创建一个空文件即可。

（6）通过初始化连续滚动微件添加 JavaScript。

```
/* site_static/site/js/list.js */
jQuery(function ($) {
    var $list = $('.item-list');
    var $loader = $('script[type="text/template"].loader');
    $list.jscroll({
        loadingHtml: $loader.html(),
        padding: 100,
        pagingSelector: '.pagination',
        nextSelector: 'a.next-page:last',
        contentSelector: '.item,.pagination'
    });
});
```

（7）添加 CSS，以便评级通过用户友好的星形（而非数字）显示。

```
/* site_static/site/css/rating.css */
.rating {
    color: #c90;
    display: block;
    position: relative;
    margin: 0;
    padding: 0;
    white-space: nowrap;
}
```

```
.rating span {
    color: #fc0;
    display: block;
    position: absolute;
    overflow: hidden;
    top: 0;
    left: 0;
    bottom: 0;
    white-space: nowrap;
}

.rating span:before,
.rating span:after {
    display: block;
    position: absolute;
    overflow: hidden;
    left: 0;
    top: 0;
    bottom: 0;
}

.rating:before {
    content: "☆☆☆☆☆";
}

.rating span:after {
    content: "★★★★★";
}
```

（8）在主站点样式的主文件中，添加一个加载指示器的样式。

```
/* site_static/site/css/style.css */
/* … */
.loading-indicator {
    display: inline-block;
    width: 45px;
    height: 45px;
}
.loading-indicator:after {
    content: "";
    display: block;
    width: 40px;
```

```
    height: 40px;
    border-radius: 50%;
    border: 5px solid rgba(0,0,0,.25);
    border-color: rgba(0,0,0,.25) transparent rgba(0,0,0,.25)
     transparent;
    animation: dual-ring 1.2s linear infinite;
}
@keyframes dual-ring {
    0% {
        transform: rotate(0deg);
    }
    100% {
        transform: rotate(360deg);
    }
}
```

4.8.3　工作方式

当在浏览器中打开位置列表视图时，页面上将显示在视图中设置为 paginate_by 的预定义条目数（即 10）。在向下滚动时，附加页面的条目和下一个分页链接将被自动加载，并附加至条目容器中。其中，分页链接使用了第 5 章中的{% modify_query %}自定义模板标签，并根据当前标签生成了调整的 URL，但会指向下一个正确的页码。如果连接速度较慢，那么当滚动至页面底部时，将会看到如图 4.5 所示的现象，直至下一个页面的条目被加载并绑定至当前列表中。

当进一步向下滚动时，将加载第 2 个、第 3 个……直至最后一个页面并绑定至底部。这一过程持续进行直至不存在加载的页面。这表示在最后一个分组中不存在任何进一步加载的分页链接。

我们采用 Cloudflare CDN URL 加载 jScroll 插件。如果选择在本地下载一个副本作为静态文件，那么可使用{% static %}查找并将脚本添加到模板中。

当加载初始页面时，基于 item-list CSS 类的元素（包含条目和分页链接）将通过 list.js 中的代码成为 jScroll 对象。实际上，这一实现是通用的，并以此支持任何遵循类似标记结构的列表显示的连续滚动行为。下列选项可用于定义其特性。

- loadingHtml：这将设置 jScroll 在加载条目新页面时将在列表结尾注入的标记。在当前示例中，这是一个动画加载指示器，并直接在标记中的包含<script type="text/template" />标签的 HTML 中进行绘制。通过给定 type 属性，浏览器不会像普通 JavaScript 那样尝试执行它，其中的内容对于用户而言仍是不可见的。

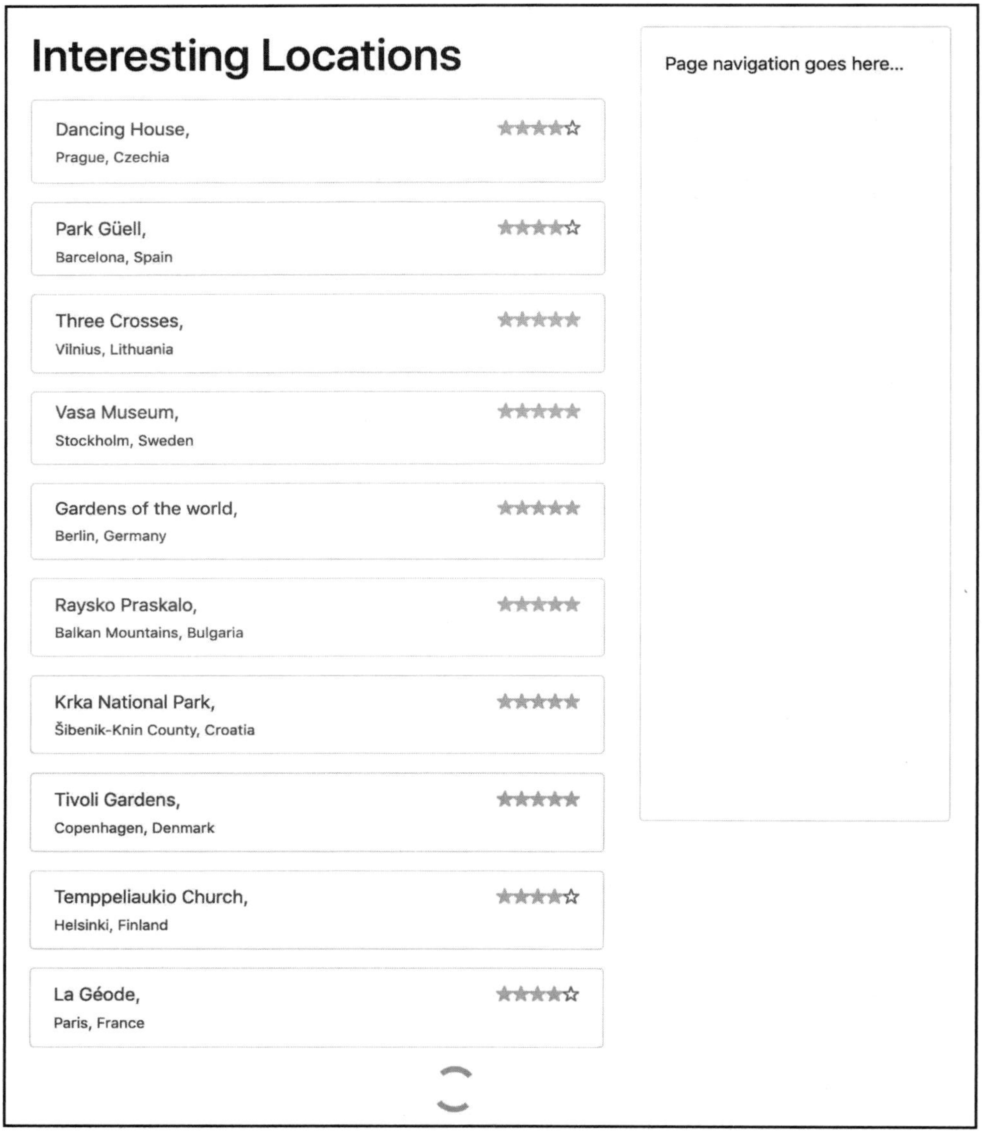

图 4.5

- padding：当页面的滚动位置位于滚动区域结束距离的范围内时，此时应加载一个新的页面。这里，我们将 padding 设置为 100 像素。
- pagingSelector：表示为一个 CSS 选择器，并指示 object_list 中的哪些 HTML 元

素是分页链接。这些内容将被隐藏在 jScroll 插件激活的浏览器中，以便连续的滚动可接管附加页面的加载，但其他浏览器中的用户仍可通过正常单击分页进行导航。

- nextSelector：这一 CSS 选择器负责查找 HTML 元素，并从中读取下一个页面的 URL。
- contentSelector：另一个 CSS 选择器，用于指定哪一个 HTML 元素应从 Ajax 加载的内容中析取并添加至容器中。
- rating.css 将插入 Unicode 星号字符，并将轮廓与填充版本重叠进而创建评级效果。采用与评级最大值（在当前示例中为 5）的百分比相等的宽度，填充后的星形将覆盖空心星形顶部上的适当空间，进而完成十进制的评级操作。在标记中，对于使用屏幕阅读器的读者，存在一个包含评级信息的 aria-label 属性。

最后，style.css 文件中的 CSS 使用 CSS 动画创建旋转的加载指示器。

4.8.4　更多内容

侧栏中包含一个用于导航的占位符。注意，当采用连续滚动机制时，条目列表之后的所有二级导航均应置于侧边栏，而不是页脚，因为访问者永远不会到达页面的结尾。

4.8.5　延伸阅读

- 第 3 章中的“过滤对象列表”示例。
- 第 3 章中的“管理分页列表”示例。
- 第 3 章中的“合成基于类的视图”示例。
- “公开 JavaScript 中的设置项”示例。
- 第 5 章中的“创建模板标签以调整请求查询参数”示例。

4.9　在模式对话框中打开对象的细节信息

在当前示例中，我们将创建一个指向位置的链接。当单击该链接时，将会打开一个 Bootstrap 模式对话框，其中包含了关于位置和 Learn more...链接相关的信息，并导航至位置详细页面，如图 4.6 所示。

其中，对话框的内容由 Ajax 加载。对于缺少 JavaScript 的访问者，细节页面将立即打开，没有中间步骤。

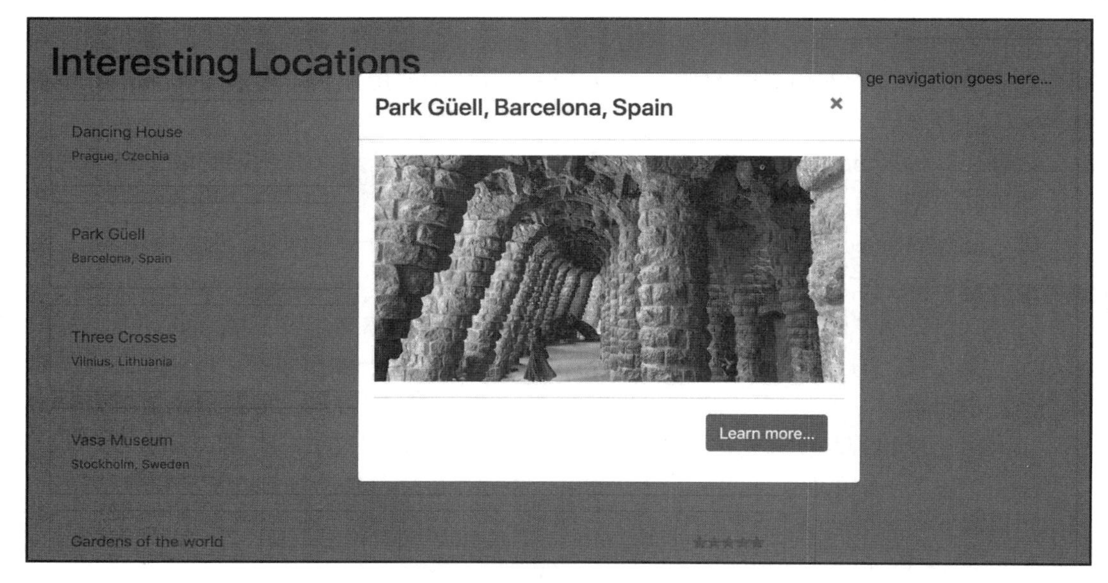

图 4.6

4.9.1 准备工作

当前示例将在前述 locations 应用程序的基础上完成。

与之前一样，确保针对位置列表和位置详细信息包含相应的视图、URL 配置和模板。

4.9.2 实现方式

逐一执行下列步骤，并作为列表视图和详细视图之间的中间步骤添加一个模式对话框。

（1）在 locations 应用程序的 URL 配置中，添加模式对话框的响应的规则。

```
# myproject/apps/locations/urls.py
from django.urls import path
from .views import LocationList, LocationDetail

urlpatterns = [
    path("", LocationList.as_view(), name="location_list"),
    path("add/", add_or_change_location, name="add_location"),
    path("<uuid:pk>/", LocationDetail.as_view(), name="location_detail"),
    path(
        "<uuid:pk>/modal/",
        LocationDetail.as_view(template_name=
```

```
        "locations/location_detail_modal.html"),
        name="location_detail_modal",
    ),
]
```

（2）创建一个模式对话框模板。

```
{# locations/location_detail_modal.html #}
{% load i18n %}
<p class="text-center">
    {% if location.picture %}
        <picture class="img-fluid">
            <source media="(max-width: 480px)"
                srcset="{{ location.picture_mobile.url }}"/>
            <source media="(max-width: 768px)"
                srcset="{{ location.picture_tablet.url }}"/>
            <img src="{{ location.picture_desktop.url }}"
                alt="{{ location.name }}"
                class="img-fluid"
            />
        </picture>
    {% endif %}
</p>
<div class="modal-footer text-right">
    <a href="{% url "locations:location_detail" pk=location.pk %}"
     class="btn btn-primary pull-right">
        {% trans "Learn more…" %}
    </a>
</div>
```

（3）在位置列表的模板中，通过添加自定义数据属性更新指向位置细节信息的链接。

```
{# locations/location_list.html #}
{# … #}
<a href="{{ location.get_url_path }}"
    data-modal-title="{{ location.get_full_address }}"
    data-modal-url="{% url 'locations:location_detail_modal'
     pk=location.pk %}"
    class="item d-block my-3">
     {# … #}
</a>
{# … #}
```

（4）在同一文件中，利用模式对话框的标记覆写 extra_body 内容。

```
{% block extra_body %}
    {# … #}
    <div id="modal" class="modal fade" tabindex="-1" role="dialog"
         aria-hidden="true" aria-labelledby="modal_title">
        <div class="modal-dialog modal-dialog-centered" role="document">
            <div class="modal-content">
                <div class="modal-header">
                    <h4 id="modal_title"
                        class="modal-title"></h4>
                    <button type="button" class="close"
                            data-dismiss="modal"
                            aria-label="{% trans 'Close' %}">
                        <span aria-hidden="true">&times;</span>
                    </button>
                </div>
                <div class="modal-body"></div>
            </div>
        </div>
    </div>
{% endblock %}
```

（5）添加一个脚本以处理模式对话框的打开和关闭，进而调整 list.js 文件。

```
/* site_static/js/list.js */
/* … */
jQuery(function ($) {
    var $list = $('.item-list');
    var $modal = $('#modal');
    $modal.on('click', '.close', function (event) {
        $modal.modal('hide');
        // do something when dialog is closed…
    });
    $list.on('click', 'a.item', function (event) {
        var $link = $(this);
        var url = $link.data('modal-url');
        var title = $link.data('modal-title');
        if (url && title) {
            event.preventDefault();
            $('.modal-title', $modal).text(title);
            $('.modal-body', $modal).load(url, function () {
                $modal.on('shown.bs.modal', function () {
                    // do something when dialog is shown…
                }).modal('show');
```

```
            });
        }
    });
});
```

4.9.3 工作方式

如果在浏览器中访问位置的列表视图，并单击其中一个位置，我们将会看到如图 4.7 所示的模式对话框。

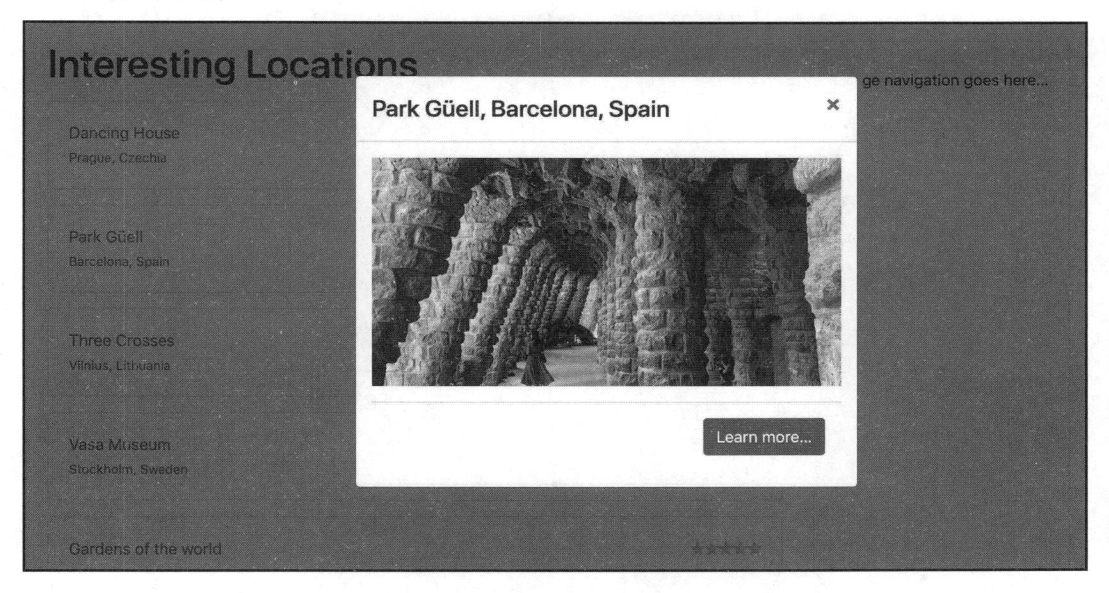

图 4.7

这里，命名为 location_detail_modal 的 URL 路径指向同一个位置详细视图，但使用了不同的模板。之前提到的模板仅包含一幅响应图像和一个包含链接 Learn more...（指向位置的正常详细页面）的模式对话框。在列表视图中，我们修改了列表条目的链接，并包含了 data-modal-title 和 data-modal-url 属性，以供后续 JavaScript 引用。第一个属性规定了完整的地址应用作标题。第二个属性规定了模式对话框主体的 HTML 应该从哪个位置获取。在列表视图结尾处，我们持有 Bootstrap 4 模式对话框的标记。该对话框包含了一个标头，其中涵盖 Close 按钮、标题和针对主要细节信息的内容区域。JavaScript 应已通过 js 块被添加。

在 JavaScript 文件中，我们通过 jQuery 框架使用较短的语法和统一的跨浏览器功能。

当加载页面时，我们针对.item-list 元素赋予了一个事件句柄 on('click')。当单击任何 a.item 时，事件将被托管至该句柄，进而将自定义数据属性分别读取和存储为 url 和 title。当这些内容被成功解析时，我们即可防止最初的单击行为（导航至完整的详细页面），并于随后设置显示的模式。

我们设置了隐藏对话框的新标题，并通过 Ajax 将模式对话框的内容加载至.modal-body 元素中。最后，模式通过 Bootstrap 4 modal() jQuery 插件向用户予以显示。

如果 JavaScript 文件无法处理源自自定义属性中的模式对话框的 URL，或者更为糟糕的是，如果 list.js 中的 JavaScript 无法完全加载或执行，那么单击位置链接将会把用户引至详细页面。对此，我们已实现了渐进式增强模式，以便用户在面临失败时也能够体验正确的模式。

4.9.4　延伸阅读

- “使用 HTML 5 数据属性”示例。
- “提供响应式图像”示例。
- “实现连续的滚动”示例。
- “实现 Like 微件”示例。

4.10　实现 Like 微件

一般的站点，最常见的是涵盖社交组件的站点，通常会集成 Facebook、Twitter 和 Google+微件来点赞和分享内容。在当前示例中，我们将指导读者构建类似的 Django 功能，并在用户点赞时在数据库中保存信息。我们应能够根据点赞内容创建特定的视图。对此，当前示例将创建一个 Like 微件，该微件包含两个状态按钮，以及一个显示点赞数量的标记。

图 4.8 显示了处于非活动状态的微件，我们单击该微件并予以激活。

图 4.9 则显示了活动状态下的微件，我们可以单击按钮以使其处于非活动状态。

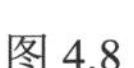
图 4.8

图 4.9

相应地，调整文件的状态由 Ajax 调用负责处理。

4.10.1　准备工作

首先创建一个 likes 应用程序，并将其添加至 INSTALLED_APPS 中。随后设置 Like 模型，它与点赞的用户存在外键关系，并与数据库中的任何对象存在通用关系。这里，我们将使用第 2 章中定义的 object_relation_base_factory。如果不打算使用混入，还可在下列模型中定义一种通用关系。

```
# myproject/apps/likes/models.py
from django.db import models
from django.utils.translation import ugettext_lazy as _
from django.conf import settings

from myproject.apps.core.models import (
    CreationModificationDateBase,
    object_relation_base_factory,
)

LikeableObject = object_relation_base_factory(is_required=True)

class Like(CreationModificationDateBase, LikeableObject):
    class Meta:
        verbose_name = _("Like")
        verbose_name_plural = _("Likes")
        ordering = ("-created",)

    user = models.ForeignKey(settings.AUTH_USER_MODEL,
     on_delete=models.CASCADE)

    def __str__(self):
        return _("{user} likes {obj}").format(user=self.user,
         obj=self.content_object)
```

此外还应确保 request 上下文预处理器在设置项中被设置。另外，对于当前登录的用户（绑定至请求中），还需要在设置项中设置身份验证中间件。

```
# myproject/settings/_base.py
# …
MIDDLEWARE = [
    # …
    "django.contrib.auth.middleware.AuthenticationMiddleware",
    # …
]
```

```
TEMPLATES = [
    {
        # …
        "OPTIONS": {
            "context_processors": [
                "django.template.context_processors.request",
                # …
            ]
        },
    }
]
```

不要忘记创建和运行迁移，并对新的 Like 模型设置数据库。

4.10.2　实现方式

具体实现方式如下列步骤所示。

（1）在 likes 应用程序中，创建包含空__init__.py 文件的 templatetags 目录，并使其生成一个 Python 模块。随后添加 likes_tags.py 文件。其中，我们定义了{% like_widget %}模板标签，如下所示。

```
# myproject/apps/likes/templatetags/likes_tags.py
from django import template
from django.contrib.contenttypes.models import ContentType
from django.template.loader import render_to_string

from ..models import Like

register = template.Library()

# TAGS

class ObjectLikeWidget(template.Node):
    def __init__(self, var):
        self.var = var

    def render(self, context):
        liked_object = self.var.resolve(context)
        ct = ContentType.objects.get_for_model(liked_object)
        user = context["request"].user

        if not user.is_authenticated:
            return ""
```

```
        context.push(object=liked_object, content_type_id=ct.pk)
        output = render_to_string("likes/includes/widget.html",
         context.flatten())
        context.pop()
        return output

@register.tag
def like_widget(parser, token):
    try:
        tag_name, for_str, var_name = token.split_contents()
    except ValueError:
        tag_name = "%r" % token.contents.split()[0]
        raise template.TemplateSyntaxError(
            f"{tag_name} tag requires a following syntax: "
            f"{{% {tag_name} for <object> %}}"
        )
    var = template.Variable(var_name)
    return ObjectLikeWidget(var)
```

（2）在同一文件中添加过滤器，以获取某个用户的 Like 状态，以及特定对象的全部 Like 数量。

```
# myproject/apps/likes/templatetags/likes_tags.py
# …
# FILTERS

@register.filter
def liked_by(obj, user):
    ct = ContentType.objects.get_for_model(obj)
    liked = Like.objects.filter(user=user, content_type=ct,
object_id=obj.pk)
    return liked.count() > 0

@register.filter
def liked_count(obj):
    ct = ContentType.objects.get_for_model(obj)
    likes = Like.objects.filter(content_type=ct, object_id=obj.pk)
    return likes.count()
```

（3）在 URL 规则中，我们需要一个规则，以处理使用 Ajax 时点赞和未点赞的视图。

```
# myproject/apps/likes/urls.py
from django.urls import path
from .views import json_set_like
```

```
urlpatterns = [
    path("<int:content_type_id>/<str:object_id>/",
          json_set_like,
          name="json_set_like")
]
```

（4）确保将 URL 映射至项目，如下所示。

```
# myproject/urls.py
from django.conf.urls.i18n import i18n_patterns
from django.urls import include, path

urlpatterns = i18n_patterns(
    # …
    path("likes/", include(("myproject.apps.likes.urls", "likes"),
     namespace="likes")),
)
```

（5）定义视图，如下所示。

```
# myproject/apps/likes/views.py
from django.contrib.contenttypes.models import ContentType
from django.http import JsonResponse
from django.views.decorators.cache import never_cache
from django.views.decorators.csrf import csrf_exempt

from .models import Like
from .templatetags.likes_tags import liked_count

@never_cache
@csrf_exempt
def json_set_like(request, content_type_id, object_id):
    """
    Sets the object as a favorite for the current user
    """
    result = {
        "success": False,
    }
    if request.user.is_authenticated and request.method == "POST":
        content_type = ContentType.objects.get(id=content_type_id)
        obj = content_type.get_object_for_this_type(pk=object_id)

        like, is_created = Like.objects.get_or_create(
            content_type=ContentType.objects.get_for_model(obj),
            object_id=obj.pk,
```

```
            user=request.user)
    if not is_created:
        like.delete()

    result = {
        "success": True,
        "action": "add" if is_created else "remove",
        "count": liked_count(obj),
    }

return JsonResponse(result)
```

（6）在任何对象的列表或详细视图的模板中，我们可添加微件的模板标签。下面向位置详细视图中添加微件，如下所示。

```
{# locations/location_detail.html #}
{% extends "base.html" %}
{% load i18n static likes_tags %}

{% block content %}
    <a href="{% url "locations:location_list" %}">{% trans
     "Interesting Locations" %}</a>
    <div class="float-right">
        {% if request.user.is_authenticated %}
            {% like_widget for location %}
        {% endif %}
    </div>
    <h1 class="map-title">{{ location.name }}</h1>
    {# … #}
{% endblock %}
```

（7）随后需要一个微件模板，如下所示。

```
{# likes/includes/widget.html #}
{% load i18n static likes_tags sekizai_tags %}
<p class="like-widget">
    <button type="button"
            class="like-button btn btn-primary{% if object|
            liked_by:request.user %} active{% endif %}"
            data-href="{% url "likes:json_set_like"
             content_type_id=content_type_id
             object_id=object.pk %}"
            data-remove-label="{% trans "Like" %}"
            data-add-label="{% trans "Unlike" %}">
        {% if object|liked_by:request.user %}
```

```
            {% trans "Unlike" %}
        {% else %}
            {% trans "Like" %}
        {% endif %}
    </button>
    <span class="like-badge badge badge-secondary">
        {{ object|liked_count }}</span>
</p>
{% addtoblock "js" %}
<script src="{% static 'likes/js/widget.js' %}"></script>
{% endaddtoblock %}
```

（8）创建 JavaScript 以处理浏览器中的点赞和非点赞动作，如下所示。

```
/* myproject/apps/likes/static/likes/js/widget.js */
(function($) {
    $(document).on("click", ".like-button", function() {
        var $button = $(this);
        var $widget = $button.closest(".like-widget");
        var $badge = $widget.find(".like-badge");

        $.post($button.data("href"), function(data) {
            if (data.success) {
                var action = data.action; // "add" or "remove"
                var label = $button.data(action + "-label");

                $button[action + "Class"]("active");
                $button.html(label);

                $badge.html(data.count);
            }
        }, "json");
    });
}(jQuery));
```

4.10.3　工作方式

目前，我们可针对站点中的任何对象使用{% like_widget for object %}模板标签，这将生成一个微件并根据当前登录用户与对象间的响应方式显示 Like 状态。

Like 按钮包含 3 个自定义的 HTML 5 数据属性。

（1）data-href 提供一个唯一的、特定于对象的 URL 以修改当前微件的状态。

（2）data-add-text 是添加 Like 关联时显示的翻译文本（Unlike）

（3）data-remove-text 类似于删除 Like 关联时的翻译文本(Like)。

当使用 django-sekizai 时，我们向页面中添加了<script src="{% static 'likes/js/widget.js' %}"></script>。注意，如果页面上存在多个 Like 微件，我们仅需包含 JavaScript 一次即可。而且，如果页面中不存在 Like 微件，那么 JavaScript 则不应包含在页面中。

在 JavaScript 文件中，Like 按钮被 like-button CSS 类识别。绑定至文档的事件监听器监视页面中任何此类按钮的单击事件，然后向 data-href 属性指定的 URL 发送一个 Ajax 调用。

指定的视图 json_set_like 接收两个参数，即内容类型 ID 和点赞对象的主键。其中，视图检查 Like 是否针对指定的对象存在。若存在，视图将移除 Like 对象，否则将添加 Like 对象。最终，该视图返回一个 JSON 响应结果，其中包含成功状态、针对 Like 对象采取的操作（添加或删除操作），以及所有用户对该对象的 Like 总数。根据返回的动作，JavaScript 将显示按钮的相应状态。

我们可在浏览器的开发人员工具（一般位于 Network 选项卡中）中调试 Ajax 响应结果。如果在开发过程中出现了任何服务器错误，并且在设置项中开启了 DEBUG，那么将会在响应预览中看到错误跟踪结果；否则将会看到返回的 JSON，如图 4.10 所示。

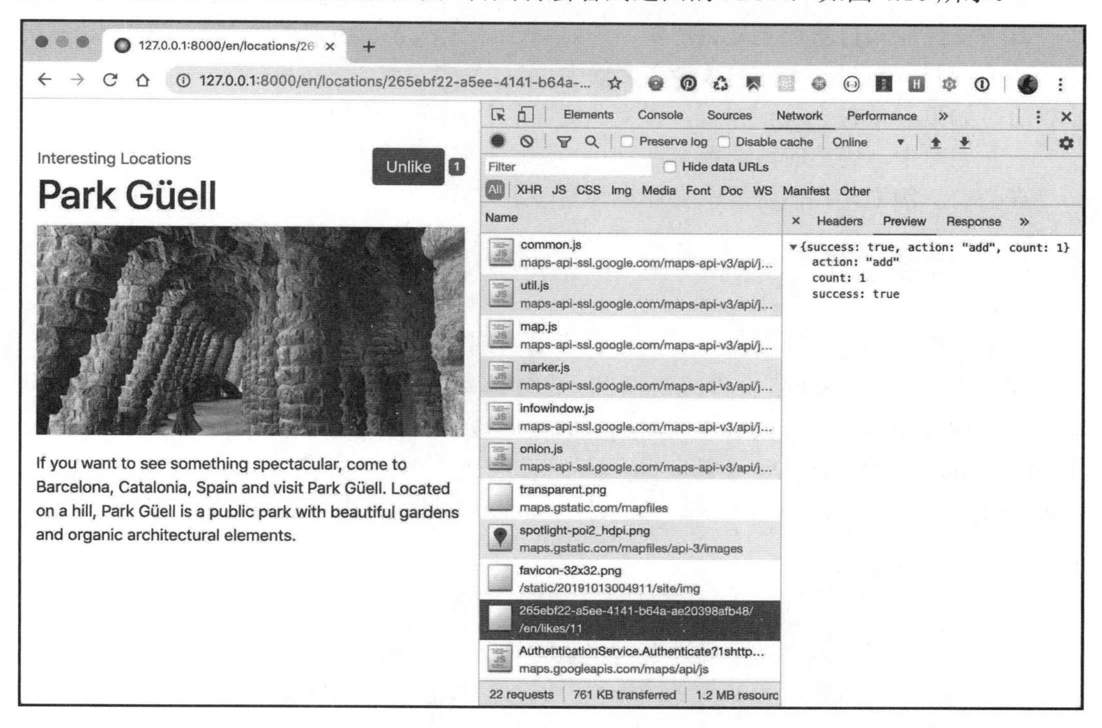

图 4.10

4.10.4 延伸阅读

- “使用 Django Sekizai”示例。
- “在模式对话框中打开对象的细节信息”示例。
- “实现连续的滚动”示例。
- “通过 Ajax 上传图像”示例。
- 第 2 章中的“创建一个模型混入以处理 Generic Relation”示例。
- 第 5 章。

4.11 通过 Ajax 上传图像

通过默认的文件输入字段，可以快速、明显地改善用户体验。

- 字段中只显示所选文件的路径，而用户希望在选择文件后看到他们所选的内容。
- 文件输入自身通常较为“狭窄”，以至于无法显示所选的大部分路径并从左端读取。因此，文件名很少在字段中可见。
- 如果表单包含验证错误，没有人希望再次选择对应的文件；但在有验证错误的表单中，仍然应该选择文件。

在当前示例中，我们将展示文件上传的改进方式。

4.11.1 准备工作

当前示例将在之前的 locations 应用程序的基础上完成。

我们自己的 JavaScript 文件将依赖于一个外部库，即 jQuery File Upload。读者可访问 https://github.com/blueimp/jQuery-File-Upload/tree/v10.2.0 下载并解压相关文件，并将其置于 site_static/site/vendor/jQuery-File-Upload-10.2.0 中。另外，该实用工具也需要使用 jquery.ui.widget.js，而后者连同其他文件一起位于 vendor/子目录夹中。

4.11.2 实现方式

下面定义位置表单，以便通过下列步骤支持 Ajax 上传。

（1）接下来创建位置模型表单，其中涵盖非必须的 picture 字段、一个隐藏的 picture_path 字段，以及 geoposition 的 latitude 和 longitude 字段。

```python
# myproject/apps/locations/forms.py
import os
from django import forms
from django.urls import reverse
from django.utils.translation import ugettext_lazy as _
from django.core.files.storage import default_storage
from crispy_forms import bootstrap, helper, layout
from .models import Location

class LocationForm(forms.ModelForm):
    picture = forms.ImageField(
        label=_("Picture"), max_length=255,
         widget=forms.FileInput(), required=False
    )
    picture_path = forms.CharField(
        max_length=255, widget=forms.HiddenInput(), required=False
    )
    latitude = forms.FloatField(
        label=_("Latitude"),
        help_text=_("Latitude (Lat.) is the angle between any point
        and the equator (north pole is at 90; south pole is at -90)."),
        required=False,
    )
    longitude = forms.FloatField(
        label=_("Longitude"),
        help_text=_("Longitude (Long.) is the angle east or west
        of an arbitrary point on Earth from Greenwich (UK),
        which is the international zero-longitude point
        (longitude=0 degrees). The anti-meridian of Greenwich is
        both 180 (direction to east) and -180 (direction to west)."),
        required=False,
    )
    class Meta:
        model = Location
        exclude = ["geoposition", "rating"]
```

（2）在该表单的__init__()方法中，我们将从模型实例中读取地理位置，并于随后定义该表单的 django-crispy-forms 布局。

```python
def __init__(self, request, *args, **kwargs):
    self.request = request
    super().__init__(*args, **kwargs)
    geoposition = self.instance.get_geoposition()
```

```
if geoposition:
    self.fields["latitude"].initial = geoposition.latitude
    self.fields["longitude"].initial = geoposition.longitude

name_field = layout.Field("name", css_class="input-block-level")
description_field = layout.Field(
    "description", css_class="input-block-level", rows="3"
)
main_fieldset = layout.Fieldset(_("Main data"), name_field,
 description_field)

picture_field = layout.Field(
    "picture",
    data_url=reverse("upload_file"),
    template="core/includes/file_upload_field.html",
)
picture_path_field = layout.Field("picture_path")

picture_fieldset = layout.Fieldset(
    _("Picture"),
    picture_field,
    picture_path_field,
    title=_("Picture upload"),
    css_id="picture_fieldset",
)

street_address_field = layout.Field(
    "street_address", css_class="input-block-level"
)
street_address2_field = layout.Field(
    "street_address2", css_class="input-block-level"
)
postal_code_field = layout.Field("postal_code",
 css_class="input-block-level")
city_field = layout.Field("city", css_class="input-block-level")
country_field = layout.Field("country", css_class="input-block-
 level")
latitude_field = layout.Field("latitude", css_class="input-block-
 level")
longitude_field = layout.Field("longitude", css_class="input-block-
 level")
address_fieldset = layout.Fieldset(
```

```
        _("Address"),
        street_address_field,
        street_address2_field,
        postal_code_field,
        city_field,
        country_field,
        latitude_field,
        longitude_field,
    )

    submit_button = layout.Submit("save", _("Save"))
    actions = bootstrap.FormActions(layout.Div(submit_button,
     css_class="col"))

    self.helper = helper.FormHelper()
    self.helper.form_action = self.request.path
    self.helper.form_method = "POST"
    self.helper.attrs = {"noValidate": "noValidate"}
    self.helper.layout = layout.Layout(main_fieldset,
     picture_fieldset, address_fieldset, actions)
```

（3）针对同一表单，添加 picture 和 picture_path 字段的验证。

```
def clean(self):
    cleaned_data = super().clean()
    picture_path = cleaned_data["picture_path"]
    if not self.instance.pk and not self.files.get("picture")
     and not picture_path:
        raise forms.ValidationError(_("Please choose an image."))
```

（4）添加表单的保存方法，并保存图像和地理位置。

```
def save(self, commit=True):
    instance = super().save(commit=False)
    picture_path = self.cleaned_data["picture_path"]
    if picture_path:
        temporary_image_path = os.path.join("temporary-uploads",
         picture_path)
        file_obj = default_storage.open(temporary_image_path)
        instance.picture.save(picture_path, file_obj, save=False)
        default_storage.delete(temporary_image_path)
    latitude = self.cleaned_data["latitude"]
    longitude = self.cleaned_data["longitude"]
    if latitude is not None and longitude is not None:
```

```
        instance.set_geoposition(longitude=longitude, latitude=latitude)
    if commit:
        instance.save()
        self.save_m2m()
    return instance
```

（5）除了之前在 locations 应用程序中定义的视图，还将添加一个 add_or_change_location 视图，如下所示。

```
# myproject/apps/locations/views.py
from django.contrib.auth.decorators import login_required
from django.shortcuts import render, redirect, get_object_or_404

from .forms import LocationForm
from .models import Location

# …

@login_required
def add_or_change_location(request, pk=None):
    location = None
    if pk:
        location = get_object_or_404(Location, pk=pk)
    if request.method == "POST":
        form = LocationForm(request, data=request.POST,
         files=request.FILES, instance=location)
        if form.is_valid():
            location = form.save()
            return redirect("locations:location_detail", pk=location.pk)
    else:
        form = LocationForm(request, instance=location)

    context = {"location": location, "form": form}
    return render(request, "locations/location_form.html", context)
```

（6）将该视图添加至 URL 配置中。

```
# myproject/apps/locations/urls.py
from django.urls import path
from .views import add_or_change_location

urlpatterns = [
    # …
    path("<uuid:pk>/change/", add_or_change_location,
```

```
    name="add_or_change_location"),
]
```

（7）在 core 应用程序的视图中，我们添加一个 upload_file 泛型函数上传图片，并可被其他带有 picture 字段的应用程序所复用。

```
# myproject/apps/core/views.py
import os
from django.core.files.base import ContentFile

from django.core.files.storage import default_storage
from django.http import JsonResponse
from django.core.exceptions import SuspiciousOperation
from django.urls import reverse
from django.views.decorators.csrf import csrf_protect
from django.utils.translation import gettext_lazy as _
from django.conf import settings
# …

@csrf_protect
def upload_file(request):
    status_code = 400
    data = {"files": [], "error": _("Bad request")}
    if request.method == "POST" and request.is_ajax() and "picture"
     in request.FILES:
        file_types = [f"image/{x}" for x in ["gif", "jpg", "jpeg", "png"]]
        file = request.FILES.get("picture")
        if file.content_type not in file_types:
            status_code = 405
            data["error"] = _("Invalid file format")
        else:
            upload_to = os.path.join("temporary-uploads", file.name)
            name = default_storage.save(upload_to,ContentFile(file.read()))
            file = default_storage.open(name)
            status_code = 200
            del data["error"]
            absolute_uploads_dir = os.path.join(
                settings.MEDIA_ROOT, "temporary-uploads"
            )
            file.filename = os.path.basename(file.name)
            data["files"].append(
                {
                    "name": file.filename,
```

```
                "size": file.size,
                "deleteType": "DELETE",
                "deleteUrl": (
                    reverse("delete_file") +
                     f"?filename={file.filename}"
                ),
                "path": file.name[len(absolute_uploads_dir) + 1 :],
            }
        )

    return JsonResponse(data, status=status_code)
```

（8）设置新上传视图的 URL 规则，如下所示。

```
# myproject/urls.py
from django.urls import path
from myproject.apps.core import views as core_views

# …

urlpatterns += [
    path(
        "upload-file/",
        core_views.upload_file,
        name="upload_file",
    ),
]
```

（9）创建位置表单的模板，如下所示。

```
{# locations/location_form.html #}
{% extends "base.html" %}
{% load i18n crispy_forms_tags %}

{% block content %}
    <div class="row">
        <div class="col-lg-8">
            <a href="{% url "locations:location_list" %}">{% trans
             "Interesting Locations" %}</a>
            <h1>
                {% if location %}
                    {% blocktrans trimmed with name = location.name %}
                        Change Location "{{ name }}"
                    {% endblocktrans %}
```

```
            {% else %}
                {% trans "Add Location" %}
            {% endif %}
        </h1>
        {% crispy form %}
    </div>
</div>
{% endblock %}
```

（10）此外还需要其他模板。为文件上传字段创建一个自定义模板，其中包含所需的 CSS 和 JavaScript。

```
{# core/includes/file_upload_field.html #}
{% load i18n crispy_forms_field static sekizai_tags %}

{% include "core/includes/picture_preview.html" %}
<{% if tag %}{{ tag }}{% else %}div{% endif %} id="div_{{
field.auto_id }}"
class="form-group{% if 'form-horizontal' in form_class %} row{%
endif %}{% if wrapper_class %} {{ wrapper_class }}{% endif %}{% if
field.css_classes %} {{ field.css_classes }}{% endif %}">
    {% if field.label and form_show_labels %}
      <label for="{{ field.id_for_label }}"
           class="col-form-label {{ label_class }}{% if field
            .field.required %} requiredField{% endif %}">
        {{ field.label|safe }}{% if field.field.required %}<span
         class="asteriskField">*</span>{% endif %}
      </label>
    {% endif %}

    <div class="{{ field_class }}">
      <span class="btn btn-success fileinput-button">
          <span>{% trans "Upload File..." %}</span>
          {% crispy_field field %}
      </span>
      {% include 'bootstrap4/layout/help_text_and_errors.html' %}
      <p class="form-text text-muted">
        {% trans "Available formats are JPG, GIF, and PNG." %}
        {% trans "Minimal size is 800 × 800 px." %}
      </p>
    </div>
</{% if tag %}{{ tag }}{% else %}div{% endif %}>
```

```
{% addtoblock "css" %}
<link rel="stylesheet" href="{% static 'site/vendor/jQuery-File-
Upload-10.2.0/css/jquery.fileupload-ui.css' %}"/>
<link rel="stylesheet" href="{% static 'site/vendor/jQuery-File-
Upload-10.2.0/css/jquery.fileupload.css' %}"/>
{% endaddtoblock %}

{% addtoblock "js" %}
<script src="{% static 'site/vendor/jQuery-File-
Upload-10.2.0/js/vendor/jquery.ui.widget.js' %}"></script>
<script src="{% static 'site/vendor/jQuery-File-
Upload-10.2.0/js/jquery.iframe-transport.js' %}"></script>
<script src="{% static 'site/vendor/jQuery-File-
Upload-10.2.0/js/jquery.fileupload.js' %}"></script>

<script src="{% static 'site/js/picture_upload.js' %}"></script>
{% endaddtoblock %}
```

（11）创建图像预览的模板。

```
{# core/includes/picture_preview.html #}
<div id="picture_preview">
   {% if form.instance.picture %}
       <img src="{{ form.instance.picture.url }}" alt=""
        class="img-fluid"/>
   {% endif %}
</div>
<div id="progress" class="progress" style="visibility: hidden">
   <div class="progress-bar progress-bar-striped
    progress-bar-animated"
       role="progressbar"
       aria-valuenow="0"
       aria-valuemax="100"
       style="width: 0%"></div>
</div>
```

（12）添加 JavaScript 以处理图像上传和预览。

```
/* site_static/site/js/picture_upload.js */
$(function() {
  $("#id_picture_path").each(function() {
    $picture_path = $(this);
    if ($picture_path.val()) {
      $("#picture_preview").html(
```

```
            '<img src="' +
              window.settings.MEDIA_URL +
              "temporary-uploads/" +
              $picture_path.val() +
              '" alt="" class="img-fluid" />'
          );
        }
    });
    $("#id_picture").fileupload({
      dataType: "json",
      add: function(e, data) {
        $("#progress").css("visibility", "visible");
        data.submit();
      },
      progressall: function(e, data) {
        var progress = parseInt((data.loaded / data.total) * 100, 10);
        $("#progress .progress-bar")
          .attr("aria-valuenow", progress)
          .css("width", progress + "%");
      },
      done: function(e, data) {
        $.each(data.result.files, function(index, file) {
          $("#picture_preview").html(
            '<img src="' +
              window.settings.MEDIA_URL +
              "temporary-uploads/" +
              file.name +
              '" alt="" class="img-fluid" />'
          );
          $("#id_picture_path").val(file.name);
        });
        $("#progress").css("visibility", "hidden");
      }
    });
});
```

4.11.3　工作方式

如果 JavaScript 无法正常运行，那么表单仍保持完全可用状态；当 JavaScript 正常运行时，我们将得到一个增强的表单，其中包含被按钮所替换的文件字段，如图 4.11 所示。

当图像通过单击 Upload File...按钮被选中后，浏览器中的对应结果如图 4.12 所示。

Picture

Picture

Upload File...

Available formats are JPG, GIF, and PNG. Minimal size is 800 × 800 px.

图 4.11

图 4.12

单击 Upload File...按钮后，该按钮将触发一个文件对话框，并询问用户选择一个文件；待选取完毕后，将立即启用 Ajax 上传处理过程。随后即可看到一个绑定的图像预览。这里，预览图上传至一个临时目录中，其文件名保存于 picture_path 隐藏字段中。当提交表单后，图像将保存至该临时位置或 picture 字段。其中，如果表单在缺少 JavaScript 的情况下被提交，或者无法加载 JavaScript，那么 picture 字段将保存一个值。如果在页面重载后其他字段存在验证错误，那么加载的预览图将基于 picture_path。

接下来深入讨论具体处理过程，并探讨其工作方式。

在 Location 模型的模型表单中，生成了非所需的 picture 字段，虽然该字段应用于模型级别。除此之外，我们还添加了 picture_path 字段，并以此执行表单的提交。在 crispy-forms 布局中，我们定义了 picture 字段的自定义模板 file_upload_field.html，并于其中设置了一个预览图、上传进度条和自定义帮助文本，其中涉及所支持的文件格式和最小尺寸。在同一模板中，我们还绑定了 jQuery File Upload 库中的 CSS 和 JavaScript 文件，以及一个自定义脚本 picture_upload.js。这里，CSS 文件将文件上传字段渲染为一个按钮，而 JavaScript 文件则负责基于 Ajax 的文件上传。

picture_upload.js 将所选文件发送至 upload_file 视图。该视图负责检查文件是否为图像类型，并于随后尝试将其保存至项目的 MEDIA_ROOT 下的 temporary-uploads/目录中。该视图返回一个 JSON，其中包含成功或失败的文件上传的详细信息。

待图像选取、上传完毕且表单提交后，将调用 LocationForm 的 save()方法。如果 picture_path 字段值存在，文件将从临时目录中获取，并复制至 Location 模型的 picture 字段中。随后删除临时目录中的图像，并保存 Location 实例。

4.11.4　更多内容

我们从模型表单中排除了 geoposition 字段，反而为地理位置数据渲染 latitude 和 longitude 字段。另外，默认的地理位置的 PointField 被渲染为 Leaflet.js 地图，且不存在定制的可能性。latitude 和 longitude 这两个字段则表现得较为灵活，我们可以使用 Google Maps API、Bing Maps API 或 Leaflet.js 在地图中显示它们；或者采用手动方式输入；或者从填写的位置地址对其进行地理编码。

出于方便考虑，这里还使用了两个帮助方法，即 get_geoposition()和 set_geoposition()，读者可参考之前的“使用 HTML 5 数据属性”示例。

4.11.5　延伸阅读

- “使用 HTML 5 数据属性”示例。
- 第 3 章中的“上传图像”示例。
- “在模式对话框中打开对象的细节信息”示例。
- “实现连续的滚动”示例。
- “实现 Like 微件”示例。
- 第 7 章中的“表单的跨站点请求伪造安全”示例。

第 5 章　自定义模板过滤器和标签

本章主要涉及下列主题。

- 遵循自定义的模板过滤器和标签规则。
- 创建一个模板过滤器并显示帖子的发布天数。
- 创建一个模板过滤器并析取第一个媒体对象。
- 创建一个模板过滤器并识别 URL。
- 创建一个模板标签并包含一个模板（如果存在）。
- 创建一个模板标签并加载模板中的 QuerySet。
- 创建一个模板标签并作为模板解析内容。
- 创建模板标签并调整请求查询参数。

5.1　简　介

Django 包含一个可扩展的模板系统，其特性包括模板继承、修改值表示的过滤器，以及表示逻辑的标签。而且，Django 还支持向应用程序中添加自定义的模板过滤器和标签，另外，自定义过滤器或标签应位于应用程序中 templatetags Python 包下的 template-tag 库文件中。随后，template-tag 库可通过{% load %}模板标签载入至任何模板中。本章将创建多个实用过滤器和标签，进而拥有对模板编辑器更多的控制。

5.2　技术需求

当与本章代码协同工作时，应安装最新稳定版本的 Python 3、MySQL 或 PostgreSQL 数据库，以及虚拟环境中的 Django 项目。

读者可访问 GitHub 储存库的 ch05 目录查看本章的全部代码，对应网址为 https://github.com/PacktPublishing/Django-3-Web-Development-Cookbook-Fourth-Edition。

5.3　遵循自定义的模板过滤器和标签规则

如果缺少相应的规则，自定义模板过滤器和标签将会令人混淆且缺乏一致性。拥有方便而灵活的模板过滤器和标签是非常必要的，它们应该尽可能地为模板编辑器服务。当增强 Django 模板系统的功能时，在当前示例中将探讨一些相关规则。

（1）当页面逻辑与视图、上下文预处理器或模型方法实现较好地匹配时，不要创建或使用自定义模板过滤器或标签。当内容特定于上下文，如对象列表或一个对象详细视图，则需要加载视图中的对象。如果需要在几乎每个页面上显示某些内容，则可创建一个上下文预处理器。当需要获取某个对象的一些属性，而该对象与模板的上下文不相关时，可使用模型的自定义方法，而非模板过滤器。

（2）利用_tags 后缀命名 template-tag 库。当 template-tag 库的命名方式与应用程序不同时，可以避免不明确的包导入问题。

（3）在新创建的库中，隔离过滤器和标签，如使用注释，如下所示。

```
# myproject/apps/core/templatetags/utility_tags.py
from django import template

register = template.Library()

""" TAGS """

# Your tags go here…

""" FILTERS """

# Your filters go here…
```

（4）当创建高级自定义模板标签时，应确保其语法易于理解，即包含遵循下列标签名称的结构。

- for [app_name.model_name]：包含该结构以使用一个特定的模型。
- using [template_name]：包含该结构以使用一个模板标签输出结果的模板。
- limit [count]：包含该结构以限制特定数量的结果。
- as [context_variable]：包含该结构以将结果存储于可被多次复用的上下文变量中。

（5）尽量避免在模板标签中多次定义位置值，除非这些值是自解释的，否则这将会使开发人员感到困惑。

（6）尽可能多地生成可解析的参数。不包含引号的字符串应被视为需要解析的上下文变量，或者是提醒 template-tag 组件结构的简短单词。

5.4　创建一个模板过滤器以显示帖子发布的天数

当谈及日期的创建和修改时，较为方便的方法是读取一个人类可读的时间差。例如，博客条目是 3 天前发布的，新闻文章是今天发布的，用户最后一次登录是昨天。在当前示例中，我们将创建一个名为 date_since 的模板过滤器，并根据日期、星期、月份和年份将日期转换为一个具有可读性的时间差。

5.4.1　准备工作

创建 core 应用程序（如果不存在），并将其置于设置项的 INSTALLED_APPS 下。随后在该应用程序中创建一个 templatetags Python 包（Python 包是指包含一个空__init__.py 文件的目录）。

5.4.2　实现方式

创建包含下列内容的 utility_tags.py 文件。

```
# myproject/apps/core/templatetags/utility_tags.py
from datetime import datetime
from django import template
from django.utils import timezone
from django.utils.translation import ugettext_lazy as _

register = template.Library()

""" FILTERS """

DAYS_PER_YEAR = 365
DAYS_PER_MONTH = 30
DAYS_PER_WEEK = 7

@register.filter(is_safe=True)
def date_since(specific_date):
    """
```

```
    Returns a human-friendly difference between today and past_date
    (adapted from https://www.djangosnippets.org/snippets/116/)
    """
    today = timezone.now().date()
    if isinstance(specific_date, datetime):
        specific_date = specific_date.date()
    diff = today - specific_date
    diff_years = int(diff.days / DAYS_PER_YEAR)
    diff_months = int(diff.days / DAYS_PER_MONTH)
    diff_weeks = int(diff.days / DAYS_PER_WEEK)
    diff_map = [
        ("year", "years", diff_years,),
        ("month", "months", diff_months,),
        ("week", "weeks", diff_weeks,),
        ("day", "days", diff.days,),
    ]

    for parts in diff_map:
        (interval, intervals, count,) = parts
        if count > 1:
            return _(f"{count} {intervals} ago")
        elif count == 1:
            return _("yesterday") \
                if interval == "day" \
                else _(f"last {interval}")
    if diff.days == 0:
        return _("today")
    else:
        # Date is in the future; return formatted date.
        return f"{specific_date:%B %d, %Y}"
```

5.4.3　工作方式

下列代码显示了用于模板中的过滤器，并渲染类似于“昨天”“上一个星期”或“5 个月前”。

```
{% load utility_tags %}
{{ object.published|date_since }}
```

我们可将这一过滤器应用于 date 和 datetime 类型值上。

每个 template-tag 库都包含一个 template.Library 类型的寄存器，并于其中收集过滤器和标签。Django 过滤器是由@register.filter 装饰器注册的函数。在当前示例中，我们传递

is_safe=True 参数以表明过滤器不会引入不安全的 HTML 标记。

默认状态下，模板系统中的过滤器的命名与函数或另一个可调用的对象保持一致。如果需要，可通过将名称传递至装饰器来设置过滤器不同的名称，如下所示。

```
@register.filter(name="humanized_date_since", is_safe=True)
def date_since(value):
    # …
```

该过滤器自身具有较好的自解释性。首先读取当前日期。如果过滤器的给定值为 datatime 类型，其 date 将被析取。随后，今日与析取值之间的差将根据 DAYS_PER_YEAR、DAYS_PER_MONTH、DAYS_PER_WEEK 或日期间隔被计算。根据计数结果，将返回不同的字符串。如果对应值表示为未来的某天，那么将显示一个格式化的日期。

5.4.4　更多内容

如果需要，还可延长时间间隔，如 20 分钟前，5 小时前，甚至 10 年前。对此，可向现有的 diff_map 集添加更大的间隔，进而显示时间差。相应地，我们需要操控 datetime 值而非 date 值。

5.4.5　延伸阅读

- "创建一个模板过滤器以析取第一个媒体对象"示例。
- "创建一个模板过滤器以识别 URL"示例。

5.5　创建一个模板过滤器以析取第一个媒体对象

假设正在开发一个博客概览页面，针对每个帖子，我们需要显示从内容中获取的该页面上的图像、音乐或视频。此时，需要从帖子的 HTML 内容中析取<figure>、<img>、<object>、<embed>、<video>、<audio>和<iframe>标签，它们存储在帖子模型的字段中。在当前示例中，我们将探讨如何使用 first_media 过滤器中的正则表达式执行这一项任务。

5.5.1　准备工作

我们将使用设置项中 INSTALLED_APPS 中设置的 core 应用程序，并在该应用程序中包含 templatetags 包。

5.5.2　实现方式

在 utility_tags.py 文件中添加下列内容。

```
# myproject/apps/core/templatetags/utility_tags.py
import re
from django import template
from django.utils.safestring import mark_safe
register = template.Library()

""" FILTERS """

MEDIA_CLOSED_TAGS = "|".join([
   "figure", "object", "video", "audio", "iframe"])
MEDIA_SINGLE_TAGS = "|".join(["img", "embed"])
MEDIA_TAGS_REGEX = re.compile(
   r"<(?P<tag>" + MEDIA_CLOSED_TAGS + ")[\S\s]+?</(?P=tag)>|" +
   r"<(" + MEDIA_SINGLE_TAGS + ")[^>]+>",
   re.MULTILINE)

@register.filter
def first_media(content):
   """
   Returns the chunk of media-related markup from the html content
   """
   tag_match = MEDIA_TAGS_REGEX.search(content)
   media_tag = ""
   if tag_match:
      media_tag = tag_match.group()
   return mark_safe(media_tag)
```

5.5.3　工作方式

如果数据库中的 HTML 内容有效，同时将下列代码置于模板中，那么这将从对象的内容（content）字段中检索媒体标签（media）；如果未查到任何媒体，则返回一个空字符串。

```
{% load utility_tags %}
{{ object.content|first_media }}
```

正则表达式是一个功能强大的特性，用于搜索或替换文本中的模式。首先，我们定义全部支持的媒体标签名称列表，并根据开放和关闭标签（MEDIA_CLOSED_TAGS）以及自关闭标签（MEDIA_SINGLE_TAGS）将其划分为多个组。在这些列表中，我们生成编译后的正则表达式 MEDIA_TAGS_REGEX。在当前示例中，我们将搜索全部可能的媒体标签，同时允许它们跨行出现。

接下来探讨正则表达式的工作方式，如下所示。

- 交替模式由管道符号（|）分隔。这些模式分为两组，首先是具有打开和关闭的正常标签（<figure>、<object>、<video>、<audio>、<iframe>和<picture>），其次是最终模式，称作自关闭或空标记（<img>和<embed>）。
- 对于多行正常标签，我们将使用[\S\s]+?模式至少与任何符号匹配一次。然而，我们应尽量减少这类操作次数，直至查找到其后的字符串。
- 因此，<figure[\S\s]+?</figure>搜索<figure>标签的开始部分及其之后的所有内容，直至找到关闭的</figure>标签。
- 类似地，当使用自闭合标签的[^>]+模式时，我们至少应搜索一次并尽可能多地搜索除直角括号（即>）之外的任何符号，直至遇到这样一个表示标签关闭的括号。

re.MULTILINE 标志确保匹配可以找到，即使它们在内容中跨越多行。随后在过滤器中，我们利用该正则表达式模式进行搜索。默认状态下，在 Django 中，任何过滤器的结果都将显示<、>和&符号，这些符号分别转义为<、>和&实体。在当前示例中，我们使用 mark_safe()函数表明当前结果是安全的，并可用于 HTML，因而任何内容都将被渲染而不会被转义。由于最初的内容是用户输入，所以在注册过滤器时，我们并未传递 is_safe=True，因为我们需要显式地证明标记是安全的。

5.5.4　更多内容

关于正则表达式，读者可参考 Python 官方文档，对应网址为 https://docs.python.org/3/library/re.html。

5.5.5　延伸阅读

- “创建一个模板过滤器以显示帖子发布的天数”示例。
- “创建一个模板过滤器以识别 URL”示例。

5.6 创建一个模板过滤器以识别 URL

Web 用户通常能够识别没有协议（http://）或尾随斜杠（/）的 URL。类似地，他们将以这种方式在地址字段中输入 URL。在当前示例中，我们将创建一个 humanize_url 过滤器，并以较短的格式向用户显示 URL，同时截断较长的地址（类似于 Twitter 对推文采取的做法）。

5.6.1 准备工作

当前示例将在前述 core 应用程序的基础上完成，后者在设置项的 INSTALLED_APPS 中被设置，并在应用程序中包含了 templatetags 包。

5.6.2 实现方式

在 core 应用程序中，在 utility_tags.py 模板库的 FILTERS 部分中，添加 humanize_url 过滤器并对其进行注册，如下所示。

```
# myproject/apps/core/templatetags/utility_tags.py
import re
from django import template

register = template.Library()

""" FILTERS """

@register.filter
def humanize_url(url, letter_count=40):
    """
    Returns a shortened human-readable URL
    """
    letter_count = int(letter_count)
    re_start = re.compile(r"^https?://")
    re_end = re.compile(r"/$")
    url = re_end.sub("", re_start.sub("", url))
    if len(url) > letter_count:
        url = f"{url[:letter_count - 1]}…"
    return url
```

5.6.3 工作方式

我们可在任意模板中使用 humanize_url 过滤器，如下所示。

```
{% load utility_tags %}
<a href="{{ object.website }}" target="_blank">
    {{ object.website|humanize_url }}
</a>
<a href="{{ object.website }}" target="_blank">
    {{ object.website|humanize_url:30 }}
</a>
```

过滤器使用正则表达式来删除开头的协议和结尾的斜杠，将 URL 缩短为给定的字母数量（默认为 40），如果完整的 URL 不符合指定的字母数量，则在截断 URL 后在末尾添加一个省略号。例如，对于 https://docs.djangoproject.com/en/3.0/howto/custom- templatetags/的 URL，40 个字符的可识别版本为 docs.djangoproject.com/en/3.0/howto/cus…。

5.6.4 延伸阅读

- “创建一个模板过滤器以显示帖子发布的天数”示例。
- “创建一个模板过滤器以析取第一个媒体对象”示例。
- “创建一个模板标签以包含一个模板”示例。

5.7 创建一个模板标签以包含一个模板

Django 提供了{% include %}模板标签，以使一个模板可渲染和包含另一个模板。然而，如果尝试包含一个文件系统中不存在的模板，该模板标签将会出现错误。在当前示例中，我们将创建一个{% try_to_include %}模板标签并可包含另一个模板（如果存在）；否则将以静默方式退出并渲染一个空字符串。

5.7.1 准备工作

当前示例将在已安装的 core 应用程序的基础上完成，并已准备了相应的自定义模板标签。

5.7.2 实现方式

执行下列步骤将创建{% try_to_include %}模板标签。

（1）创建相关函数并解析模板标签参数，如下所示。

```
# myproject/apps/core/templatetags/utility_tags.py
from django import template
from django.template.loader import get_template

register = template.Library()

""" TAGS """

@register.tag
def try_to_include(parser, token):
    """
    Usage: {% try_to_include "some_template.html" %}

    This will fail silently if the template doesn't exist.
    If it does exist, it will be rendered with the current context.
    """
    try:
        tag_name, template_name = token.split_contents()
    except ValueError:
        tag_name = token.contents.split()[0]
        raise template.TemplateSyntaxError(
            f"{tag_name} tag requires a single argument")
    return IncludeNode(template_name)
```

（2）在同一文件中需要一个自定义 IncludeNode 类，该类扩展自 template.Node 基类。下面将其插入至 try_to_include()函数之前，如下所示。

```
class IncludeNode(template.Node):
    def __init__(self, template_name):
        self.template_name = template.Variable(template_name)

    def render(self, context):
        try:
            # Loading the template and rendering it
            included_template = self.template_name.resolve(context)
            if isinstance(included_template, str):
```

```
            included_template = get_template(included_template)
        rendered_template = included_template.render(
            context.flatten()
        )
    except (template.TemplateDoesNotExist,
            template.VariableDoesNotExist,
            AttributeError):
        rendered_template = ""
    return rendered_template


@register.tag
def try_to_include(parser, token):
    # …
```

5.7.3 工作方式

高级自定义模板标签由以下两项内容构成。

（1）解析模板标签参数的函数。

（2）负责模板标签逻辑和输出的 Node 类。

{% try_to_include %}模板标签接收一个参数，即 template_name。因此，在 try_to_include()函数中，我们尝试仅将标记的划分内容赋予 tag_name 变量（即 try_to_include）和 template_name 变量。如果该过程无法正常工作，则会出现一个 TemplateSyntaxError 错误。该函数返回 IncludeNode 对象，获取 template_name 字段并将其存储于一个模板 Variable 对象中以供后续使用。

在 IncludeNode 的 render()方法中，我们解析了 template_name 变量。如果上下文变量被传递至模板标签中，其值将用于 template_name。如果一个带引号的字符串被传递至模板标签，那么引号中的内容将用于 included_template，而对应于上下文变量的字符串将被解析为与其等价的字符串。

最后尝试通过解析后的 included_template 字符串加载模板，并利用当前模板上下文渲染模板。如果该过程无法正常工作，那么将返回一个空字符串。

上述模板标签的应用场合至少包含以下两种。

（1）当包含一个模板，其路径定义于模型中，如下所示。

```
{% load utility_tags %}
{% try_to_include object.template_path %}
```

（2）当包含一个模板时，其路径是利用{% with %}模板标签在模板上下文变量的作用域中定义的。当需要在 Django CMS 模板的占位符中为插件创建自定义布局时，这将十分有用。

```
{# templates/cms/start_page.html #}
{% load cms_tags %}
{% with editorial_content_template_path=
"cms/plugins/editorial_content/start_page.html" %}
   {% placeholder "main_content" %}
{% endwith %}
```

稍后，占位符可被 editorial_content 插件填充，editorial_content_template_path 上下文变量将被读取，并且模板可被安全地包含进来。

```
{# templates/cms/plugins/editorial_content.html #}
{% load utility_tags %}
{% if editorial_content_template_path %}
   {% try_to_include editorial_content_template_path %}
{% else %}
   <div>
      <!-- Some default presentation of
      editorial content plugin -->
   </div>
{% endif %}
```

5.7.4　更多内容

我们可以使用{% try_to_include %}标签和默认的{% include %}标签的任意组合来包含扩展了其他模板的模板。

这对于大规模 Web 平台十分有用。其中包含了不同种类的列表，在这些列表中，复杂的条目拥有与微件相同的结构，但却包含不同的数据源。

例如，在艺术家列表模板中，可以包含 artist_item 模板，如下所示。

```
{% load utility_tags %}
{% for object in object_list %}
   {% try_to_include "artists/includes/artist_item.html" %}
{% endfor %}
```

该模板将从条目库中扩展，如下所示。

```
{# templates/artists/includes/artist_item.html #}
{% extends "utils/includes/item_base.html" %}
{% block item_title %}
```

```
    {{ object.first_name }} {{ object.last_name }}
{% endblock %}
```

条目库定义了任何条目的标记，同时还包含了一个 Like 微件，如下所示。

```
{# templates/utils/includes/item_base.html #}
{% load likes_tags %}
<h3>{% block item_title %}{% endblock %}</h3>
{% if request.user.is_authenticated %}
    {% like_widget for object %}
{% endif %}
```

5.7.5　延伸阅读

- ❑ 第 4 章中的“实现 Like 微件”示例。
- ❑ “创建一个模板标签以加载模板中的 QuerySet”示例。
- ❑ “创建一个模板标签以作为模板解析内容”示例。
- ❑ “创建模板标签以调整请求查询参数”示例。

5.8　创建一个模板标签以加载模板中的 QuerySet

一般来讲，显示于 Web 页面上的内容将通过视图在上下文中被定义。如果对应内容显示于每个页面，从逻辑上讲，应创建一个上下文处理器以使其全局有效。另一种情况是，当需要显示附加内容时，如在某些页面（如开始页面或某个对象的详细页面）上显示最新消息或随机报价。此时，可利用一个自定义的{% load_objects %}模板标签加载所需的内容，当前示例将对此加以解释。

5.8.1　准备工作

当前示例将在 core 应用程序的基础上完成，其中安装并准备好了自定义模板标签。

除此之外，当展示相关概念时，还需要创建一个包含 Article 模型的 news 应用程序，如下所示。

```
# myproject/apps/news/models.py
from django.db import models
from django.urls import reverse
from django.utils.translation import ugettext_lazy as _
```

```
from myproject.apps.core.models import CreationModificationDateBase,
UrlBase

class ArticleManager(models.Manager):
    def random_published(self):
        return self.filter(
            publishing_status=self.model.PUBLISHING_STATUS_PUBLISHED,
        ).order_by("?")

class Article(CreationModificationDateBase, UrlBase):
    PUBLISHING_STATUS_DRAFT, PUBLISHING_STATUS_PUBLISHED = "d", "p"
    PUBLISHING_STATUS_CHOICES = (
        (PUBLISHING_STATUS_DRAFT, _("Draft")),
        (PUBLISHING_STATUS_PUBLISHED, _("Published")),
    )
    title = models.CharField(_("Title"), max_length=200)
    slug = models.SlugField(_("Slug"), max_length=200)

    content = models.TextField(_("Content"))
    publishing_status = models.CharField(
        _("Publishing status"),
        max_length=1,
        choices=PUBLISHING_STATUS_CHOICES,
        default=PUBLISHING_STATUS_DRAFT,
    )

    custom_manager = ArticleManager()

    class Meta:
        verbose_name = _("Article")
        verbose_name_plural = _("Articles")

    def __str__(self):
        return self.title

    def get_url_path(self):
        return reverse("news:article_detail", kwargs={"slug": self.slug})
```

其中，Article 模型的 custom_manager 值得关注。管理器可用于列出随机发布的文章。

当使用前述示例时，可利用 URL 配置、视图、模板和管理设置完成应用程序。随后，可利用管理表单将某些文章添加至数据库中。

5.8.2　实现方式

高级自定义模板标签由一个函数（解析传递至标签的参数）和一个 Node 类（渲染标签的输出结果或调整模板上下文）构成。执行下列步骤并创建{% load_objects %}模板标签。

（1）创建对应函数并处理模板标签参数的解析工作，如下所示。

```
# myproject/apps/core/templatetags/utility_tags.py
from django import template
from django.apps import apps

register = template.Library()

""" TAGS """

@register.tag
def load_objects(parser, token):
    """
    Gets a queryset of objects of the model specified by app and
    model names

    Usage:
        {% load_objects [<manager>.]<method>
                        from <app_name>.<model_name>
                        [limit <amount>]
                        as <var_name> %}

    Examples:
        {% load_objects latest_published from people.Person
                        limit 3 as people %}
        {% load_objects site_objects.all from news.Article
                        as articles %}
        {% load_objects site_objects.all from news.Article
                        limit 3 as articles %}
    """
    limit_count = None
    try:
        (tag_name, manager_method,
         str_from, app_model,
         str_limit, limit_count,
         str_as, var_name) = token.split_contents()
    except ValueError:
```

```
    try:
        (tag_name, manager_method,
         str_from, app_model,
         str_as, var_name) = token.split_contents()
    except ValueError:
        tag_name = token.contents.split()[0]
        raise template.TemplateSyntaxError(
            f"{tag_name} tag requires the following syntax: "
            f"{{% {tag_name} [<manager>.]<method> from "
            "<app_name>.<model_name> [limit <amount>] "
            "as <var_name> %}")
try:
    app_name, model_name = app_model.split(".")
except ValueError:
    raise template.TemplateSyntaxError(
        "load_objects tag requires application name "
        "and model name, separated by a dot")
model = apps.get_model(app_name, model_name)
return ObjectsNode(
    model, manager_method, limit_count, var_name
)
```

（2）在同一文件中创建自定义 ObjectsNode 类，该类扩展自 template.Node 基类。接下来将其插入至 load_objects()函数前，如下所示。

```
class ObjectsNode(template.Node):
    def __init__(self, model, manager_method, limit, var_name):
        self.model = model
        self.manager_method = manager_method
        self.limit = template.Variable(limit) if limit else None
        self.var_name = var_name

    def render(self, context):
        if "." in self.manager_method:
            manager, method = self.manager_method.split(".")
        else:
            manager = "_default_manager"
            method = self.manager_method

        model_manager = getattr(self.model, manager)
        fallback_method = self.model._default_manager.none
        qs = getattr(model_manager, method, fallback_method)()
        limit = None
```

```
        if self.limit:
            try:
                limit = self.limit.resolve(context)
            except template.VariableDoesNotExist:
                limit = None
        context[self.var_name] = qs[:limit] if limit else qs
        return ""

@register.tag
def load_objects(parser, token):
    # …
```

5.8.3 工作方式

{% load_objects %}模板标签从指定的应用程序和模型加载一个由管理器的方法定义的 QuerySet，将结果限制到指定的数量，并将结果保存到给定的上下文变量中。

下列代码展示了如何使用刚刚创建的模板标签。该过程将全部新闻文章加载至任意模板中，如下所示。

```
{% load utility_tags %}
{% load_objects all from news.Article as all_articles %}
<ul>
    {% for article in all_articles %}
        <li><a href="{{ article.get_url_path }}">
         {{ article.title }}</a></li>
    {% endfor %}
</ul>
```

其中使用了 Article 模型的默认 objects 管理器的 all()方法，并通过模型 Meta 类中定义的 ordering 属性对文章进行排序。

接下来的示例使用了基于自定义方法的自定义管理器并从数据库中查询对象。这里，管理器是一个接口，并提供了数据库查询操作模型。

默认状态下，每个模型至少包含一个名为 objects 的管理器。对于 Article 模型，我们添加了一个额外的 custom_manager 管理器，其中包含了一个 random_published()方法。下列代码展示了如何将该方法与{% load_objects%}模板标签结合使用，进而加载随机发布的文章。

```
{% load utility_tags %}
{% load_objects custom_manager.random_published from news.Article
limit 1 as random_published_articles %}
```

```
<ul>
    {% for article in random_published_articles %}
        <li><a href="{{ article.get_url_path }}">
         {{ article.title }}</a></li>
    {% endfor %}
</ul>
```

接下来研究{% load_objects %}模板标签代码。在解析函数中，该标签存在两个表单，即基于 limit 和缺少 limit 的表单。这里，字符串经解析后，如果格式被识别，那么模板标签组件将被传递至 ObjectsNode 类中。

在 Node 类的 render()方法中，我们检查管理器的名称及其方法的名称。如果未指定管理器，那么将使用_default_manager。这是一个任何模型的自动属性，并由 Django 注入，同时指向第一个有效的 models.Manager()实例。在大多数时候，_default_manager 将会是 objects 管理器。随后，我们调用管理器的方法，如果对应的方法不存在，则回退至一个空的 QuerySet。如果 limit 已被定义，那么我们将解析其值并相应地限制 QuerySet 的数量。最后，我们将结果 QuerySet 存储至上下文变量中，此处由 var_name 给定。

5.8.4 延伸阅读

- 第 2 章中的“利用与 URL 相关的方法创建一个模型混入”示例。
- 第 2 章中的“创建一个模型混入以处理日期的创建和修改”示例。
- “创建一个模板标签以包含一个模板”示例。
- “创建一个模板标签以作为模板解析内容”示例。
- “创建模板标签以调整请求查询参数”示例。

5.9 创建一个模板标签以作为模板解析内容

在当前示例中，我们将创建{% parse %}模板标签，进而将模板片段置于数据库中。对于已认证或非认证用户，当需要提供不同的内容，或者打算包含个性化的问候，抑或不希望硬编码数据库中的媒体路径时，这将十分有用。

5.9.1 准备工作

同样，当前示例在前述 core 应用程序的基础上完成，同时自定义模板标签应处于可用状态。

5.9.2　实现方式

高级的自定义模板标签由函数（解析传递至标签中的参数）和一个 Node 类（渲染标签的输出结果或调整模板上下文）构成。下列步骤将创建{% parse %}模板标签。

（1）创建函数并解析模板的参数，如下所示。

```
# myproject/apps/core/templatetags/utility_tags.py
from django import template

register = template.Library()

""" TAGS """

@register.tag
def parse(parser, token):
    """
    Parses a value as a template and prints or saves to a variable

    Usage:
        {% parse <template_value> [as <variable>] %}

    Examples:
        {% parse object.description %}
        {% parse header as header %}
        {% parse "{{ MEDIA_URL }}js/" as js_url %}
    """
    bits = token.split_contents()
    tag_name = bits.pop(0)
    try:
        template_value = bits.pop(0)
        var_name = None
        if len(bits) >= 2:
            str_as, var_name = bits[:2]
    except ValueError:
        raise template.TemplateSyntaxError(
            f"{tag_name} tag requires the following syntax: "
            f"{{% {tag_name} <template_value> [as <variable>] %}}")
    return ParseNode(template_value, var_name)
```

（2）在同一文件中创建自定义 ParseNode 类，该类扩展自 template.Node 基类，如下列代码片段所示（置于 parse()函数之前）。

```
class ParseNode(template.Node):
    def __init__(self, template_value, var_name):
        self.template_value = template.Variable(template_value)
        self.var_name = var_name

    def render(self, context):
        template_value = self.template_value.resolve(context)
        t = template.Template(template_value)
        context_vars = {}
        for d in list(context):
            for var, val in d.items():
                context_vars[var] = val
        req_context = template.RequestContext(
            context["request"], context_vars
        )
        result = t.render(req_context)
        if self.var_name:
            context[self.var_name] = result
            result = ""
        return result

@register.tag
def parse(parser, token):
    # …
```

5.9.3　工作方式

{% parse %}模板标签允许我们将一个值解析为模板，并立即对其进行渲染，或将其存储于一个上下文变量中。

如果持有一个包含描述字段的对象，可包含模板变量或逻辑，那么可通过下列代码对象进行解析和渲染。

```
{% load utility_tags %}
{% parse object.description %}
```

此外，也可使用引号括起来的字符串定义一个值以供解析，如下所示。

```
{% load static utility_tags %}
{% get_static_prefix as STATIC_URL %}
{% parse "{{ STATIC_URL }}site/img/" as image_directory %}
<img src="{{ image_directory }}logo.svg" alt="Logo" />
```

接下来看一看{% parse %}模板标签代码。其间，解析函数逐位地检查模板标签参数。首先是解析名称和模板值。如果标记中还有更多位，则需要使用可选的 as 单词后跟上下文变量的组合。模板值和可选的变量名将被传递至 ParseNode 类中。

该类的 render()方法首先解析模板变量值，并从中创建一个模板对象。另外，context_vars 被复制，并生成一个请求上下文，即模板渲染器。如果变量名被定义，对应结果将存储于其中，同时渲染一个空字符串，否则将立即显示渲染后的模板。

5.9.4　延伸阅读

- “创建一个模板标签以包含一个模板”示例。
- “创建一个模板标签以加载模板中的 QuerySet”示例。
- “创建模板标签以调整请求查询参数”示例。

5.10　创建模板标签以调整请求查询参数

Django 提供了一个方便灵活的系统，只要在 URL 配置文件中添加正则表达式规则，即可创建规范且干净的 URL。但是，目前尚缺少一种内建技术管理查询参数。搜索或可筛选对象列表的视图需要接收查询参数，以便使用另一个参数钻取过滤后的结果，或转到另一个页面。

在当前示例中，我们将创建{% modify_query %}、{% add_to_query %}和{% remove_from_query %}模板标签，进而可添加、修改或移除当前查询的参数。

5.10.1　准备工作

当前示例将在 core 应用程序的基础上完成，该程序在 INSTALLED_APPS 中被设置，并包含了 templatetags 包。

另外，确保 request 上下文预处理器已被添加至 OPTIONS 下的 TEMPLATES 设置项的 context_processors 列表中，如下所示。

```
# myproject/settings/_base.py
TEMPLATES = [
    {
        "BACKEND": "django.template.backends.django.DjangoTemplates",
        "DIRS": [os.path.join(BASE_DIR, "myproject", "templates")],
        "APP_DIRS": True,
```

```
        "OPTIONS": {
            "context_processors": [
                "django.template.context_processors.debug",
                "django.template.context_processors.request",
                "django.contrib.auth.context_processors.auth",
                "django.contrib.messages.context_processors.messages",
                "django.template.context_processors.media",
                "django.template.context_processors.static",
                "myproject.apps.core.context_processors.website_url",
            ]
        },
    }
]
```

5.10.2　实现方式

对于这些模板标签，我们将使用@simple_tag 装饰器，并解析组件以及反定义渲染函数，如下所示。

（1）添加一个帮助方法，将每个标签输出的查询字符串整合在一起。

```
# myproject/apps/core/templatetags/utility_tags.py
from urllib.parse import urlencode

from django import template
from django.utils.encoding import force_str
from django.utils.safestring import mark_safe

register = template.Library()

""" TAGS """

def construct_query_string(context, query_params):
    # empty values will be removed
    query_string = context["request"].path
    if len(query_params):
        encoded_params = urlencode([
            (key, force_str(value))
            for (key, value) in query_params if value
        ]).replace("&", "&amp;")
        query_string += f"?{encoded_params}"
    return mark_safe(query_string)
```

（2）创建{% modify_query %}模板标签。

```
@register.simple_tag(takes_context=True)
def modify_query(context, *params_to_remove, **params_to_change):
    """Renders a link with modified current query parameters"""
    query_params = []
    for key, value_list in context["request"].GET.lists():
        if not key in params_to_remove:
            # don't add key-value pairs for params_to_remove
            if key in params_to_change:
                # update values for keys in params_to_change
                query_params.append((key, params_to_change[key]))
                params_to_change.pop(key)
            else:
                # leave existing parameters as they were
                # if not mentioned in the params_to_change
                for value in value_list:
                    query_params.append((key, value))
                    # attach new params
    for key, value in params_to_change.items():
        query_params.append((key, value))
    return construct_query_string(context, query_params)
```

（3）创建{% add_to_query %}模板标签。

```
@register.simple_tag(takes_context=True)
def add_to_query(context, *params_to_remove, **params_to_add):
    """Renders a link with modified current query parameters"""
    query_params = []
    # go through current query params..
    for key, value_list in context["request"].GET.lists():
        if key not in params_to_remove:
            # don't add key-value pairs which already
            # exist in the query
            if (key in params_to_add
                    and params_to_add[key] in value_list):
                params_to_add.pop(key)
            for value in value_list:
                query_params.append((key, value))
    # add the rest key-value pairs
    for key, value in params_to_add.items():
        query_params.append((key, value))
    return construct_query_string(context, query_params)
```

（4）创建{% remove_from_query %}模板标签。

```
@register.simple_tag(takes_context=True)
def remove_from_query(context, *args, **kwargs):
    """Renders a link with modified current query parameters"""
    query_params = []
    # go through current query params..
    for key, value_list in context["request"].GET.lists():
        # skip keys mentioned in the args
        if key not in args:
            for value in value_list:
                # skip key-value pairs mentioned in kwargs
                if not (key in kwargs and
                        str(value) == str(kwargs[key])):
                    query_params.append((key, value))
    return construct_query_string(context, query_params)
```

5.10.3　工作方式

上述 3 个生成的模板标签其行为类似。首先，这些模板标签将当前查询参数从 request.GET 字典 QueryDict 对象读取至一个新的（键、值）query_params 元组列表中。随后，对应值根据位置参数和关键字参数更新。最后，新的查询字符串通过之前定义的帮助方法创建。在该处理过程中，所有的空格和特殊字符均为 URL 编码，且连接查询参数的&字符被转义。这个新的查询字符串将被返回至模板中。

注意：

关于 QueryDict 对象的更多内容，读者可查看 Django 官方文档，对应网址为 https://docs.djangoproject.com/en/3.0/ref/request-response/#querydict-objects。

接下来探讨{% modify_query %}模板标签的应用方式。其中，模板标签中的位置参数定义了查询参数中被移除的参数；而关键字参数则定义了当前查询中更新的查询参数。如果当前 URL 为 http://127.0.0.1:8000/artists/?category=fine-art&page=5，我们可以使用下列模板标签渲染访问下一个页面的链接。

```
{% load utility_tags %}
<a href="{% modify_query page=6 %}">6</a>
```

下列代码片段表示为使用上述模板标签后渲染的输出结果。

```
<a href="/artists/?category=fine-art&amp;page=6">6</a>
```

此外，还可使用以下示例渲染一个链接，该链接重置分页并访问另一个分类 sculpture，如下所示。

```
{% load utility_tags %}
<a href="{% modify_query "page" category="sculpture" %}">
   Sculpture
</a>
```

因此，采用上述代码片段渲染的输出结果如下所示。

```
<a href="/artists/?category=sculpture">
   Sculpture
</a>
```

当使用{% add_to_query %}模板标签时，我们可以逐步添加具有相同名称的参数。例如，如果当前的 URL 表示为 http://127.0.0.1:8000/artists/?category=fine-art，我们可在下列代码片段的帮助下添加另一个分类 Sculpture。

```
{% load utility_tags %}
<a href="{% add_to_query category="sculpture" %}">
   + Sculpture
</a>
```

这将在模板内被渲染，如下所示。

```
<a href="/artists/?category=fine-art&amp;category=sculpture">
   + Sculpture
</a>
```

最后，借助{% remove_from_query %}模板标签，我们可逐步移除包含相同名称的参数。例如，如果当前 URL 表示为 http://127.0.0.1:8000/artists/?category=fine-art&category=sculpture，那么我们可以借助下列代码片段移除 Sculpture 分类。

```
{% load utility_tags %}
<a href="{% remove_from_query category="sculpture" %}">
   - Sculpture
</a>
```

这将在模板中渲染，如下所示。

```
<a href="/artists/?category=fine-art">
   - Sculpture
</a>
```

5.10.4 延伸阅读

- 第 3 章中的“过滤对象列表”示例。
- “创建一个模板标签以包含一个模板”示例。
- “创建一个模板标签以加载模板中的 QuerySet”示例。
- “创建一个模板标签以作为模板解析内容”示例。

第6章 模型管理

本章主要涉及下列主题。

- 自定义修改列表页上的列。
- 创建可排序的内联。
- 生成管理动作。
- 开发修改列表过滤器。
- 修改第三方应用程序的应用程序标记。
- 创建自定义账户应用程序。
- 创建用户头像。
- 向修改表单中插入一幅地图。

6.1 简　　介

Django 框架包含数据模型的内建管理系统，我们可方便地设置可过滤的、可搜索的和可排序的列表以浏览模型。同时，我们还可以配置表单并添加和管理数据。本章将探讨一些高级技术并通过实际用例自定义管理机制。

6.2 技术需求

当与本章代码协同工作时，需要安装最新稳定版本的 Python、MySQL 或 PostgreSQL 数据库，以及基于虚拟环境的 Django 项目。

读者可查看 GitHub 储存库的 chapter 06 目录查看本章的代码，对应网址为 https://github.com/PacktPublishing/Django-3-Web-Development-Cookbook-Fourth-Edition。

6.3 自定义修改列表页面上的列

在默认的 Django 管理系统中，修改列表视图提供了特定模型所有实例的概览。默认状态下，list_display 模型管理属性控制着不同列表中显示的字段。除此之外，我们还可实现自定义的管理方法，并返回关系中的数据或显示自定义 HTML。在当前示例中，我们将结合 list_display 属性创建一个特殊的函数，进而在某个列表视图列中显示一幅图像。通过添加 list_editable 设置项，还可直接在列表视图中编辑某个字段。

6.3.1 准备工作

当前示例需要使用 Pillow 和 django-imagekit 库。下面在虚拟环境中使用下列命令对其进行安装。

```
(env)$ pip install Pillow
(env)$ pip install django-imagekit
```

确保 django.contrib.admin 和 imagekit 位于设置项的 INSTALLED_APPS 中。

```
# myproject/settings/_base.py
INSTALLED_APPS = [
    # …
    "django.contrib.admin",
    "imagekit",
]
```

在 URL 配置中连接管理站点，如下所示。

```
# myproject/urls.py
from django.contrib import admin
from django.conf.urls.i18n import i18n_patterns
from django.urls import include, path

urlpatterns = i18n_patterns(
    # …
    path("admin/", admin.site.urls),
)
```

随后，创建一个新的 products 应用程序，并将其置于 INSTALLED_APPS 下。该应用程序包含 Product 和 ProductPhoto 模型。这里，一件商品可包含多幅照片。针对当前示例，我们还可使用第 2 章中定义的 UrlMixin。

接下来在 models.py 文件中创建 Product 和 ProductPhoto，如下所示。

```
# myproject/apps/products/models.py
import os
```

```
from django.urls import reverse, NoReverseMatch
from django.db import models
from django.utils.timezone import now as timezone_now
from django.utils.translation import ugettext_lazy as _

from ordered_model.models import OrderedModel

from myproject.apps.core.models import UrlBase

def product_photo_upload_to(instance, filename):
    now = timezone_now()
    slug = instance.product.slug
    base, ext = os.path.splitext(filename)
    return f"products/{slug}/{now:%Y%m%d%H%M%S}{ext.lower()}"

class Product(UrlBase):
    title = models.CharField(_("title"), max_length=200)
    slug = models.SlugField(_("slug"), max_length=200)
    description = models.TextField(_("description"), blank=True)
    price = models.DecimalField(
        _("price (EUR)"), max_digits=8, decimal_places=2,
         blank=True, null=True
    )

    class Meta:
        verbose_name = _("Product")
        verbose_name_plural = _("Products")

    def get_url_path(self):
        try:
            return reverse("product_detail", kwargs={"slug": self.slug})
        except NoReverseMatch:
            return ""

    def __str__(self):
        return self.title

class ProductPhoto(models.Model):
    product = models.ForeignKey(Product, on_delete=models.CASCADE)
    photo = models.ImageField(_("photo"),
```

```
    upload_to=product_photo_upload_to)

    class Meta:
        verbose_name = _("Photo")
        verbose_name_plural = _("Photos")

    def __str__(self):
        return self.photo.name
```

6.3.2　实现方式

在当前示例中，我们将创建一个简单的 Product 模型的管理系统，它将以内联的方式将 ProductPhoto 模型的实例绑定至产品上。

在 list_display 属性中，我们将包含模型管理的 first_photo()方法，该方法用于显示多对一关系中的第一张照片。

（1）创建包含下列代码的 admin.py 文件。

```
# myproject/apps/products/admin.py
from django.contrib import admin
from django.template.loader import render_to_string
from django.utils.html import mark_safe
from django.utils.translation import ugettext_lazy as _

from .models import Product, ProductPhoto

class ProductPhotoInline(admin.StackedInline):
    model = ProductPhoto
    extra = 0
    fields = ["photo"]
```

（2）在同一文件中，添加产品的管理。

```
@admin.register(Product)
class ProductAdmin(admin.ModelAdmin):
    list_display = ["first_photo", "title", "has_description", "price"]
    list_display_links = ["first_photo", "title"]
    list_editable = ["price"]

    fieldsets = ((_("Product"), {"fields": ("title", "slug",
     "description", "price")}),)
    prepopulated_fields = {"slug": ("title",)}
```

```
    inlines = [ProductPhotoInline]

def first_photo(self, obj):
        project_photos = obj.productphoto_set.all()[:1]
        if project_photos.count() > 0:
        photo_preview = render_to_string(
            "admin/products/includes/photo-preview.html",
            {"photo": project_photos[0], "product": obj},
            )
            return mark_safe(photo_preview)
        return ""

    first_photo.short_description = _("Preview")

def has_description(self, obj):
return bool(obj.description)

    has_description.short_description = _("Has description?")
    has_description.admin_order_field = "description"
    has_description.boolean = True
```

（3）创建模板，用于生成 photo-preview，如下所示。

```
{# admin/products/includes/photo-preview.html #}
{% load imagekit %}
{% thumbnail "120x120" photo.photo -- alt=
"{{ product.title }} preview" %}
```

6.3.3　工作方式

当添加包含照片的多个产品，并在浏览器中查看产品管理列表时，对应结果如图 6.1 所示。

list_display 属性通常用于定义字段，以便在管理列表视图中对其进行显示。例如，TITLE 和 PRICE 表示为 Product 模型的字段。除了普通的字段名，list_display 属性还接收下列内容。

- 一个函数或另一个可调用的函数。
- 模型管理类的属性名。
- 模型的属性名。

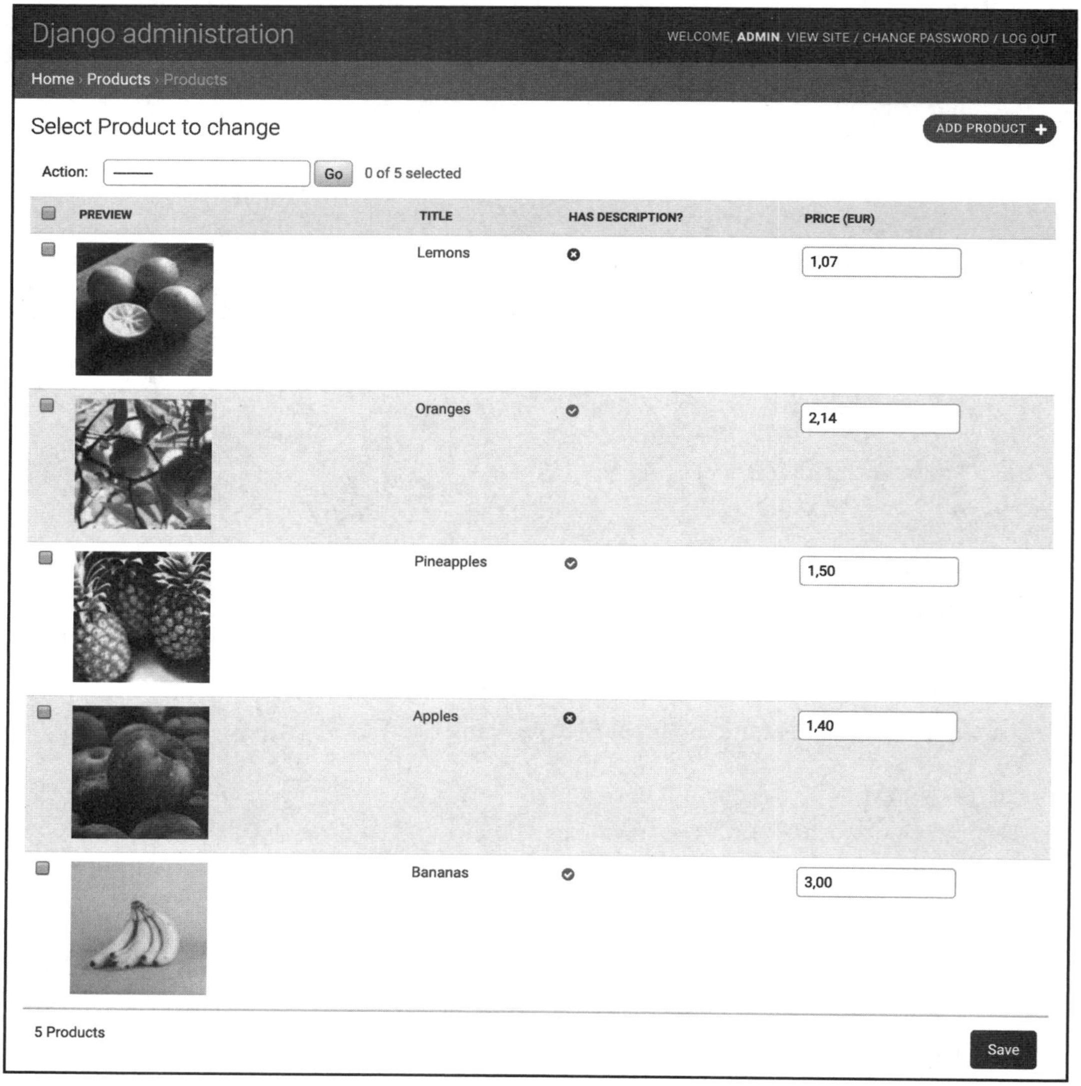

图 6.1

当在 list_display 中使用可调用的函数时，每个函数将获得作为第一个参数传递的模型实例。因此，在当前示例中，我们在模型管理类中定义了 get_photo()方法，该方法以 obj 形式接收 Product 实例，并尝试从多对一关系中获取第一个 ProductPhoto 对象。如果存在，该方法返回生成自 include 模板并带有<img>标签的 HTML。通过设置 list_display_ links，我们针对 Product 模型生成了链接至管理修改表单的照片和标题。

对于 list_display 中所使用的可调用函数，我们还可设置多个属性。

- 可调用函数的 short_description 属性定义了在列表上方显示的 TITLE。
- 默认状态下，返回自可调用函数的值在管理中并未呈排序状态，但是可以通过设置 admin_order_field 属性来定义应该根据哪个数据库字段对生成的列进行排序。另外，也可以在字段前加上连字符，以表示逆置的排序顺序。
- 通过设置 boolean = True，可以显示 True 或 False 值的图标。

最后，如果将 PRICE 字段包含至 list_editable 设置项中，那么该字段将成为一个可编辑的字段。由于存在可编辑的字段，那么底部将出现一个 Save 按钮，以便保存整个产品列表。

6.3.4　延伸阅读

- 第 2 章中的“利用与 URL 相关的方法创建一个模型混入”示例。
- “创建管理动作”示例。
- “开发修改列表过滤器”示例。

6.4　创建可排序的内联

我们可能需要根据生成日期、出现日期或字母顺序排序数据库中的大多数模型。但有些时候，用户还应能够显示自定义排序顺序中的条目。这适用于分类、图片库、策划表或类似的情形。在当前示例中，我们将讨论如何使用 django-ordered-model，以支持管理中的自定义排序。

6.4.1　准备工作

当前示例将在前述已定义的 products 应用程序示例的基础上完成，具体步骤如下所示。

（1）在虚拟环境中安装 django-ordered-model。

```
(env)$ pip install django-ordered-model
```

（2）将 ordered_model 添加至设置项的 INSTALLED_APPS 中。

（3）调整之前定义的 products 应用程序中的 ProductPhoto 模型，如下所示。

```
# myproject/apps/products/models.py
```

```
from django.db import models
from django.utils.translation import ugettext_lazy as _

from ordered_model.models import OrderedModel

# …

class ProductPhoto(OrderedModel):
    product = models.ForeignKey(Product, on_delete=models.CASCADE)
    photo = models.ImageField(_("photo"),
     upload_to=product_photo_upload_to)

order_with_respect_to = "product"

    class Meta(OrderedModel.Meta):
        verbose_name = _("Photo")
        verbose_name_plural = _("Photos")

def __str__(self):
return self.photo.name
```

OrderedModel 类引入了一个 order 字段。生成并运行迁移，并将 ProductPhoto 的新 order 字段添加至数据库中。

6.4.2　实现方式

当设置可排序的产品照片时，我们需要调整 products 应用程序的模型管理。

（1）调整管理文件中的 ProductPhotoInline，如下所示。

```
# myproject/apps/products/admin.py
from django.contrib import admin
from django.template.loader import render_to_string
from django.utils.html import mark_safe
from django.utils.translation import ugettext_lazy as _
from ordered_model.admin import OrderedTabularInline,
OrderedInlineModelAdminMixin

from .models import Product, ProductPhoto

class ProductPhotoInline(OrderedTabularInline):
    model = ProductPhoto
```

```
    extra = 0
    fields = ("photo_preview", "photo", "order", "move_up_down_links")
    readonly_fields = ("photo_preview", "order", "move_up_down_links")
    ordering = ("order",)

    def get_photo_preview(self, obj):
        photo_preview = render_to_string(
            "admin/products/includes/photo-preview.html",
            {"photo": obj, "product": obj.product},
        )
        return mark_safe(photo_preview)

    get_photo_preview.short_description = _("Preview")
```

（2）调整 ProductAdmin，如下所示。

```
@admin.register(Product)
class ProductAdmin(OrderedInlineModelAdminMixin, admin.ModelAdmin):
    # …
```

6.4.3　工作方式

当打开 Change Product 表单时，对应结果如图 6.2 所示。

在当前模型中，我们设置了 order_with_respect_to 属性，以确保排序独立于每件产品，而不是排序产品照片的整个列表。

在 Django administration 中，产品照片可通过遵循产品自身细节并作为表格内联进行编辑。在第一列中，我们包含了一个照片预览图，并通过前述示例中使用的同一 photo-preview.html 模板生成该预览图。在第二列中，存在一个修改照片的字段。随后是 ORDER 字段列，后面是一个带有箭头的按钮的列，以便可在一侧以手动方式重新排列照片。其中，箭头按钮源自 move_up_down_links 方法。最后是一个复选框列，以便可删除内联。

readonly_fields 属性通知 Django 的某些字段或方法为只读。如果打算使用另一种方法显示修改表单中的内容，则需要将相关方法置于 readonly_fields 列表中。在当前示例中，get_photo_preview 和 move_up_down_links 就是此类方法。

move_up_down_links 被定义在 OrderedTabularInline 中，我们扩展的是 OrderedTabularInline，而不是 adminStackedInline 或 adminTabularInline。这将渲染箭头按钮，以便在产品照片中切换位置。

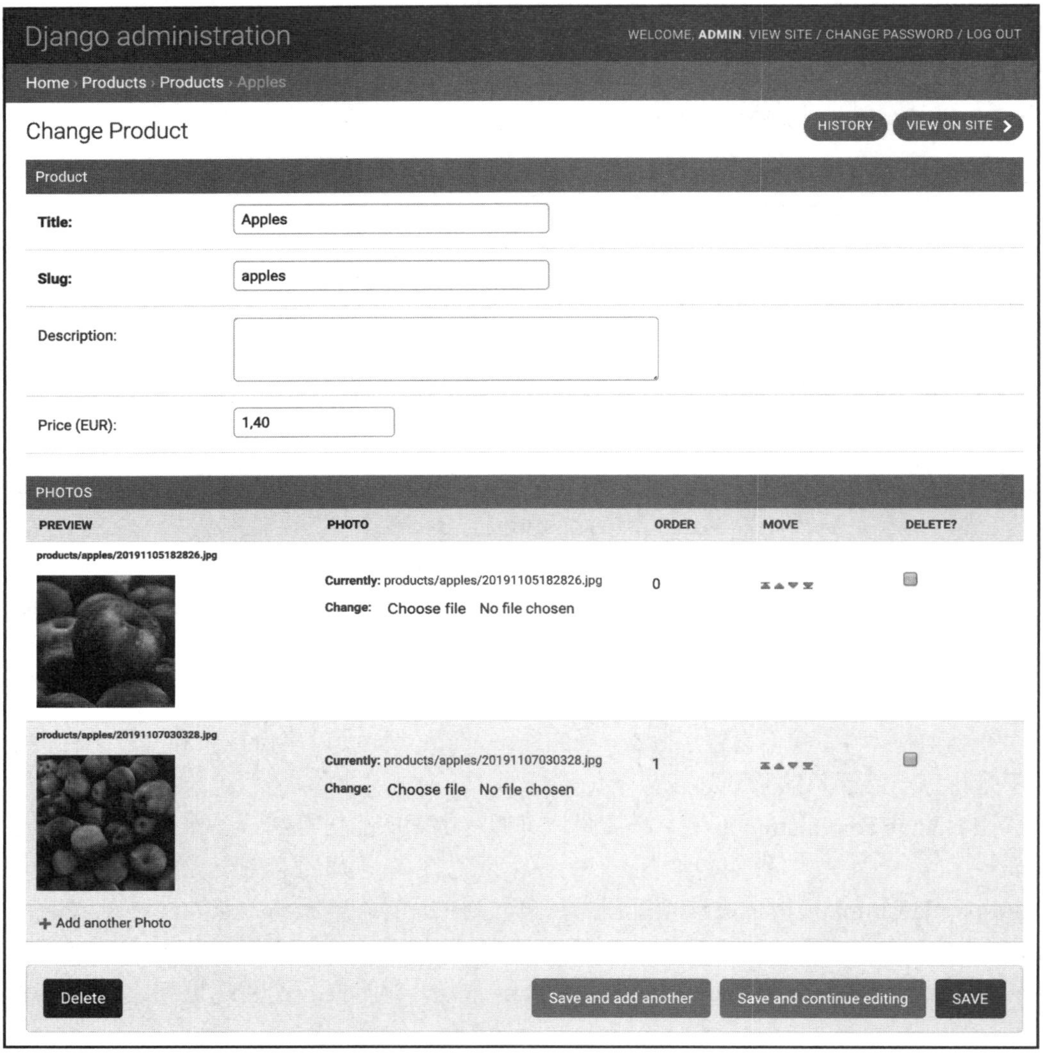

Django administration

WELCOME, **ADMIN**. VIEW SITE / CHANGE PASSWORD / LOG OUT

Home › Products › Products › Apples

Change Product

HISTORY　VIEW ON SITE ›

Product

Title: Apples

Slug: apples

Description:

Price (EUR): 1,40

PHOTOS

PREVIEW	PHOTO	ORDER	MOVE	DELETE?
products/apples/20191105182826.jpg	**Currently:** products/apples/20191105182826.jpg **Change:** Choose file No file chosen	0		
products/apples/20191107030328.jpg	**Currently:** products/apples/20191107030328.jpg **Change:** Choose file No file chosen	1		

+ Add another Photo

Delete　Save and add another　Save and continue editing　SAVE

图 6.2

6.4.4 延伸阅读

- “自定义修改列表页面上的列”示例。
- “创建管理动作”示例。
- “开发修改列表过滤器”示例。

6.5　创建管理动作

Django 管理系统提供了多个动作，并可以执行选择列表中的条目。默认状态下，仅提供一个动作，用于删除所选的实例。在当前示例中，我们将针对 Product 模型列表创建一个附加动作，这将允许管理员将所选产品导出至 Excel 电子表格中。

6.5.1　准备工作

当前示例将在 products 应用程序的基础上完成。确保在虚拟环境中安装了 openpyxl 模块，以创建 Excel 电子表格，如下所示。

```
(env)$ pip install openpyxl
```

6.5.2　实现方式

管理动作为接收 3 个参数的函数，如下所示。

（1）当前 ModelAdmin 值。

（2）当前 HttpRequest 值。

（3）QuerySet 值，其中包含了所选的条目。

执行下列步骤创建一个自定义管理动作并导出一个电子表格。

（1）在 products 应用程序中的 admin.py 文件中，创建一个电子表格列配置的 ColumnConfig 类，如下所示。

```
# myproject/apps/products/admin.py
from openpyxl import Workbook
from openpyxl.styles import Alignment, NamedStyle, builtins
from openpyxl.styles.numbers import FORMAT_NUMBER
from openpyxl.writer.excel import save_virtual_workbook

from django.http.response import HttpResponse
from django.utils.translation import ugettext_lazy as _
from ordered_model.admin import OrderedTabularInline,
OrderedInlineModelAdminMixin

# other imports…
```

```
class ColumnConfig:
    def __init__(
            self,
            heading,
            width=None,
            heading_style="Headline 1",
            style="Normal Wrapped",
            number_format=None,
        ):
        self.heading = heading
        self.width = width
        self.heading_style = heading_style
        self.style = style
        self.number_format = number_format
```

（2）在同一文件中，创建 export_xlsx()函数。

```
def export_xlsx(modeladmin, request, queryset):
    wb = Workbook()
    ws = wb.active
    ws.title = "Products"

    number_alignment = Alignment(horizontal="right")
    wb.add_named_style(
        NamedStyle(
            "Identifier", alignment=number_alignment,
            number_format=FORMAT_NUMBER
        )
    )
    wb.add_named_style(
        NamedStyle("Normal Wrapped",
         alignment=Alignment(wrap_text=True))
    )

    column_config = {
        "A": ColumnConfig("ID", width=10, style="Identifier"),
        "B": ColumnConfig("Title", width=30),
        "C": ColumnConfig("Description", width=60),
        "D": ColumnConfig("Price", width=15, style="Currency",
            number_format="#,##0.00 €"),
        "E": ColumnConfig("Preview", width=100, style="Hyperlink"),
    }

    # Set up column widths, header values and styles
```

```
    for col, conf in column_config.items():
        ws.column_dimensions[col].width = conf.width

        column = ws[f"{col}1"]
        column.value = conf.heading
        column.style = conf.heading_style

    # Add products
    for obj in queryset.order_by("pk"):
        project_photos = obj.productphoto_set.all()[:1]
        url = ""
        if project_photos:
            url = project_photos[0].photo.url
        data = [obj.pk, obj.title, obj.description, obj.price, url]
        ws.append(data)

        row = ws.max_row
        for row_cells in ws.iter_cols(min_row=row, max_row=row):
            for cell in row_cells:
                conf = column_config[cell.column_letter]
                cell.style = conf.style
                if conf.number_format:
                    cell.number_format = conf.number_format

    mimetype = "application/vnd.openxmlformatsofficedocument.
    spreadsheetml.sheet"
    charset = "utf-8"
    response = HttpResponse(
        content=save_virtual_workbook(wb),
        content_type=f"{mimetype}; charset={charset}",
        charset=charset,
    )
    response["Content-Disposition"] = "attachment;
    filename=products.xlsx"
    return response

export_xlsx.short_description = _("Export XLSX")
```

（3）将 actions 设置项添加至 ProductAdmin 中，如下所示。

```
@admin.register(Product)
class ProductAdmin(OrderedInlineModelAdminMixin, admin.ModelAdmin):
    # …
```

```
    actions = [export_xlsx]
    # ...
```

6.5.3 工作方式

当在浏览器中查看产品管理列表页时，我们将看到一个名为 Export XLSX 的新动作，连同默认的 Delete selected Products 动作，如图 6.3 所示。

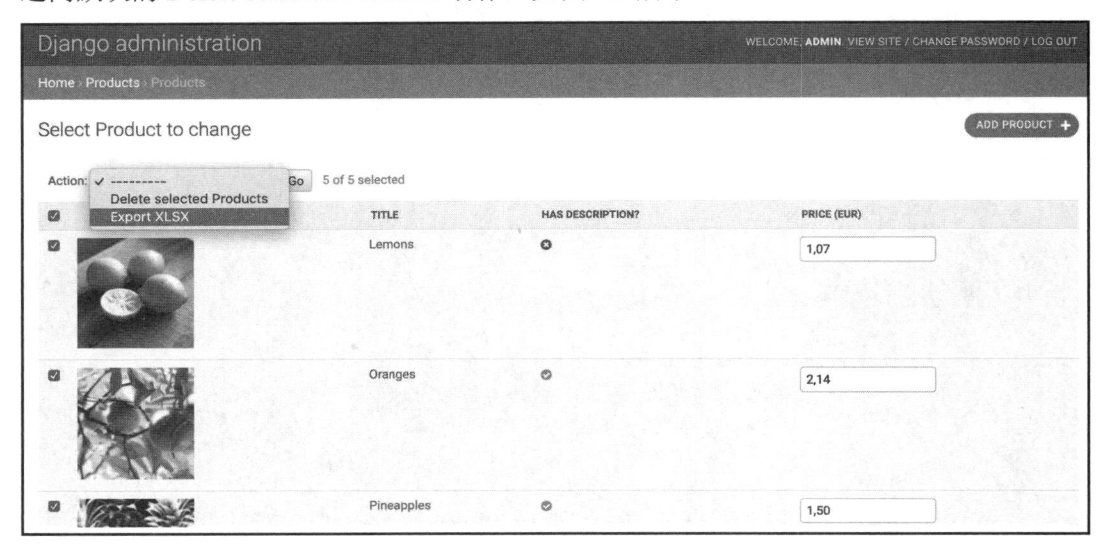

图 6.3

我们使用 openpyxl Python 模块创建 OpenOffice XML 文件，该文件兼容于 Excel 和其他电子表格软件。

首先创建一个工作簿，并选择活动的工作表，同时将其标题设置为 Products。由于存在一些在整个工作表中使用的通用样式，所以这些样式被设置为命名样式，以便可以根据名称应用至每个单元。这些样式、列标题和列宽度存储为 Config 对象，column_config 字典将列字母键映射到对象。随后执行遍历操作并设置标题和列宽度。

我们使用工作表的 append()方法在 QuerySet 中添加每件所选产品的内容，并按 ID 排序，包括当照片可用时产品的第一张照片的 URL。接下来，通过遍历刚刚添加的行中的每个单元格分别设置产品数据的样式，再次引用 column_config 来一致地应用样式。

默认情况下，管理动作对 QuerySet 执行一些操作，并将管理员重定向回修改列表页面。然而，对于更加复杂的动作，可返回 HttpResponse。export_xlsx()函数将工作簿的虚拟副本保存至 HttpResponse 中，其内容类型和字符集适合于 Office Open XML (OOXML)

电子表格。当使用 Content-Disposition 标头时，我们设置响应结果，以便它可作为 products.xlsx 文件下载。最终的电子表格可在 Open Office 中打开，如图 6.4 所示。

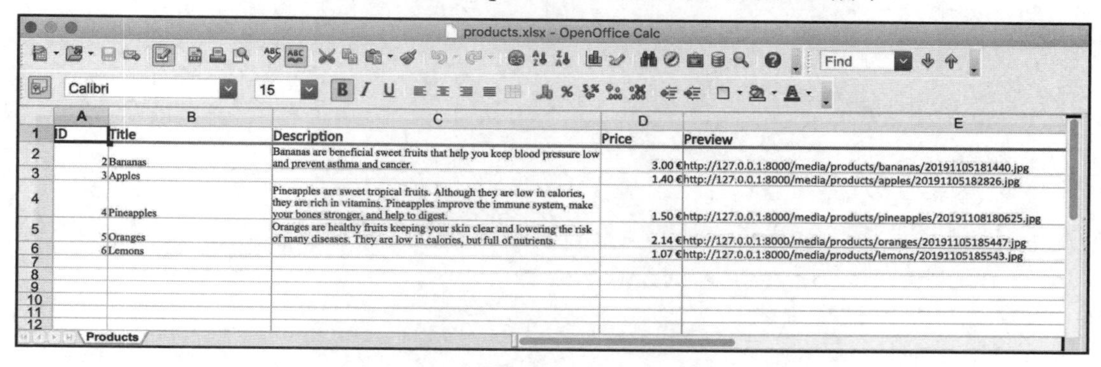

ID	Title	Description	Price	Preview
2	Bananas	Bananas are beneficial sweet fruits that help you keep blood pressure low and prevent asthma and cancer.	3.00 €	http://127.0.0.1:8000/media/products/bananas/20191105181440.jpg
3	Apples		1.40 €	http://127.0.0.1:8000/media/products/apples/20191105182826.jpg
4	Pineapples	Pineapples are sweet tropical fruits. Although they are low in calories, they are rich in vitamins. Pineapples improve the immune system, make your bones stronger, and help to digest.	1.50 €	http://127.0.0.1:8000/media/products/pineapples/20191108180625.jpg
5	Oranges	Oranges are healthy fruits keeping your skin clear and lowering the risk of many diseases. They are low in calories, but full of nutrients.	2.14 €	http://127.0.0.1:8000/media/products/oranges/20191105185447.jpg
6	Lemons		1.07 €	http://127.0.0.1:8000/media/products/lemons/20191105185543.jpg

图 6.4

6.5.4　延伸阅读

- “自定义修改列表页面上的列表”示例。
- “开发修改列表过滤器”示例。
- 第 9 章。

6.6　开发修改列表过滤器

如果希望管理员能够按照日期、关系或字段选择过滤修改列表，则需要使用管理模型的 list_filter 属性。除此之外，还可能存在定制的过滤器。在当前示例中，我们将添加一个过滤器并按照与其绑定的照片数量选择产品。

6.6.1　准备工作

当前示例将在之前的 products 示例的基础上完成。

6.6.2　实现方式

具体实现方式如下列步骤所示。

（1）在 admin.py 文件中，创建一个 PhotoFilter 类，该类从 SimpleListFilter 类扩展而来的。

```
# myproject/apps/products/admin.py
from django.contrib import admin
from django.db import models
from django.utils.translation import ugettext_lazy as _

# other imports…

ZERO = "zero"
ONE = "one"
MANY = "many"

class PhotoFilter(admin.SimpleListFilter):
    # Human-readable title which will be displayed in the
    # right admin sidebar just above the filter options.
    title = _("photos")

    # Parameter for the filter that will be used in the
    # URL query.
    parameter_name = "photos"

    def lookups(self, request, model_admin):
        """
        Returns a list of tuples, akin to the values given for
        model field choices. The first element in each tuple is the
        coded value for the option that will appear in the URL
        query. The second element is the human-readable name for
        the option that will appear in the right sidebar.
        """
        return (
            (ZERO, _("Has no photos")),
            (ONE, _("Has one photo")),
            (MANY, _("Has more than one photo")),
        )

    def queryset(self, request, queryset):
        """
        Returns the filtered queryset based on the value
        provided in the query string and retrievable via
        `self.value()`.

        """
        qs = queryset.annotate(num_photos=
         models.Count("productphoto"))
```

```
        if self.value() == ZERO:
            qs = qs.filter(num_photos=0)
        elif self.value() == ONE:
            qs = qs.filter(num_photos=1)
        elif self.value() == MANY:
            qs = qs.filter(num_photos__gte=2)
        return qs
```

（2）将一个列表过滤器添加至 ProductAdmin，如下所示。

```
@admin.register(Product)
class ProductAdmin(OrderedInlineModelAdminMixin, admin.ModelAdmin):
    # …
    list_filter = [PhotoFilter]
    # …
```

6.6.3 工作方式

根据已创建的自定义字段，列表过滤器显示于产品列表的侧栏，如图 6.5 所示。

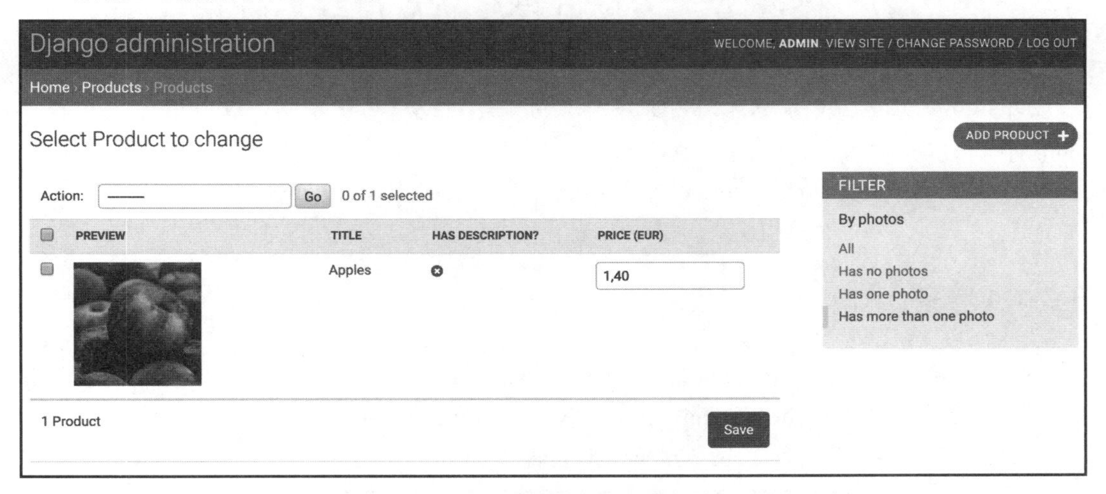

图 6.5

PhotoFilter 类包含一个可翻译的标题和查询参数名作为属性，此外还包含两种方法，如下所示。

（1）lookups()方法定义了过滤器的选择方案。

（2）queryset()方法定义了选取特定值时如何过滤 QuerySet 对象。

在 lookups()方法中，我们定义了 3 种选择方案，如下所示。

（1）no photos。

（2）one photo。

（3）绑定的 more than one photo。

在 queryset()方法中，我们使用 QuerySet 的 annotate()方法选择每件产品的照片数量。随后，该计数结果根据选取方案进行过滤。

关于聚合函数（如 annotate()）的更多内容，读者可参考 Django 官方文档，对应网址为 https://docs.djangoproject.com/en/3.0/topics/db/aggregation/。

6.6.4　延伸阅读

- ❑ “自定义修改列表页面上的列”示例。
- ❑ “创建管理动作”示例。
- ❑ “创建一个自定义账户应用程序”示例。

6.7　修改第三方应用程序的应用程序标记

Django 框架包含大量的第三方应用程序可供用户在项目中使用，读者可访问 https://djangopackages.org/进行查看和比较。在当前示例中，我们将讨论如何在管理中重命名 python-social-auth 应用程序的标记。类似地，我们还将修改 Django 第三方应用程序的标记。

6.7.1　准备工作

遵循 https://python-social-auth.readthedocs.io/en/latest/configuration/django.html 中的步骤，并在项目中安装 Python Social Auth。Python Social Auth 允许用户登录社交网络账户或其 Open ID。随后将显示如图 6.6 所示的管理页面。

PYTHON SOCIAL AUTH		
Associations	+ Add	Change
Nonces	+ Add	Change
User social auths	+ Add	Change

图 6.6

6.7.2 实现方式

首先，将 PYTHON SOCIAL AUTH 标记修改为用户友好的名称，如 SOCIAL AUTHENTICATION。具体步骤如下所示。

（1）创建一个名为 accounts 的应用程序。在 apps.py 文件中，添加下列内容。

```
# myproject/apps/accounts/apps.py
from django.apps import AppConfig
from django.utils.translation import ugettext_lazy as _

class AccountsConfig(AppConfig):
    name = "myproject.apps.accounts"
    verbose_name = _("Accounts")

    def ready(self):
        pass

class SocialDjangoConfig(AppConfig):
    name = "social_django"
    verbose_name = _("Social Authentication")
```

（2）设置 Python Social Auth 的步骤之一是将"social_django"应用程序添加至 INSTALLED_APPS。随后，将当前应用程序替换为"myproject.apps.accounts.apps.SocialDjangoConfig"。

```
# myproject/settings/_base.py
# …
INSTALLED_APPS = [
    # …
    #"social_django",
    "myproject.apps.accounts.apps.SocialDjangoConfig",
    # …
]
```

6.7.3 工作方式

当查看管理索引页面时，对应结果如图 6.7 所示。

INSTALLED_APPS 设置项接收应用程序路径或应用程序配置路径。相应地，我们可传递一个应用程序配置，而非默认的应用程序路径。这里，我们改变了应用程序的详细名称，甚至可通过一些信号处理程序或对应用程序执行一些其他的初始设置。

SOCIAL AUTHENTICATION		
Associations	+ Add	Change
Nonces	+ Add	Change
User social auths	+ Add	Change

图 6.7

6.7.4 延伸阅读

- “创建一个自定义账户应用程序”示例。
- “获取用户头像”示例。

6.8 创建一个自定义账户应用程序

Django 包含了一个 django.contrib.auth 应用程序，并可用于身份验证，进而可通过用户名和密码登录，并使用管理特性。该应用程序可通过自身的功能进行扩展。在当前示例中，我们将创建一个自定义用户和角色模型，并对其设置管理。相应地，我们将能够通过电子邮件和密码进行登录，而非用户名和密码。

6.8.1 准备工作

创建一个 accounts 应用程序，并将该应用程序置于设置项中的 INSTALLED_APPS 中。

```
# myproject/apps/_base.py
INSTALLED_APPS = [
    # …
    "myproject.apps.accounts",
]
```

6.8.2 实现方式

下列步骤将覆写用户和分组模型。

（1）在 accounts 应用程序中创建 models.py 文件，如下所示。

```
# myproject/apps/accounts/models.py
import uuid

from django.contrib.auth.base_user import BaseUserManager
from django.db import models
from django.contrib.auth.models import AbstractUser, Group
from django.utils.translation import ugettext_lazy as _

class Role(Group):
    class Meta:
        proxy = True
        verbose_name = _("Role")
        verbose_name_plural = _("Roles")

    def __str__(self):
        return self.name

class UserManager(BaseUserManager):
    def create_user(self, username="", email="", password="",
     **extra_fields):
        if not email:
            raise ValueError("Enter an email address")
        email = self.normalize_email(email)
        user = self.model(username=username, email=email, **extra_fields)
        user.set_password(password)
        user.save(using=self._db)
        return user

    def create_superuser(self, username="", email="", password=""):
        user = self.create_user(email=email, password=password,
         username=username)
        user.is_superuser = True
        user.is_staff = True
        user.save(using=self._db)
        return user

class User(AbstractUser):
    uuid = models.UUIDField(primary_key=True, default=None,
     editable=False)
    # change username to non-editable non-required field
    username = models.CharField(
        _("username"), max_length=150, editable=False, blank=True
```

```
    )
    # change email to unique and required field
    email = models.EmailField(_("email address"), unique=True)

    USERNAME_FIELD = "email"
    REQUIRED_FIELDS = []

    objects = UserManager()

    def save(self, *args, **kwargs):
        if self.pk is None:
            self.pk = uuid.uuid4()
        super().save(*args, **kwargs)
```

（2）利用 User 模型的管理配置在 accounts 应用程序中创建 admin.py 文件。

```
# myproject/apps/accounts/admin.py
from django.contrib import admin
from django.contrib.auth.admin import UserAdmin, Group, GroupAdmin
from django.urls import reverse
from django.contrib.contenttypes.models import ContentType
from django.http import HttpResponse
from django.shortcuts import get_object_or_404, redirect
from django.utils.encoding import force_bytes
from django.utils.safestring import mark_safe
from django.utils.translation import ugettext_lazy as _
from django.contrib.auth.forms import UserCreationForm

from .helpers import download_avatar
from .models import User, Role

class MyUserCreationForm(UserCreationForm):
    def save(self, commit=True):
        user = super().save(commit=False)
        user.username = user.email
        user.set_password(self.cleaned_data["password1"])
        if commit:
            user.save()
        return user

@admin.register(User)
class MyUserAdmin(UserAdmin):
    save_on_top = True
```

```
list_display = [
    "get_full_name",
    "is_active",
    "is_staff",
    "is_superuser",
]
list_display_links = [
    "get_full_name",
]
search_fields = ["email", "first_name", "last_name", "id", "username"]
ordering = ["-is_superuser", "-is_staff", "last_name", "first_name"]

fieldsets = [
    (None, {"fields": ("email", "password")}),

    (_("Personal info"), {"fields": ("first_name", "last_name")}),
    (
        _("Permissions"),
        {
            "fields": (
                "is_active",
                "is_staff",
                "is_superuser",
                "groups",
                "user_permissions",
            )
        },
    ),
    (_("Important dates"), {"fields":("last_login","date_joined")}),
]
add_fieldsets = (
    (None, {"classes": ("wide",), "fields": ("email",
     "password1", "password2")}),
)
add_form = MyUserCreationForm

def get_full_name(self, obj):
    return obj.get_full_name()

get_full_name.short_description = _("Full name")
```

（3）在同一文件中，添加 Role 模型的配置。

```
admin.site.unregister(Group)

@admin.register(Role)
class MyRoleAdmin(GroupAdmin):
    list_display = ("__str__", "display_users")
    save_on_top = True

    def display_users(self, obj):
        links = []
        for user in obj.user_set.all():
            ct = ContentType.objects.get_for_model(user)
            url = reverse(
                "admin:{}_{}_change".format(ct.app_label,
                    ct.model), args=(user.pk,)
            )
            links.append(
                """<a href="{}" target="_blank">{}</a>""".format(
                    url,
                    user.get_full_name() or user.username,
                )
            )
        return mark_safe(u"<br />".join(links))

    display_users.short_description = _("Users")
```

6.8.3 工作方式

默认的用户管理列表如图 6.8 所示。

默认的分组管理列表如图 6.9 所示。

在当前示例中，我们创建了两个模型。

（1）Role 模型表示为来自 django.contrib.auth 应用程序中 Group 模型中的一个代理。创建 Role 模型是为了将 Group 的详细名称重命名为 Role。

（2）User 模型扩展了和 django.contrib.auth 中的 User 模型相同的抽象 AbstractUser 类。创建 User 模型是为了用 UUIDField 替换主键，并允许我们通过电子邮件和密码而不是用户名和密码登录。

管理类 MyUserAdmin 和 MyRoleAdmin 扩展了已贡献的 UserAdmin 和 GroupAdmin 类，并覆写了某些属性。随后，我们为 User 和 Group 模型取消注册现有的管理类，并注册新的修改过的管理类。

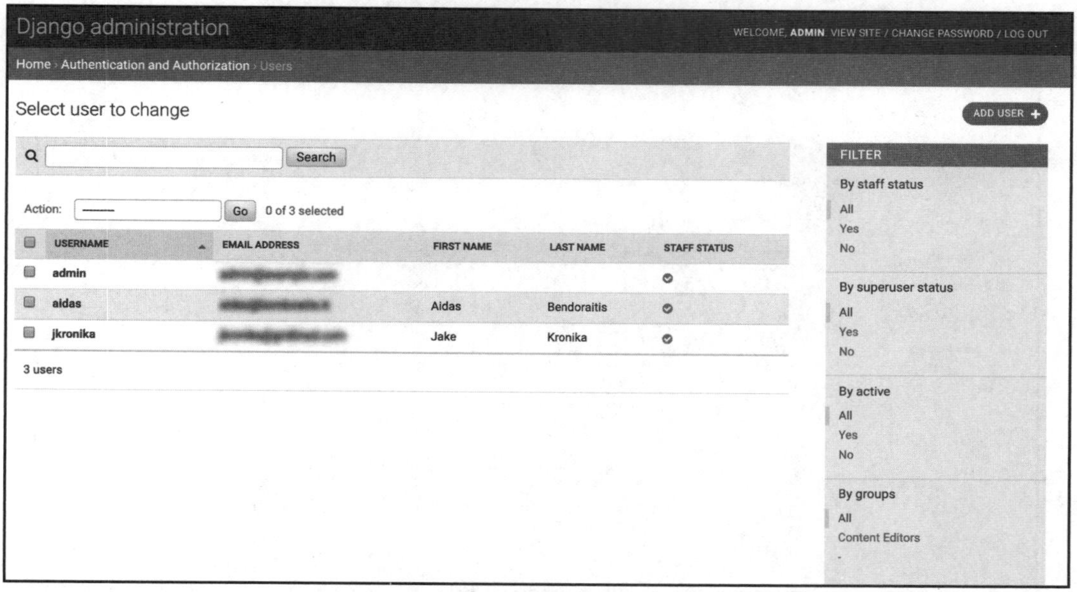

图 6.8

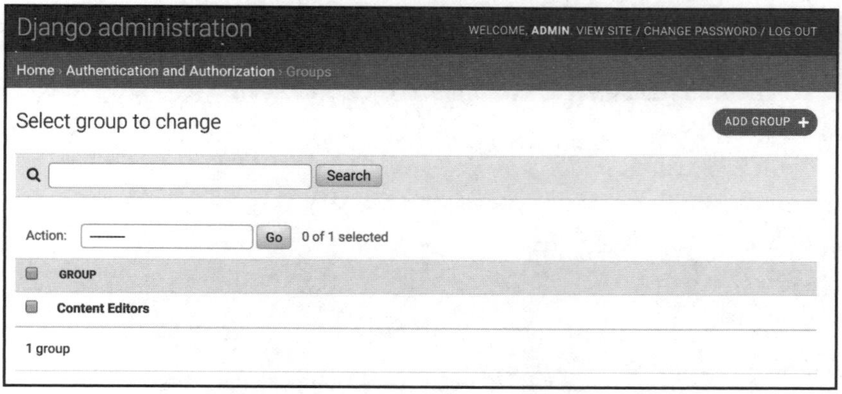

图 6.9

当前，用户管理如图 6.10 所示。

修改后的用户管理设置项在列表视图中显示了比默认设置项更多的字段、附加的过滤器和排序选项。

在新的分组管理设置项的修改列表中，我们将显示分配至特定分组的用户，在浏览器中，如图 6.11 所示。

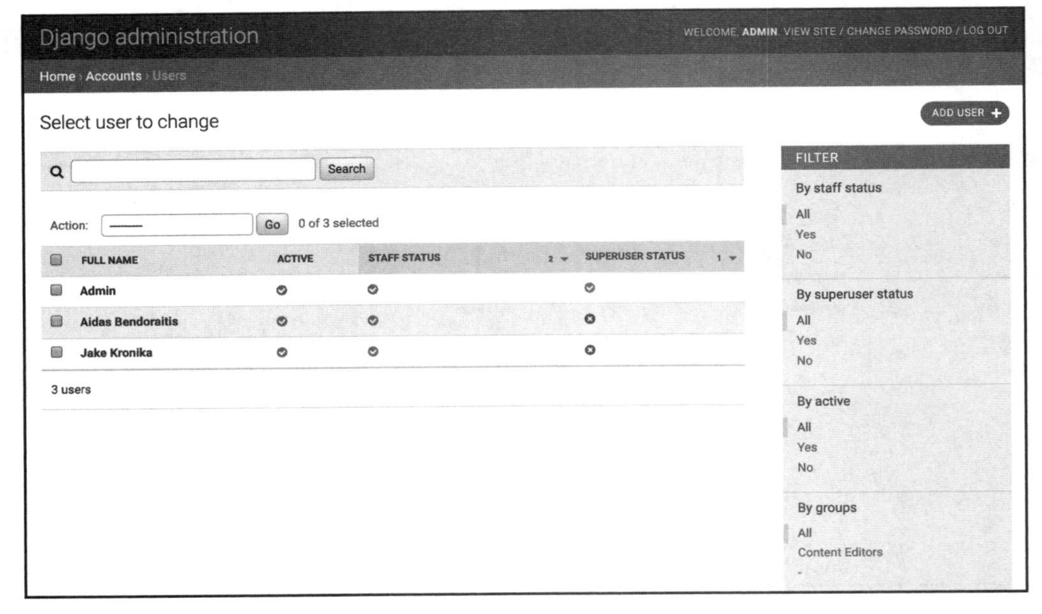

图 6.10

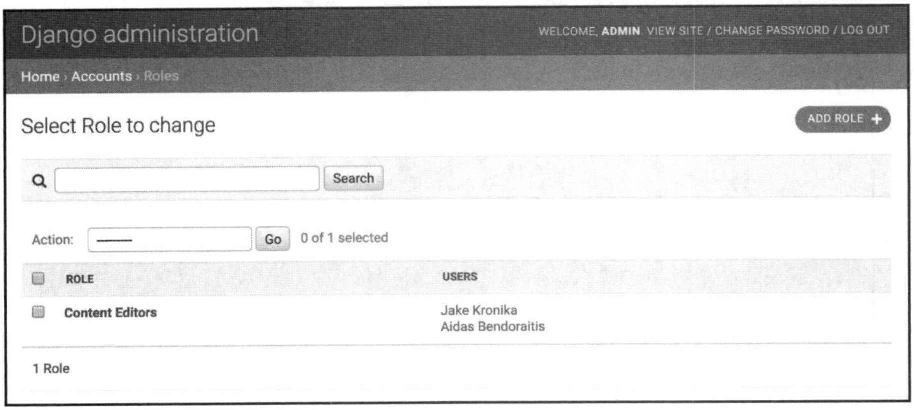

图 6.11

6.8.4　延伸阅读

- “自定义修改列表页面上的列”示例。
- “将一幅地图插入至修改表单中”示例。

6.9　获取用户头像

前述内容采用一个自定义的 User 模型进行身份验证，通过添加更多有用的字段，进一步增强相关功能。在当前示例中，我们将添加一个 avatar 字段，并且能够从 Gravatar 服务（https://en.gravatar.com/）中下载用户的头像。另外，该服务的用户还可上传头像并将其分配至电子邮件中。据此，不同的评论系统和社交平台可根据用户的电子邮件哈希值显示来自 Gravatar 的头像。

6.9.1　准备工作

当前示例将在之前创建的 accounts 应用程序的基础上完成。

6.9.2　实现方式

下列步骤用于增强 accounts 应用程序中的 User 模型。

（1）将 avatar 字段和 django-imagekit 缩略图规范添加到用户模型中。

```
# myproject/apps/accounts/models.py
import os

from imagekit.models import ImageSpecField
from pilkit.processors import ResizeToFill
from django.utils import timezone

# …

def upload_to(instance, filename):
    now = timezone.now()
    filename_base, filename_ext = os.path.splitext(filename)
    return "users/{user_id}/{filename}{ext}".format(
        user_id=instance.pk,
        filename=now.strftime("%Y%m%d%H%M%S"),
        ext=filename_ext.lower(),
    )

class User(AbstractUser):
    # …
```

```
    avatar = models.ImageField(_("Avatar"), upload_to=upload_to,
    blank=True)
    avatar_thumbnail = ImageSpecField(
        source="avatar",
        processors=[ResizeToFill(60, 60)],
        format="JPEG",
        options={"quality": 100},
    )

    # ...
```

（2）加入一些方法以下载并向 MyUserAdmin 类显示 Gravatar。

```
# myprojects/apps/accounts/admin.py
from django.contrib import admin
from django.contrib.auth.admin import UserAdmin, Group, GroupAdmin
from django.urls import reverse
from django.contrib.contenttypes.models import ContentType
from django.http import HttpResponse
from django.shortcuts import get_object_or_404
from django.utils.encoding import force_bytes
from django.utils.safestring import mark_safe
from django.utils.translation import ugettext_lazy as _
from django.contrib.auth.forms import UserCreationForm

from .helpers import download_avatar
from .models import User, Role

class MyUserCreationForm(UserCreationForm):
    def save(self, commit=True):
        user = super().save(commit=False)
        user.username = user.email
        user.set_password(self.cleaned_data["password1"])
        if commit:
            user.save()
        return user

@admin.register(User)
class MyUserAdmin(UserAdmin):
    save_on_top = True
    list_display = [
```

```
        "get_avatar",
        "get_full_name",
        "download_gravatar",
        "is_active",
        "is_staff",
        "is_superuser",
    ]
    list_display_links = [
        "get_avatar",
        "get_full_name",
    ]
    search_fields = ["email", "first_name", "last_name", "id", "username"]
    ordering = ["-is_superuser", "-is_staff", "last_name", "first_name"]

    fieldsets = [
        (None, {"fields": ("email", "password")}),
        (_("Personal info"), {"fields": ("first_name", "last_name")}),
        (
            _("Permissions"),
            {
                "fields": (
                    "is_active",
                    "is_staff",
                    "is_superuser",
                    "groups",
                    "user_permissions",
                )
            },
        ),
        (_("Avatar"), {"fields": ("avatar",)}),
        (_("Important dates"), {"fields": ("last_login", "date_joined")}),
    ]
    add_fieldsets = (
        (None, {"classes": ("wide",), "fields": ("email",
        "password1", "password2")}),
    )
    add_form = MyUserCreationForm

    def get_full_name(self, obj):
        return obj.get_full_name()

    get_full_name.short_description = _("Full name")
```

```
def get_avatar(self, obj):
    from django.template.loader import render_to_string
    html = render_to_string("admin/accounts
     /includes/avatar.html", context={
        "obj": obj
    })
    return mark_safe(html)

get_avatar.short_description = _("Avatar")

def download_gravatar(self, obj):
    from django.template.loader import render_to_string
    info = self.model._meta.app_label,
     self.model._meta.model_name
    gravatar_url = reverse("admin:%s_%s_download_gravatar" %
     info, args=[obj.pk])
    html = render_to_string("admin/accounts
     /includes/download_gravatar.html", context={
        "url": gravatar_url
    })
    return mark_safe(html)

download_gravatar.short_description = _("Gravatar")

def get_urls(self):
    from functools import update_wrapper
    from django.conf.urls import url

    def wrap(view):
        def wrapper(*args, **kwargs):
            return self.admin_site.admin_view(view)(*args, **kwargs)

        wrapper.model_admin = self
        return update_wrapper(wrapper, view)

    info = self.model._meta.app_label,
     self.model._meta.model_name

    urlpatterns = [
        url(
            r"^(.+)/download-gravatar/$",
```

```
            wrap(self.download_gravatar_view),
            name="%s_%s_download_gravatar" % info,
        )
    ] + super().get_urls()

    return urlpatterns

def download_gravatar_view(self, request, object_id):
    if request.method != "POST":
        return HttpResponse(
            "{} method not allowed.".format(request.method),
             status=405
        )
    from .models import User

    user = get_object_or_404(User, pk=object_id)
    import hashlib

    m = hashlib.md5()
    m.update(force_bytes(user.email))
    md5_hash = m.hexdigest()
    # d=404 ensures that 404 error is raised if gravatar is not
    # found instead of returning default placeholder

    url = "https://www.gravatar.com/avatar
     /{md5_hash}?s=800&d=404".format(
        md5_hash=md5_hash
    )
    download_avatar(object_id, url)
    return HttpResponse("Gravatar downloaded.", status=200)
```

（3）利用下列内容将 helpers.py 文件添加至 accounts 应用程序中。

```
# myproject/apps/accounts/helpers.py

def download_avatar(user_id, image_url):
    import tempfile
    import requests
    from django.contrib.auth import get_user_model
    from django.core.files import File

    response = requests.get(image_url, allow_redirects=True,
```

```
    stream=True)
   user = get_user_model().objects.get(pk=user_id)

   if user.avatar: # delete the old avatar
      user.avatar.delete()

   if response.status_code != requests.codes.ok:
      user.save()
      return

   file_name = image_url.split("/")[-1]

   image_file = tempfile.NamedTemporaryFile()

   # Read the streamed image in sections
   for block in response.iter_content(1024 * 8):
      # If no more file then stop
      if not block:
         break
      # Write image block to temporary file
      image_file.write(block)

   user.avatar.save(file_name, File(image_file))
   user.save()
```

（4）在管理文件中创建一个头像模板。

```
{# admin/accounts/includes/avatar.html #}
{% if obj.avatar %}
   <img src="{{ obj.avatar_thumbnail.url }}" alt=""
    width="30" height="30" />
{% endif %}
```

（5）创建一个 button 模板以下载 Gravatar。

```
{# admin/accounts/includes/download_gravatar.html #}
{% load i18n %}
<button type="button" data-url="{{ url }}" class="button
js_download_gravatar download-gravatar">
   {% trans "Get Gravatar" %}
</button>
```

（6）利用 JavaScript 创建一个用户修改列表管理的模板，并处理 Get Gravatar 按钮上的鼠标单击事件。

```
{# admin/accounts/user/change_list.html #}
{% extends "admin/change_list.html" %}
{% load static %}

{% block footer %}
{{ block.super }}
<style nonce="{{ request.csp_nonce }}">
.button.download-gravatar {
   padding: 2px 10px;
}
</style>
<script nonce="{{ request.csp_nonce }}">
django.jQuery(function($) {
   $('.js_download_gravatar').on('click', function(e) {
       e.preventDefault();
       $.ajax({
           url: $(this).data('url'),
           cache: 'false',
           dataType: 'json',
           type: 'POST',
           data: {},
           beforeSend: function(xhr) {
               xhr.setRequestHeader('X-CSRFToken',
                '{{ csrf_token }}');
           }
       }).then(function(data) {
           console.log('Gravatar downloaded.');
           document.location.reload(true);
       }, function(data) {
           console.log('There were problems downloading the Gravatar.');
           document.location.reload(true);
       });
   })
})

</script>
{% endblock %}
```

当查看用户修改列表管理时，对应结果如图 6.12 所示。

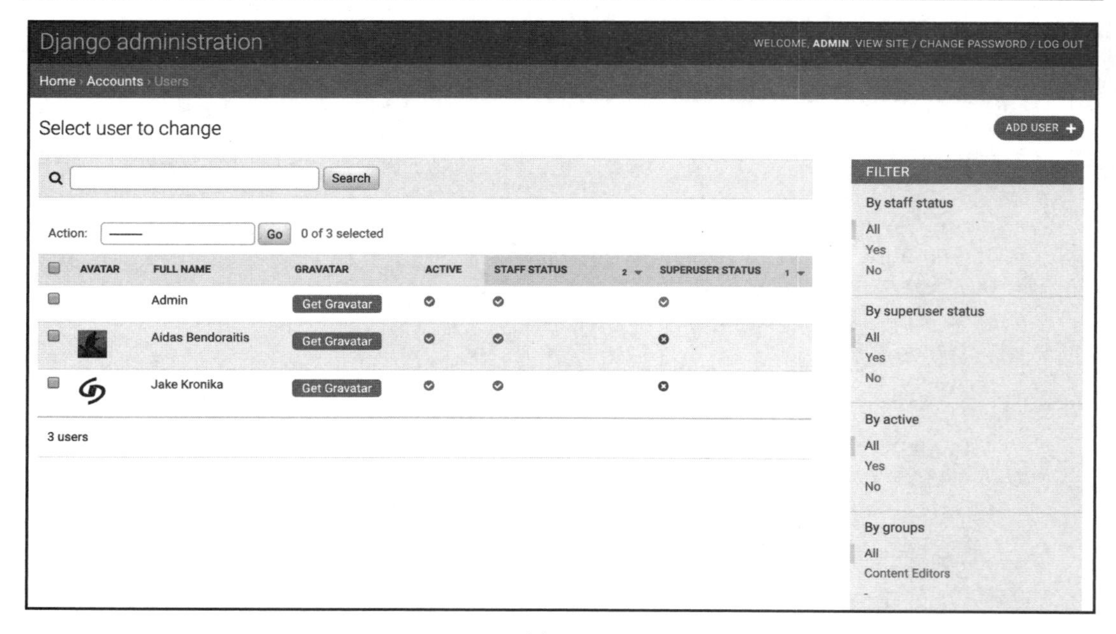

图 6.12

其中，对应列依次为用户的 AVATAR、FULL NAME 和获取头像的按钮。当用户单击 Get Gravatar 按钮时，JavaScript onclick 事件处理程序将向 download_gravatar_view 生成一个 POST 请求。该视图针对用户的头像生成一个 URL，这取决于用户电子邮件的 MD5 哈希值，随后将调用一个帮助函数并下载用户图像，同时将其链接至 avatar 字段。

6.9.3　更多内容

Gravatar 图像较小且易于下载。当从不同的服务中下载较大的图像时，读者可使用 Celery 或 Huey 任务队列在后台检索图像。关于 Celery，读者可访问 https://docs.celeryproject.org/en/latest/django/first-steps-with-django.html；关于 Huey，读者可访问 https://huey.readthedocs.io/en/0.4.9/django.html。

6.9.4　延伸阅读

- “修改第三方应用程序的应用程序标记”示例。
- “创建一个自定义账户应用程序”示例。

6.10　将一幅地图插入至修改表单中

Google Maps 提供了一个 JavaScript API，并且可以将地图插入至站点中。在当前示例中，我们将利用 Location 模型创建一个 locations 应用程序，并扩展修改表单的模板，以添加一幅地图。其中，管理员可发现和标记某个位置的地理坐标。

6.10.1　准备工作

注册一个 Google Maps API 密钥，并将其公开至模板中。类似于第 4 章中的“使用 HTML5 数据属性”示例。注意，针对当前示例，在 Google Cloud Platform 控制台中，用户需要激活 Maps JavaScript API 和 Geocoding API。另外，为了使这些 API 能够正常工作，还需要设置计费数据。

下面继续创建一个 locations 应用程序。

（1）将应用程序置于设置项的 INSTALLED_APPS 中。

```
# myproject/settings/_base.py
INSTALLED_APPS = [
    # …
    "myproject.apps.locations",
]
```

（2）利用名称、描述、地址、地理位置坐标和图像创建一个 Location 模型，如下所示。

```
# myproject/apps/locations/models.py
import os
import uuid
from collections import namedtuple

from django.contrib.gis.db import models
from django.urls import reverse
from django.conf import settings
from django.utils.translation import gettext_lazy as _
from django.utils.timezone import now as timezone_now

from myproject.apps.core.models import
CreationModificationDateBase, UrlBase
```

```
COUNTRY_CHOICES = getattr(settings, "COUNTRY_CHOICES", [])

Geoposition = namedtuple("Geoposition", ["longitude", "latitude"])

def upload_to(instance, filename):
    now = timezone_now()
    base, extension = os.path.splitext(filename)
    extension = extension.lower()
    return f"locations/{now:%Y/%m}/{instance.pk}{extension}"

class Location(CreationModificationDateBase, UrlBase):
    uuid = models.UUIDField(primary_key=True, default=None,
     editable=False)
    name = models.CharField(_("Name"), max_length=200)
    description = models.TextField(_("Description"))
    street_address = models.CharField(_("Street address"),
     max_length=255, blank=True)
    street_address2 = models.CharField(
        _("Street address (2nd line)"), max_length=255, blank=True
    )
    postal_code = models.CharField(_("Postal code"),
     max_length=255, blank=True)
    city = models.CharField(_("City"), max_length=255, blank=True)
    country = models.CharField(
        _("Country"), choices=COUNTRY_CHOICES, max_length=255,
        blank=True
    )
    geoposition = models.PointField(blank=True, null=True)
    picture = models.ImageField(_("Picture"), upload_to=upload_to)

    class Meta:
        verbose_name = _("Location")
        verbose_name_plural = _("Locations")

    def __str__(self):
        return self.name

    def get_url_path(self):
        return reverse("locations:location_detail",
         kwargs={"pk": self.pk})
```

```
    def save(self, *args, **kwargs):
        if self.pk is None:
            self.pk = uuid.uuid4()
        super().save(*args, **kwargs)

    def delete(self, *args, **kwargs):
        if self.picture:
            self.picture.delete()
        super().delete(*args, **kwargs)

    def get_geoposition(self):
        if not self.geoposition:
            return None
        return Geoposition(self.geoposition.coords[0],
         self.geoposition.coords[1])

    def set_geoposition(self, longitude, latitude):
        from django.contrib.gis.geos import Point
        self.geoposition = Point(longitude, latitude, srid=4326)
```

（3）针对 PostgreSQL 数据库安装 PostGIS 扩展。对此，最为简单的方式是运行 dbshell 管理命令，并执行下列命令。

```
> CREATE EXTENSION postgis;
```

（4）利用地理位置（稍后将对此进行修改）创建模型默认的管理。

```
# myproject/apps/locations/admin.py
from django.contrib.gis import admin
from .models import Location

@admin.register(Location)
class LocationAdmin(admin.OSMGeoAdmin):
    pass
```

针对来自贡献的 gis 模块的地理 Point 字段，默认的 Django 管理采用了 Leaflet.js JavaScript 映射库。其中，单元图源自 Open Street Maps，对应管理如图 6.13 所示。

注意，在默认的设置中，我们无法通过手动方式输入纬度和经度，同时也不能从地址信息中对地理位置进行地理编码。我们将在当前示例中实现这些功能。

Django administration　WELCOME, **ADMIN**. VIEW SITE / CHANGE PASSWORD / LOG OUT

Home › Locations › Locations › Park Güell

Change Location　HISTORY　VIEW ON SITE ›

Name: Park Güell

Description: If you want to see something spectacular, come to Barcelona, Catalonia, Spain and visit Park Güell. Located on a hill, Park Güell is a public park with beautiful gardens and organic architectural elements.

Street address: Park Güell

Street address (2nd line):

Postal code: 08024

City: Barcelona

Country: Spain

Geoposition:

Delete all Features

Picture: **Currently:** locations/2019/11/c6dc6d86-0345-4135-9ef3-34a6f36b1df4.jpg

Change: Choose file　No file chosen

Delete　Save and add another　Save and continue editing　SAVE

图 6.13

6.10.2　实现方式

Location 模型的管理将与多个文件整合使用。执行下列步骤将创建 Location 模型。

（1）创建 Location 模型的管理配置。注意，我们还将创建一个自定义模型表单，以生成独立的 latitude 和 longitude 字段。

```
# myproject/apps/locations/admin.py
from django.contrib import admin
from django import forms
from django.conf import settings
from django.template.loader import render_to_string
from django.utils.translation import ugettext_lazy as _

from .models import Location

LATITUDE_DEFINITION = _(
    "Latitude (Lat.) is the angle between any point and the "
    "equator (north pole is at 90°; south pole is at -90°)."
)

LONGITUDE_DEFINITION = _(
    "Longitude (Long.) is the angle east or west of a point "
    "on Earth at Greenwich (UK), which is the international "
    "zero-longitude point (longitude = 0°). The anti-meridian "
    "of Greenwich (the opposite side of the planet) is both "
    "180° (to the east) and -180° (to the west)."
)

class LocationModelForm(forms.ModelForm):
    latitude = forms.FloatField(
        label=_("Latitude"), required=False,
help_text=LATITUDE_DEFINITION
    )
    longitude = forms.FloatField(
        label=_("Longitude"), required=False,
help_text=LONGITUDE_DEFINITION
    )

    class Meta:
        model = Location
        exclude = ["geoposition"]

    def __init__(self, *args, **kwargs):
```

```
        super().__init__(*args, **kwargs)
        if self.instance:
            geoposition = self.instance.get_geoposition()
            if geoposition:
                self.fields["latitude"].initial = geoposition.latitude
                self.fields["longitude"].initial = geoposition.longitude

    def save(self, commit=True):
        cleaned_data = self.cleaned_data
        instance = super().save(commit=False)
        instance.set_geoposition(
            longitude=cleaned_data["longitude"],
            latitude=cleaned_data["latitude"],
        )
        if commit:
            instance.save()
            self.save_m2m()
        return instance

@admin.register(Location)
class LocationAdmin(admin.ModelAdmin):
    form = LocationModelForm
    save_on_top = True
    list_display = ("name", "street_address", "description")
    search_fields = ("name", "street_address", "description")

    def get_fieldsets(self, request, obj=None):
        map_html = render_to_string(
            "admin/locations/includes/map.html",
            {"MAPS_API_KEY": settings.GOOGLE_MAPS_API_KEY},
        )
        fieldsets = [
            (_("Main Data"), {"fields": ("name", "description")}),
            (
                _("Address"),
                {
                    "fields": (
                        "street_address",
                        "street_address2",
                        "postal_code",
                        "city",
                        "country",
                        "latitude",
                        "longitude",
```

```
                )
            },
        ),
        (_("Map"), {"description": map_html, "fields": []}),
        (_("Image"), {"fields": ("picture",)}),
    ]
    return fieldsets
```

（2）当创建一个自定义修改表单模板时，需要将 admin/locations/location/下的 change_form.html 新文件添加至模板目录中。该模板扩展自默认的 admin/change_form.html 模板，并将覆写 extrastyle 和 field_sets 块，如下所示。

```
{# admin/locations/location/change_form.html #}
{% extends "admin/change_form.html" %}
{% load i18n static admin_modify admin_urls %}

{% block extrastyle %}
   {{ block.super }}
   <link rel="stylesheet" type="text/css"
      href="{% static 'site/css/location_map.css' %}" />
{% endblock %}

{% block field_sets %}
   {% for fieldset in adminform %}
      {% include "admin/includes/fieldset.html" %}
   {% endfor %}
   <script src="{% static 'site/js/location_change_form.js'
    %}"></script>
{% endblock %}
```

（3）必须创建一个地图模板，随后将其插入至 Map 字段集中，如下所示。

```
{# admin/locations/includes/map.html #}
{% load i18n %}
<div class="form-row map js_map">
   <div class="canvas">
      <!-- THE GMAPS WILL BE INSERTED HERE DYNAMICALLY -->
   </div>
   <ul class="locations js_locations"></ul>
   <div class="btn-group">
      <button type="button"
             class="btn btn-default locate-address
              js_locate_address">
          {% trans "Locate address" %}
```

```
        </button>
        <button type="button"
              class="btn btn-default remove-geo js_remove_geo">
           {% trans "Remove from map" %}
        </button>
    </div>
</div>
<script src="https://maps-api-ssl.google.com/maps/api/js?key={{
MAPS_API_KEY }}"></script>
```

（4）默认状态下，地图并未样式化，因而需要添加一些 CSS，如下所示。

```
/* site_static/site/css/location_map.css */
.map {
   box-sizing: border-box;
   width: 98%;
}
.map .canvas,
.map ul.locations,
.map .btn-group {
   margin: 1rem 0;
}
.map .canvas {
   border: 1px solid #000;
   box-sizing: padding-box;
   height: 0;
   padding-bottom: calc(9 / 16 * 100%); /* 16:9 aspect ratio */
   width: 100%;
}
.map .canvas:before {
   color: #eee;
   color: rgba(0, 0, 0, 0.1);
   content: "map";
   display: block;
   font-size: 5rem;
   line-height: 5rem;
   margin-top: -25%;
   padding-top: calc(50% - 2.5rem);
   text-align: center;
}
.map ul.locations {
   padding: 0;
}
.map ul.locations li {
   border-bottom: 1px solid #ccc;
```

```
    list-style: none;
}
.map ul.locations li:first-child {
    border-top: 1px solid #ccc;
}
.map .btn-group .btn.remove-geo {
    float: right;
}
```

（5）下面创建一个 location_change_form.js JavaScript 文件。这里，我们并不打算使用全局变量“污染”环境。因此，我们将从闭包开始，进而为变量和函数创建一个私有作用域。我们将在该文件中使用 jQuery（jQuery 与贡献的管理系统一起被提供，且兼具简单性和跨浏览器功能），如下所示。

```
/* site_static/site/js/location_change_form.js */
(function ($, undefined) {
    var gettext = window.gettext || function (val) {
        return val;
    };
    var $map, $foundLocations, $lat, $lng, $street, $street2,
        $city, $country, $postalCode, gMap, gMarker;
    // …this is where all the further JavaScript functions go…
}(django.jQuery));
```

（6）创建一些 JavaScript 函数并将其逐一添加至 location_change_form.js 文件中。其中，getAddress4search()函数收集地址字段中的地址字符串，以供后续地理编码使用，如下所示。

```
function getAddress4search() {
    var sStreetAddress2 = $street2.val();
    if (sStreetAddress2) {
        sStreetAddress2 = " " + sStreetAddress2;
    }

    return [
        $street.val() + sStreetAddress2,
        $city.val(),
        $country.val(),
        $postalCode.val()
    ].join(", ");
}
```

（7）updateMarker()函数接收 latitude 和 longitude 参数，并在地图上绘制或移动一个标记。除此之外，该函数还可上传可拖曳的标记，如下所示。

```
function updateMarker(lat, lng) {
    var point = new google.maps.LatLng(lat, lng);

    if (!gMarker) {
        gMarker = new google.maps.Marker({
            position: point,
            map: gMap
        });
    }

    gMarker.setPosition(point);
    gMap.panTo(point, 15);
    gMarker.setDraggable(true);

    google.maps.event.addListener(gMarker, "dragend",
        function() {
            var point = gMarker.getPosition();
            updateLatitudeAndLongitude(point.lat(), point.lng());
        }
    );
}
```

（8）updateLatitudeAndLongitude()函数（引用于前述 dragend 事件监听器）接收 latitude 和 longitude 参数，并利用 id_latitude 和 id_longitude ID 更新字段值，如下所示。

```
function updateLatitudeAndLongitude(lat, lng) {
    var precision = 1000000;
    $lat.val(Math.round(lat * precision) / precision);
    $lng.val(Math.round(lng * precision) / precision);
}
```

（9）autocompleteAddress()函数从 Google Maps 地理编码中获取结果，并将对应结果列在地图之下，以选择正确的结果。如果仅存在一个结果，这将更新地理位置和地址字段，如下所示。

```
function autocompleteAddress(results) {
    var $item = $('<li/>');
    var $link = $('<a href="#"/>');

    $foundLocations.html("");
    results = results || [];

    if (results.length) {
        results.forEach(function (result, i) {
            $link.clone()
```

```
            .html(result.formatted_address)
            .click(function (event) {
                event.preventDefault();
                updateAddressFields(result.address_components);

                var point = result.geometry.location;
                updateLatitudeAndLongitude(
                    point.lat(), point.lng());
                updateMarker(point.lat(), point.lng());
                $foundLocations.hide();
            })
            .appendTo($item.clone()
            .appendTo($foundLocations));
    });
    $link.clone()
        .html(gettext("None of the above"))
        .click(function(event) {
            event.preventDefault();
            $foundLocations.hide();
        })
        .appendTo($item.clone().appendTo($foundLocations));
    $foundLocations.show();
  } else {
    $foundLocations.hide();
  }
}
```

（10）updateAddressFields()函数接收一个嵌套的字典，并以地址组件作为参数，同时填充所有地址字段，如下所示。

```
function updateAddressFields(addressComponents) {
  var streetName, streetNumber;
  var typeActions = {
    "locality": function(obj) {
      $city.val(obj.long_name);
    },
    "street_number": function(obj) {
      streetNumber = obj.long_name;
    },
    "route": function(obj) {
      streetName = obj.long_name;
    },
    "postal_code": function(obj) {
      $postalCode.val(obj.long_name);
```

```
        },
        "country": function(obj) {
            $country.val(obj.short_name);
        }
    };

    addressComponents.forEach(function(component) {
        var action = typeActions[component.types[0]];
        if (typeof action === "function") {
            action(component);
        }
    });

    if (streetName) {
        var streetAddress = streetName;
        if (streetNumber) {
            streetAddress += " " + streetNumber;
        }
        $street.val(streetAddress);
    }
}
```

（11）最后一个是初始化函数，并在加载页面时调用该函数。其间将 onclick 事件处理程序绑定至按钮上，创建一个 Google Map，并在初始化状态下标记 latitude 和 longitude 字段中定义的地理位置。

```
$(function(){
    $map = $(".map");

    $foundLocations = $map.find("ul.js_locations").hide();
    $lat = $("#id_latitude");
    $lng = $("#id_longitude");
    $street = $("#id_street_address");
    $street2 = $("#id_street_address2");
    $city = $("#id_city");
    $country = $("#id_country");
    $postalCode = $("#id_postal_code");

    $map.find("button.js_locate_address")
        .click(function(event) {
            var geocoder = new google.maps.Geocoder();
            geocoder.geocode(
                {address: getAddress4search()},
```

```
                function (results, status) {
                    if (status === google.maps.GeocoderStatus.OK) {
                        autocompleteAddress(results);
                    } else {
                        autocompleteAddress(false);
                    }
                }
            );
        });

    $map.find("button.js_remove_geo")
        .click(function() {
            $lat.val("");
            $lng.val("");
            gMarker.setMap(null);
            gMarker = null;
        });

    gMap = new google.maps.Map($map.find(".canvas").get(0), {
        scrollwheel: false,
        zoom: 16,
        center: new google.maps.LatLng(51.511214, -0.119824),
        disableDoubleClickZoom: true
    });

    google.maps.event.addListener(gMap, "dblclick", function(event)
    {
        var lat = event.latLng.lat();
        var lng = event.latLng.lng();
        updateLatitudeAndLongitude(lat, lng);
        updateMarker(lat, lng);
    });

    if ($lat.val() && $lng.val()) {
        updateMarker($lat.val(), $lng.val());
    }
});
```

6.10.3　工作方式

当查看浏览器中的 Change Location 表单时，可以看到字段集中显示的 Map，随后是包含地址字段的字段集，如图 6.14 所示。

Django administration
WELCOME, ADMIN. VIEW SITE / CHANGE PASSWORD / LOG OUT
Home › Locations › Locations › Park Güell
Change Location
HISTORY
VIEW ON SITE
Delete
Save and add another
Save and continue editing
SAVE
Main Data
Name: Park Güell
Description: If you want to see something spectacular, come to Barcelona, Catalonia, Spain and visit Park Güell. Located on a hill, Park Güell is a public park with beautiful gardens and organic architectural elements.
Address
Street address: Park Güell
Street address (2nd line):
Postal code: 08024
City: Barcelona
Country: Spain
Latitude:
Latitude (Lat.) is the angle between any point and the equator (north pole is at 90°; south pole is at -90°).
Longitude:
Longitude (Long.) is the angle east or west of a point on Earth at Greenwich (UK), which is the international zero-longitude point (longitude = 0°). The anti-meridian of Greenwich (the opposite side of the planet) is both 180° (to the east) and -180° (to the west).
Map
Map
Satellite
Club London
Peacock Theatre
Economics and Political...
City Lit
McDonald's
Cambridge Theatre
Prince Edward Theatre
Sarastro
Royal Courts of Justice
Apex Temple Court
Masala Zone Covent Garden
Temple Church
Royal Opera House
Covent Garden
Aldwych Theatre
Temple Brew House
Dishoom
Novello Theatre
Arts Theatre
Five Guys
London Transport Museum
The Courtauld Gallery
King's College London
Two Temple Place
Honourable Society Of Middle Temple
Leicester Square
Porterhouse Covent Garden
Strand Palace
Somerset House
Middle Temple Gardens
ODEON Luxe Leicester Square
Temple
TKTS
Hotel The Savoy
Victoria Embankment
National Portrait Gallery
St Martin-in-the-Fields
The National Gallery
Waterloo Bridge
Trafalgar Square
Charing Cross
Heaven
Oxo Tower
Gabriel's Wharf
Sofitel St James
Mulberry Bush
Embankment
Google
Map data ©2019 Google
Terms of Use
Locate address
Remove from map
Image
Picture: Currently: locations/2019/11/c6dc6d86-0345-4135-9ef3-34a6f36b1df4.jpg
Change: Choose file No file chosen
Delete
Save and add another
Save and continue editing
SAVE

图 6.14

地图下方存在两个按钮，即 Locate address 和 Remove from map 按钮。

当单击 Locate address 按钮时，地理编码将被调用以搜索输入地址的地理坐标。执行地理编码的结果表示为以嵌套字典格式列出的一个或多个地址。我们将把地址表示为一个可单击的链接的列表，如图 6.15 所示。

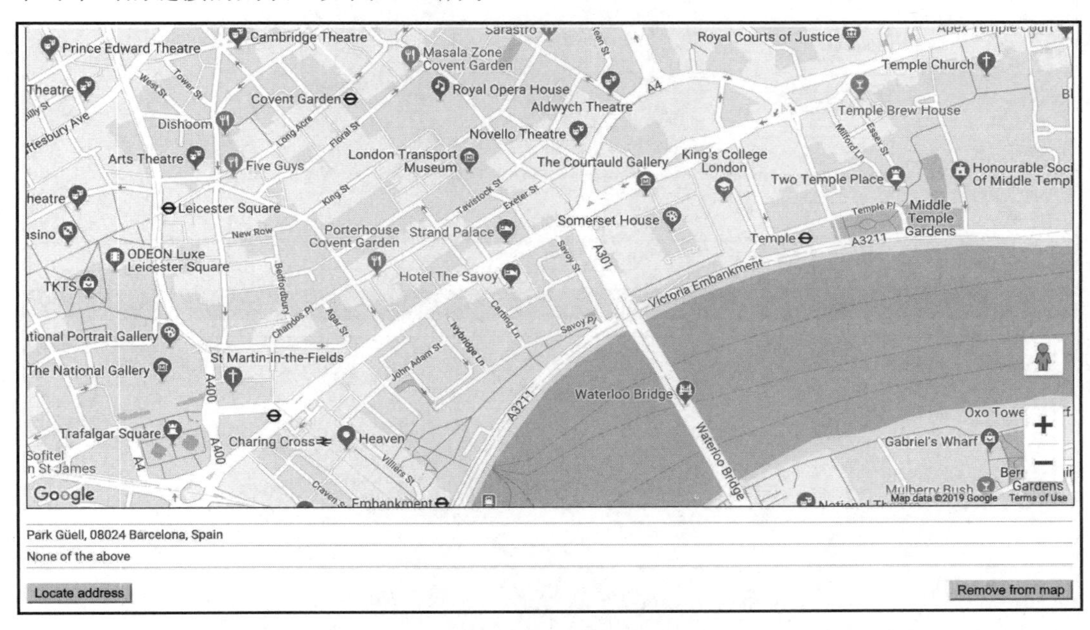

图 6.15

当在开发工具的控制台中查看嵌套字典的结构时，可将下列代码行置于 autocompleteAddress()函数的开始处。

```
console.log(JSON.stringify(results, null, 4));
```

当单击某一部分时，标记将呈现于地图上并显示准确的地理位置。此时，Latitude 和 Longitude 字段的填写结果如图 6.16 所示。

随后，管理员可通过拖曳方式移动地图上的标记。另外，双击地图上的任何一处均可更新地理坐标和标记位置。

最后，如果单击 Remove from map 按钮，地理坐标将被清除，且标记也随之被移除。

管理使用自定义 LocationModelForm（不包含 geoposition 字段）、添加 Latitude 和 Longitude 字段，并处理对应值的保存和加载操作。

Address

Street address: Park Güell

Street address (2nd line):

Postal code: 08024

City: Barcelona

Country: Spain

Latitude: 41,414495

Latitude (Lat.) is the angle between any point and the equator (north pole is at 90°; south pole is at -90°).

Longitude: 2,152695

Longitude (Long.) is the angle east or west of a point on Earth at Greenwich (UK), which is the international zero-longitude point (longitude = 0°). The anti-meridian of Greenwich (the opposite side of the planet) is both 180° (to the east) and -180° (to the west).

Map

Map Satellite

Centre Cívic El Coll-La Bruguera
Escola Virolai
Delicias
Parc de La Creueta del Coll
Placeta Montserrat
El Mirador
Poliesportiu Creueta del Coll
Mirador de Joan Sales - Mirador del Virolai
Plaça de Raimon Casellas
Plaça de Can Baró
18 Torres Marsanas
Club Natació Catalunya
Alberg Barcelona Xanascat
Escola Montseny
Jesuïtes Gràcia. Col·legi Kostka
Parc Güell
La Torreta del Parc Güell
Vall de Baldomar
Delfos
Casa del Guarda
Plaça Mons
Jardins de Maria Baldó
Gaudí Experiència
Club Tennis de La Salut
Holmes Place
Vallcarca
la República Argentina
Maps Informática Industrial
Escola Súnion
Sant Josep de la Muntanya
Hospital de l'Esperança (Consorci Mar Parc de...
Comerç Vallcarca
Institut FP Sanitària
Mercat de l'Est
Map data ©2019 Inst. Geogr. Nacional Terms of Use

Locate address　　Remove from map

图 6.16

6.10.4　延伸阅读

读者还可参考第 4 章中的内容。

第 7 章　安全和性能

本章主要涉及下列主题。

- 表单的跨站点请求伪造（CSRF）安全。
- 基于内容安全政策（CSP）的请求安全。
- 使用 django-admin-honeypot。
- 实现密码验证。
- 下载授权的文件。
- 向图像中添加动态水印。
- 基于 Auth0 的身份验证。
- 缓存方法返回值。
- 使用 Memcached 缓存 Django 视图。
- 使用 Redis 缓存 Django 视图。

7.1　简　　介

如果软件不适宜地暴露敏感信息、让用户忍受漫长的等待时间，或者占用了大量的硬件，那么此类软件永远不会持久地存在。作为开发人员，我们有责任保证应用程序的安全性和性能。在本章中，我们将研究在操作 Django 应用程序的同时保证用户处于安全状态的许多方法。随后，我们将研究一些缓存选项，这些选项可以减少处理过程，并以较低的金钱和时间成本向用户提供数据。

7.2　技 术 需 求

当与本章代码协同工作时，我们需要安装最新稳定版本的 Python、MySQL 或 PostgreSQL 数据库，以及基于虚拟环境的 Django 项目。

读者还可访问 GitHub 储存库的 ch07 目录中查看本章代码，对应网址为 https://github.com/PacktPublishing/Django-3-Web-Development-Cookbook-Fourth-Edition。

7.3　表单的跨站点请求伪造安全

如果缺少适当的预防措施，恶意站点可能会调用网站请求，从而导致服务器出现意外变化，如可能会影响到用户的身份验证或者未经用户允许修改相关内容。Django 绑定了一个系统，可用于防止此类 CSRF 攻击。

7.3.1　准备工作

当前示例将在第 3 章示例的基础上完成。

7.3.2　实现方式

下列步骤将启用 Django 中的 CSRF 防护措施。

（1）确保 CsrfViewMiddleware 包含在项目的设置项中，如下所示。

```
# myproject/settings/_base.py
MIDDLEWARE = [
    "django.middleware.security.SecurityMiddleware",
    "django.contrib.sessions.middleware.SessionMiddleware",
    "django.middleware.common.CommonMiddleware",
    "django.middleware.csrf.CsrfViewMiddleware",
    "django.contrib.auth.middleware.AuthenticationMiddleware",
    "django.contrib.messages.middleware.MessageMiddleware",
    "django.middleware.clickjacking.XFrameOptionsMiddleware",
    "django.middleware.locale.LocaleMiddleware",
]
```

（2）确保表单视图通过请求上下文被渲染。例如，在现有的 ideas 应用程序中，我们持有：

```
# myproject/apps/ideas/views.py
from django.contrib.auth.decorators import login_required
from django.shortcuts import render

@login_required
def add_or_change_idea(request, pk=None):
    # …
    return render(request, "ideas/idea_form.html", context)
```

（3）在表单模板中，确保使用 POST 方法并包含{% csrf_token %}标签。

```
{# ideas/idea_form.html #}
{% extends "base.html" %}
{% load i18n crispy_forms_tags static %}

{% block content %}
    <h1>
        {% if idea %}
            {% blocktrans trimmed with title=idea
             .translated_title %}
                Change Idea "{{ title }}"
            {% endblocktrans %}
        {% else %}
            {% trans "Add Idea" %}
        {% endif %}
    </h1>
    <form action="{{ request.path }}" method="post">
        {% csrf_token %}
        {{ form.as_p }}
        <p>
            <button type="submit">{% trans "Save" %}</button>
        </p>
    </form>
{% endblock %}
```

（4）如果针对表单布局使用 django-crispy-forms，默认状态下将包含 CSRF 令牌。

```
{# ideas/idea_form.html #}
{% extends "base.html" %}
{% load i18n crispy_forms_tags static %}

{% block content %}
    <h1>
        {% if idea %}
            {% blocktrans trimmed with title=idea
             .translated_title %}
                Change Idea "{{ title }}"
            {% endblocktrans %}
        {% else %}
            {% trans "Add Idea" %}
        {% endif %}
    </h1>
    {% crispy form %}
{% endblock %}
```

7.3.3　工作方式

Django 采用一种隐藏字段方法来防止 CSRF 攻击。在服务器上令牌根据特定于请求的随机信息的组合而被生成。通过 CsrfViewMiddleware，该令牌通过请求上下文自动生效。虽然并不建议禁用该中间件，但却可以通过@csrf_protect 装饰器标记各个视图，以获取相同的行为。

```
from django.views.decorators.csrf import csrf_protect

@csrf_protect
def my_protected_form_view():
    # …
```

类似地，通过使用@csrf_exempt 装饰器，我们可以从 CSRF 检查中排除各个视图，即使上述中间件被启用。

```
from django.views.decorators.csrf import csrf_exempt

@csrf_exempt
def my_unsecured_form_view():
    # …
```

内建的{% csrf_token %}标签生成提供令牌的隐藏输入字段，如下所示。

```
<input type="hidden" name="csrfmiddlewaretoken"
value="29sQH3UhogpseHH60eEaTq0xKen9TvbKe5lpT9xs30cR01dy5QVAtATWmAHvUZFk">
```

包含通过 GET、HEAD、OPTIONS 或 TRACE 方法提交请求的表单的令牌被视为无效的，因为使用这些方法的请求首先不应产生任何副作用。在大多数情况下，需要 CSRF 防护的 Web 表单一般是 POST 表单。

当受保护的方法利用不安全的方法且在缺少所需令牌的情况下被提交时，Django 的内建表单验证机制将对此进行识别并直接拒绝请求。相应地，仅包含基于有效值的令牌的提交才会被处理。最终，外部站点将无法修改服务器，因为它们将无法知道并包含当前有效的令牌值。

7.3.4　更多内容

在许多场合下，我们希望增强表单，以便该表单可通过 Ajax 提交。该过程需要采用 CSRF 令牌进行保护。虽然可将令牌作为附加数据注入至每个请求中，但这种方法需要开发人员对每个 POST 请求均执行此类操作。对此，存在一种使用 CSRF 令牌头的替代方案可大大提升效率。

首先需要检索令牌值，具体做法取决于 CSRF_USE_SESSIONS 设置值。当 CSRF_USE_SESSIONS 为 True 时，令牌存储于会话而非 cookie 中，因而须使用{% csrf_token %}标签将其包含在 DOM 中。随后，可读取该元素检索 JavaScript 中的数据。

```
var input = document.querySelector('[name="csrfmiddlewaretoken"]');
var csrfToken = input && input.value;
```

默认状态下，当 CSRF_USE_SESSIONS 设置项为 False 时，那么首选令牌值的源为 csrftoken cookie。虽然可运行 cookie 操作方法，但是存在一些实用程序可简化这一过程。例如，可采用 js-cookie API（https://github.com/js-cookie/js-cookie）并通过名称轻松地析取令牌，如下所示。

```
var csrfToken = Cookies.get('crsftoken');
```

一旦析取了令牌，则需要将其设置为 XmlHttpRequest 的 CSRF 令牌标头值。虽然该过程可针对每个请求分别进行，但这与为每个请求将数据添加至请求参数中具有相同的缺点。相反，我们可采用 jQuery 并在数据发送之前将其自动绑定至所有请求上，如下所示。

```
var CSRF_SAFE_METHODS = ['GET', 'HEAD', 'OPTIONS', 'TRACE'];
$.ajaxSetup({
    beforeSend: function(xhr, settings) {
        if (CSRF_SAFE_METHODS.indexOf(settings.type) < 0
            && !this.crossDomain) {
            xhr.setRequestHeader("X-CSRFToken", csrfToken);
        }
    }
});
```

7.3.5　延伸阅读

- 第 3 章中的“利用 CRUDL 函数创建一个应用程序”示例。
- “实现密码验证”示例。
- “下载授权文件”示例。
- “基于 Auth0 的身份验证”示例。

7.4　基于内容安全政策的请求安全

动态多用户站点通常允许用户添加源自不同媒体类型的各种数据，如图像、视频、音频、HTML、JavaScript 代码片段等。其间，用户也可能会向站点添加恶意代码、窃取 cookies 或其他个人信息、在后台中调用有害的 Ajax 请求，或者执行其他破坏任务。现代浏览器支持一个附加的安全层，并将媒体源列入白名单中，即 CSP。在当前示例中，我们将展示如何在 Django 站点中使用 CSP。

7.4.1　准备工作

当前示例将在第 3 章 ideas 应用程序的基础上完成。

7.4.2　实现方式

下列步骤将利用 CSP 保护项目。

（1）将 django-csp 安装至虚拟环境中。

```
(env)$ pip install django-csp==3.6
```

（2）在设置项中添加 CSPMiddleware。

```
# myproject/settings/_base.py
MIDDLEWARE = [
    "django.middleware.security.SecurityMiddleware",
    "django.contrib.sessions.middleware.SessionMiddleware",
    "django.middleware.common.CommonMiddleware",
    "django.middleware.csrf.CsrfViewMiddleware",
    "django.contrib.auth.middleware.AuthenticationMiddleware",
    "django.contrib.messages.middleware.MessageMiddleware",
    "django.middleware.clickjacking.XFrameOptionsMiddleware",
    "django.middleware.locale.LocaleMiddleware",
    "csp.middleware.CSPMiddleware",
]
```

（3）在同一设置文件中，添加 django-csp 设置项，并将所信任的媒体源添加至白名单中，如 jQuery 和 Bootstrap 的 CDN（稍后将对此予以解释）。

```
# myproject/settings/_base.py
```

```
CSP_DEFAULT_SRC = [
    "'self'",
    "https://stackpath.bootstrapcdn.com/",
]
CSP_SCRIPT_SRC = [
    "'self'",
    "https://stackpath.bootstrapcdn.com/",
    "https://code.jquery.com/",
    "https://cdnjs.cloudflare.com/",
]
CSP_IMG_SRC = ["*", "data:"]
CSP_FRAME_SRC = ["*"]
```

（4）如果模板中包含内联脚本或样式，可使用密码 nonce 将其加入白名单中，如下所示。

```
<script nonce="{{ request.csp_nonce }}">
    window.settings = {
        STATIC_URL: '{{ STATIC_URL }}',
        MEDIA_URL: '{{ MEDIA_URL }}',
    }
</script>
```

7.4.3　工作方式

CSP 指示符可添加至头部分或响应头的元标签中。

- meta 元标签语法如下所示。

```
<meta http-equiv="Content-Security-Policy" content="img-src *
data:; default-src 'self' https://stackpath.bootstrapcdn.com/
'nonce-WWNu7EYqfTcVVZDs'; frame-src *; script-src 'self'
https://stackpath.bootstrapcdn.com/ https://code.jquery.com/
https://cdnjs.cloudflare.com/">
```

- 我们选择的 django-csp 模块使用了 response headers 创建了打算加载至站点中的列表源。我们可检查浏览器查看器 Network 中的头，如下所示。

```
Content-Security-Policy: img-src * data:; default-src 'self'
https://stackpath.bootstrapcdn.com/ 'nonce-WWNu7EYqfTcVVZDs';
frame-src *; script-src 'self' https://stackpath.bootstrapcdn.com/
https://code.jquery.com/ https://cdnjs.cloudflare.com/
```

CSP 允许定义资源类型，同时还允许资源彼此邻接。其中，可使用的主指示符包括：

- default-src 用作所有未设置源的回退，并在 Django 设置项中由 CSP_DEFAULT_SRC 控制。
- script-src 用于<script>标签，并在 Django 设置项中由 CSP_DEFAULT_SRC 控制。
- style-src 用于<style>、<link rel="stylesheet">标签以及 CSS @import 语句，并由 CSP_STYLE_SRC 设置项控制。
- img-src 用于<img>标签，并由 CSP_IMG_SRC 设置项控制。
- frame-src 用于<frame>和<iframe>标签，并由 CSP_FRAME_SRC 设置项控制。
- media-src 用于<audio>、<video>和<track>标签，并由 CSP_MEDIA_SRC 设置项控制。
- font-src 用于 Web 字体，并由 CSP_FONT_SRC 设置项控制。
- connect-src 用于 JavaScript 加载的资源，并由 CSP_CONNECT_SRC 设置项控制。

注意：

读者可访问 https://developer.mozilla.org/en-US/docs/Web/HTTP/Headers/Content-Security-Policy 和 https://django-csp.readthedocs.io/en/latest/configuration.html 查看资源类型和类似设置的完整列表。

每个指示符的值可以是下列列表中的一个或多个值。

- *：支持全部资源。
- 'none'：禁用全部资源。
- 'self'：支持同一个域中的资源。
- 协议，如 https:或 data:。
- 域，如 example.com 或*.example.com。
- 站点 URL，如 https://example.com。
- 'unsafe-inline'：支持内联<script>或<style>标签。
- 'unsafe-eval'：允许使用 eval()函数执行脚本。
- 'nonce-<b64-value>'：允许使用加密 nonces 指定特殊的标签。
- 'sha256-...'::允许通过源哈希值使用资源。

配置 django-csp 并不存在通用方式，且通常是一个试错的过程，下面列出了几项原则。

（1）针对已有的工作项目添加 CSP。过早的限制只会让网站的开发变得更加困难。

（2）检查所有的脚本、样式、字体和静态文件，它们被硬编码至模板并加入至白名单中。

（3）如果允许媒体嵌入至博客帖子或其他动态内容，那么应支持图像、多媒体和帧等全部资源，如下所示。

```
# myproject/settings/_base.py
CSP_IMG_SRC = ["*"]
CSP_MEDIA_SRC = ["*"]
CSP_FRAME_SRC = ["*"]
```

（4）当采用内联脚本或样式时，可将 nonce="{{ request.csp_nonce }}"添加于其中。

（5）避免'unsafe-inline'和'unsafe-eval' CSP 值，除非在站点中输入 HTML 的唯一方法是在模板中对其进行硬编码。

（6）浏览站点并搜索任何未经正确加载的内容。如果在开发人员控制台中看到如下消息，则意味着 CSP 限制了内容。

```
Refused to execute inline script because it violates the following
Content Security Policy directive: "script-src 'self'
https://stackpath.bootstrapcdn.com/
https://code.jquery.com/ https://cdnjs.cloudflare.com/". Either
the 'unsafeinline' keyword, a hash ('sha256-
P1v4zceJ/oPr/yp20lBqDnqynDQhHf76lljlXUxt7NI='), or a nonce
('nonce-...') is required to enable inline execution.
```

这些错误十分常见，因为一些第三方工具（如 django-cms、Django Debug Toolbar 和 Google Analytics）尝试通过 JavaScript 包含资源。我们可利用错误消息中看到的资源哈希值（如'sha256-P1v4zceJ/oPr/yp20lBqDnqynDQhHf76lljlXUxt7NI='）将这些资源加入白名单中。

（7）当开发现代增强型 Web 应用程序（Progressive Web Apps，PWAs）时，应考虑检查由 CSP_MANIFEST_SRC 和 CSP_WORKER_SRC 设置项控制的清单和 Web worker。

7.4.4　延伸阅读

读者还可参考“表单的跨站点请求伪造安全”示例。

7.5　使用 django-admin-honeypot

如果保持 Django 站点的默认管理路径，那么黑客则有可能运用蛮力攻击，并尝试使用不同的密码登录。对此，django-admin-honeypot 应用程序可伪造登录界面并检测那些蛮力攻击。在当前示例中，我们将学习如何使用 django-admin-honeypot 应用程序。

7.5.1 准备工作

当前示例将在扩展前述项目的基础上完成。

7.5.2 实现方式

下列步骤将设置 django-admin-honeypot。

（1）在虚拟环境中安装模块。

```
(env)$ pip install django-admin-honeypot==1.1.0
```

（2）在设置项中向 INSTALLED_APPS 中添加"admin_honeypot"。

```
# myproject/settings/_base.py
INSTALLED_APPS = (
    # …
    "admin_honeypot",
)
```

（3）调整 URL 规则。

```
# myproject/urls.py
from django.contrib import admin
from django.conf.urls.i18n import i18n_patterns
from django.urls import include, path

urlpatterns = i18n_patterns(
    # …
    path("admin/", include("admin_honeypot.urls",
    namespace="admin_honeypot")),
    path("management/", admin.site.urls),
)
```

7.5.3 工作方式

当访问默认的管理 URL（http://127.0.0.1:8000/en/admin/）时，将会看到如图 7.1 所示的登录界面，但输入内容将会被描述为无效的密码。

相应地，真正的站点管理位于 http://127.0.0.1:8000/en/management/，其中可以看到源自 honeypot 的跟踪登录行为。

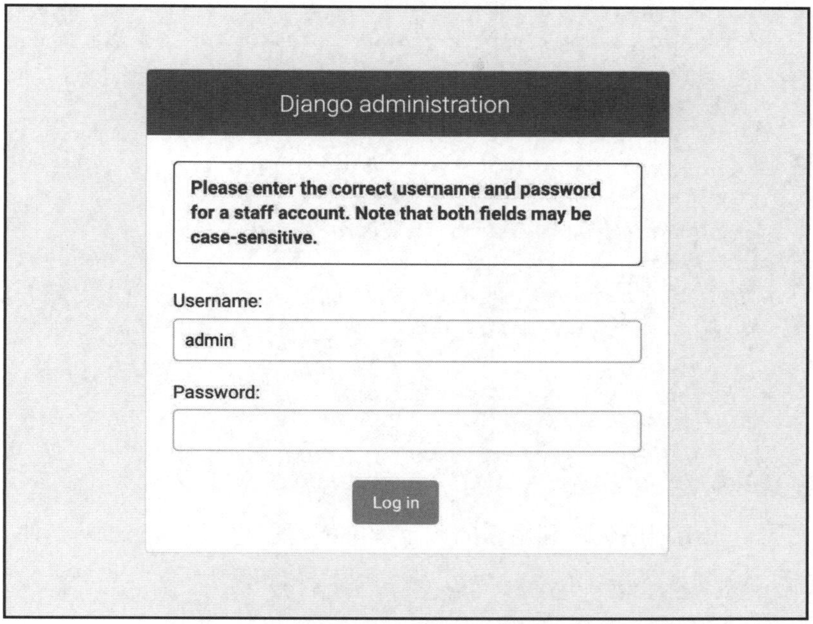

图 7.1

7.5.4　更多内容

在本书编写时，django-admin-honeypot 在 Django 3.0 中尚有缺陷，即管理界面在渲染处会转义 HTML。在 django-admin-honeypot 更新并发布新版本之前，我们可对此稍作改进，如下所示。

（1）在 admin.py 文件中创建一个 admin_honeypot_fix 应用程序。

```
# myproject/apps/admin_honeypot_fix/admin.py
from django.contrib import admin

from admin_honeypot.admin import LoginAttemptAdmin
from admin_honeypot.models import LoginAttempt
from django.utils.safestring import mark_safe
from django.utils.translation import gettext_lazy as _

admin.site.unregister(LoginAttempt)
```

```
@admin.register(LoginAttempt)
class FixedLoginAttemptAdmin(LoginAttemptAdmin):
    def get_session_key(self, instance):
        return mark_safe('<a href="?session_key=
        %(key)s">%(key)s</a>' % {'key': instance.session_key})
    get_session_key.short_description = _('Session')

    def get_ip_address(self, instance):
        return mark_safe('<a href="?ip_address=%(ip)s">%(ip)s</a>'
         % {'ip': instance.ip_address})
    get_ip_address.short_description = _('IP Address')

    def get_path(self, instance):
        return mark_safe('<a href="?path=%(path)s">%(path)s</a>'
         % {'path': instance.path})
    get_path.short_description = _('URL')
```

（2）在同一应用程序中，利用新的应用程序配置创建一个 apps.py 文件。

```
# myproject/apps/admin_honeypot_fix/apps.py
from django.apps import AppConfig
from django.utils.translation import gettext_lazy as _

class AdminHoneypotConfig(AppConfig):
    name = "admin_honeypot"
    verbose_name = _("Admin Honeypot")

    def ready(self):
        from .admin import FixedLoginAttemptAdmin
```

（3）在设置项的 INSTALLED_APPS 中，利用新的应用程序配置替换"admin_honeypot"。

```
# myproject/settings/_base.py
INSTALLED_APPS = [
    # …
    #"admin_honeypot",
    "myproject.apps.admin_honeypot_fix.apps.AdminHoneypotConfig",
]
```

图 7.2 显示了 honeypot 的登录行为。

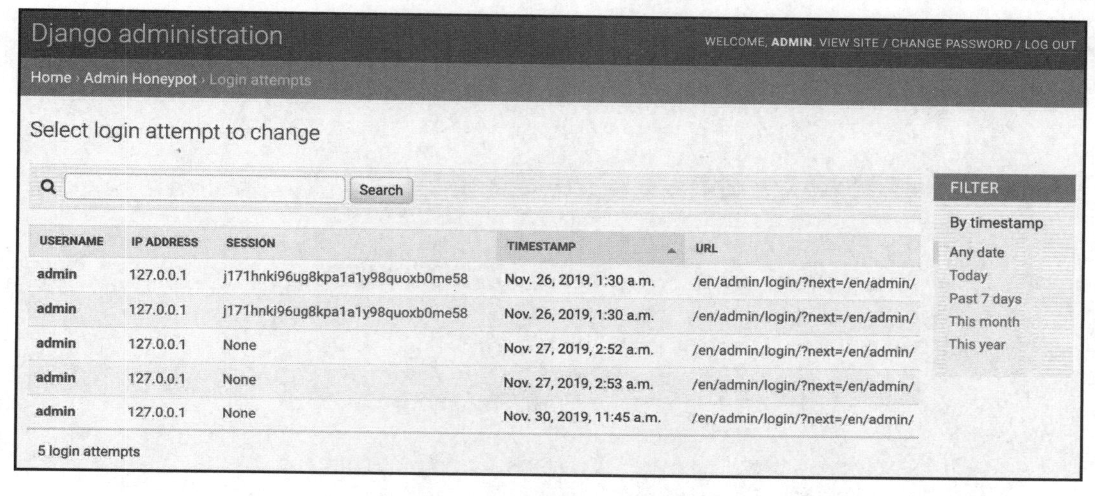

图 7.2

7.5.5　延伸阅读

- “实现密码验证”示例。
- “基于 Auth0 的身份验证”示例。

7.6　实现密码验证

在软件安全问题列表中，排在首位的是用户选择了不安全的密码。在当前示例中，我们将学习如何通过内置和自定义密码验证器强制执行最低密码要求，从而引导用户设置更安全的身份验证机制。

7.6.1　准备工作

打开项目的设置项文件并定位在 AUTH_PASSWORD_VALIDATORS 设置处。此外，还需要创建一个包含 password_validation.py 文件的新的 auth_extra 应用程序。

7.6.2　实现方式

下列步骤将构建更加强大的密码验证机制。

（1）添加一些选项以定制 Django 中包含的验证器设置项。

```
# myproject/settings/_base.py
AUTH_PASSWORD_VALIDATORS = [
    {
        "NAME": "django.contrib.auth.password_validation."
        "UserAttributeSimilarityValidator",
        "OPTIONS": {"max_similarity": 0.5},
    },
    {
        "NAME": "django.contrib.auth.password_validation."
        "MinimumLengthValidator",
        "OPTIONS": {"min_length": 12},
    },
    {"NAME": "django.contrib.auth.password_validation."
    "CommonPasswordValidator"},
    {"NAME": "django.contrib.auth.password_validation."
    "NumericPasswordValidator"},
]
```

（2）在新的 auth_extra 应用程序中，将 MaximumLengthValidator 类添加至 password_validation.py 文件中，如下所示。

```
# myproject/apps/auth_extra/password_validation.py
from django.core.exceptions import ValidationError
from django.utils.translation import gettext as _

class MaximumLengthValidator:
    def __init__(self, max_length=24):
        self.max_length = max_length

    def validate(self, password, user=None):
        if len(password) > self.max_length:
            raise ValidationError(
                self.get_help_text(pronoun="this"),
                code="password_too_long",
                params={'max_length': self.max_length},
            )

    def get_help_text(self, pronoun="your"):
        return _(f"{pronoun.capitalize()} password must contain "
                 f"no more than {self.max_length} characters")
```

（3）在同一文件中创建 SpecialCharacterInclusionValidator 类。

```
class SpecialCharacterInclusionValidator:
```

```
    DEFAULT_SPECIAL_CHARACTERS = ('$', '%', ':', '#', '!')

    def __init__(self, special_chars=DEFAULT_SPECIAL_CHARACTERS):
        self.special_chars = special_chars

    def validate(self, password, user=None):
        has_specials_chars = False
        for char in self.special_chars:
            if char in password:
                has_specials_chars = True
                break
        if not has_specials_chars:
            raise ValidationError(
                self.get_help_text(pronoun="this"),
                code="password_missing_special_chars"
            )

    def get_help_text(self, pronoun="your"):
        return _(f"{pronoun.capitalize()} password must contain at"
                 " least one of the following special characters: "
                 f"{', '.join(self.special_chars)}")
```

（4）向设置项中添加新的验证器。

```
# myproject/settings/_base.py
from myproject.apps.auth_extra.password_validation import (
    SpecialCharacterInclusionValidator,
)

AUTH_PASSWORD_VALIDATORS = [
    {
        "NAME": "django.contrib.auth.password_validation."
        "UserAttributeSimilarityValidator",
        "OPTIONS": {"max_similarity": 0.5},
    },
    {
        "NAME": "django.contrib.auth.password_validation."
        "MinimumLengthValidator",
        "OPTIONS": {"min_length": 12},
    },
    {"NAME": "django.contrib.auth.password_validation."
    "CommonPasswordValidator"},
    {"NAME": "django.contrib.auth.password_validation."
    "NumericPasswordValidator"},
```

```
    {
        "NAME": "myproject.apps.auth_extra.password_validation."
        "MaximumLengthValidator",
        "OPTIONS": {"max_length": 32},
    },
    {
        "NAME": "myproject.apps.auth_extra.password_validation."
        "SpecialCharacterInclusionValidator",
        "OPTIONS": {
            "special_chars": ("{", "}", "^", "&")
            + SpecialCharacterInclusionValidator
             .DEFAULT_SPECIAL_CHARACTERS
        },
    },
]
```

7.6.3　工作方式

Django 包含了一组默认的密码验证器。

- UserAttributeSimilarityValidator 确保所选择的任何密码不会与用户的某些属性过于相似。默认情况下，相似性比率设置为 0.7，检查的属性包括用户名、姓氏和名以及电子邮件地址。如果这些属性包含多个单词，那么每个单词将被独立地查看。
- MinimumLengthValidator 检查输入的密码应包含最小字符长度。默认状态下，密码应为 8 个或更多个字符。
- CommonPasswordValidator 引用一个包含经常使用的密码列表的文件，因此是不安全的。默认状态下，Django 使用的列表包含 1000 个密码。
- NumericPasswordValidator 验证输入的密码是否完全由数字组成。

当使用 startproject 管理命令创建一个新项目时，这些项目与默认选项一起作为初始验证集。在当前示例中，我们将根据项目需求调整选项，并将密码的最小长度提升至 12 个字符。

对于 UserAttributeSimilarityValidator，我们还将 max_similarity 减至 0.5，这意味着，与默认行为相比，密码与用户属性之间存在巨大差异。

password_validation.py 则包含两个新验证器。

（1）MaximumLengthValidator 与内建的最小长度验证器类似，并确保密码不超过默认的 24 个字符。

（2）SpecialCharacterInclusionValidator 检查密码中是否包含一个或多个特殊字符，

默认状态下定义为$、%、:、#和!字符。

每个验证器类均包含两个所需的方法。

（1）validate()方法执行与 password 参数之间的真正的检查。作为可选项，当用户经过身份验证后，将传递第 2 个 user 参数。

（2）需提供 get_help_text()方法，该方法返回一个字符串，用于表述用户的验证要求。

最后，我们向设置项中添加了一个新的验证器，覆写默认方法以支持最大 32 个字符的密码长度，并能够向默认的特殊字符列表中加入{、}、^和&字符。

7.6.4　更多内容

对于 createsuperuser 和 changepassword 管理命令，以及用于更改或重置密码的内置表单，将自动执行 AUTH_PASSWORD_VALIDATORS 中提供的验证器。有些时候，我们需要针对自定义密码管理代码采用相同的验证，Django 提供了这种集成级别的函数，我们可以在 django.contrib.auth.password_validation 模块中查看 Django auth 应用程序的详细信息。

7.6.5　延伸阅读

- “下载授权文件”示例。
- “基于 Auth0 的身份验证”示例。

7.7　下载授权文件

某些时候，我们可能只允许特定的人从站点中下载知识产权内容，如音乐、视频、文学作品或其他艺术作品，这些内容仅供付费用户访问。在当前示例中，我们将学习如何利用 Django auth 应用程序限制图像的下载权限（仅针对授权用户）。

7.7.1　准备工作

当前示例将在第 3 章 ideas 应用程序的基础上完成。

7.7.2　实现方式

具体实现方式如下列步骤所示。

（1）创建一个视图，且需要经过身份验证才可下载文件，如下所示。

```
# myproject/apps/ideas/views.py
import os

from django.contrib.auth.decorators import login_required
from django.http import FileResponse, HttpResponseNotFound
from django.shortcuts import get_object_or_404
from django.utils.text import slugify

from .models import Idea

@login_required
def download_idea_picture(request, pk):
    idea = get_object_or_404(Idea, pk=pk)
    if idea.picture:
        filename, extension =
        os.path.splitext(idea.picture.file.name)
        extension = extension[1:] # remove the dot
        response = FileResponse(
            idea.picture.file, content_type=f"image/{extension}"
        )
        slug = slugify(idea.title)[:100]
        response["Content-Disposition"] = (
            "attachment; filename="
            f"{slug}.{extension}"
        )
    else:
        response = HttpResponseNotFound(
            content="Picture unavailable"
        )
    return response
```

（2）向 URL 配置中添加下载视图。

```
# myproject/apps/ideas/urls.py
from django.urls import path

from .views import download_idea_picture

urlpatterns = [
    # ...
    path(
        "<uuid:pk>/download-picture/",
```

```
        download_idea_picture,
        name="download_idea_picture",
    ),
]
```

（3）在项目 URL 配置中设置登录视图。

```
# myproject/urls.py
from django.conf.urls.i18n import i18n_patterns
from django.urls import include, path

urlpatterns = i18n_patterns(
    # …
    path("accounts/", include("django.contrib.auth.urls")),
    path("ideas/", include(("myproject.apps.ideas.urls", "ideas"),
     namespace="ideas")),
)
```

（4）创建一个登录表单模板，如下所示。

```
{# registration/login.html #}
{% extends "base.html" %}
{% load i18n %}

{% block content %}
    <h1>{% trans "Login" %}</h1>
    <form action="{{ request.path }}" method="POST">
        {% csrf_token %}
        {{ form.as_p }}
        <button type="submit" class="btn btn-primary">{% trans
         "Log in" %}</button>
    </form>
{% endblock %}
```

（5）在 idea 详细模板中，向下载内容添加一个链接。

```
{# ideas/idea_detail.html #}
{% extends "base.html" %}
{% load i18n %}

{% block content %}
…
    <a href="{% url 'ideas:download_idea_picture' pk=idea.pk %}"
     class="btn btn-primary">{% trans "Download picture" %}</a>
{% endblock %}
```

对于绕开 Django 直接下载受限文件的用户，我们应该予以限制。对此，在 Apache Web 服务器上，如果用户运行的是 Apache 2.4，则可将一个.htaccess 文件置于 media/ideas 目录中，如下所示。

```
# media/ideas/.htaccess
Require all denied
```

注意：

当使用 django-imagekit 时，生成的镜像版本将被存储在 media/CACHE 目录中，且不会受到.htaccess 配置的影响。

7.7.3　工作方式

download_idea_picture 视图从一个特定的 idea 流式化上传的原始图像。被设置为 Content-Disposition 的 attachment 的头使得文件可下载，而非在浏览器中立即显示。另外，文件的文件名也需要在头中被设置，类似于 gamified-donation-platform.jpg。如果 idea 的图像无效，那么将显示一个 404 页面，其中包含一条简单的消息：Picture unavailable。

如果未登录用户尝试访问可下载的文件，@login_required 装饰器将访问用户重定向至登录页面。默认状态下，该登录页面如图 7.3 所示。

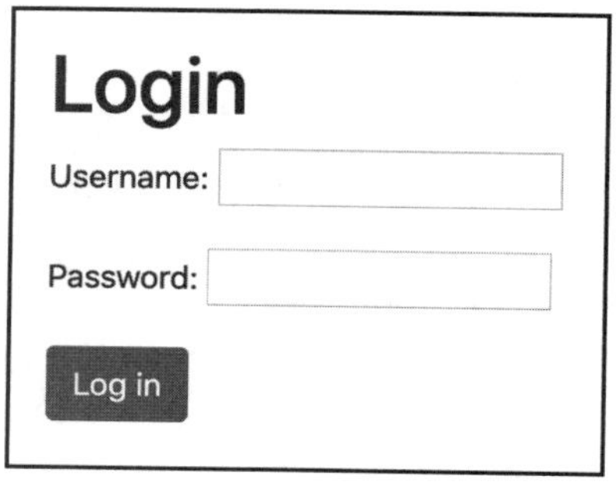

Login

Username:

Password:

Log in

图 7.3

7.7.4　延伸阅读

- 第 3 章中的“上传图像”示例。
- 第 3 章中的“利用自定义模板创建一个表单布局”示例。
- 第 3 章中的“利用 django-crispy-forms 创建一个表单布局”示例。
- 第 4 章中的“安排 bese.html 模板”示例。

- “实现密码验证”示例。
- “向图像中添加动态水印”示例。

7.8　向图像中添加动态水印

考虑到知识产权和艺术版权等问题，某些图像仅可供人们查看，且不可进行非法传播。在当前示例中，我们将学习如何在站点的图像上添加水印。

7.8.1　准备工作

当前示例将在第 3 章 core 和 ideas 应用程序的基础上完成。

7.8.2　实现方式

下列步骤将水印添加至所显示的图像上。

（1）在虚拟环境中安装 django-imagekit。

```
(env)$ pip install django-imagekit==4.0.2
```

（2）在设置项中，将"imagekit"置于 INSTALLED_APPS 中。

```
# myproject/settings/_base.py
INSTALLED_APPS = [
    # …
    "imagekit",
]
```

（3）在 core 应用程序中，利用 WatermarkOverlay 类创建一个名为 processors.py 的文件，如下所示。

```
# myproject/apps/core/processors.py
from pilkit.lib import Image

class WatermarkOverlay(object):
    def __init__(self, watermark_image):
        self.watermark_image = watermark_image

    def process(self, img):
        original = img.convert('RGBA')
```

```
        overlay = Image.open(self.watermark_image)
        img = Image.alpha_composite(original,
        overlay).convert('RGB')
        return img
```

（4）在 Idea 模型中，在 picture 字段一侧添加 watermarked_picture_large 规范，如下所示。

```
# myproject/apps/ideas/models.py
import os

from imagekit.models import ImageSpecField
from pilkit.processors import ResizeToFill

from django.db import models
from django.conf import settings
from django.utils.translation import gettext_lazy as _
from django.utils.timezone import now as timezone_now

from myproject.apps.core.models import
CreationModificationDateBase, UrlBase
from myproject.apps.core.processors import WatermarkOverlay

def upload_to(instance, filename):
    now = timezone_now()
    base, extension = os.path.splitext(filename)
    extension = extension.lower()
    return f"ideas/{now:%Y/%m}/{instance.pk}{extension}"

class Idea(CreationModificationDateBase, UrlBase):
    # …
    picture = models.ImageField(
        _("Picture"), upload_to=upload_to
    )
    watermarked_picture_large = ImageSpecField(
        source="picture",
        processors=[
            ResizeToFill(800, 400),
            WatermarkOverlay(
                watermark_image=os.path.join(settings.STATIC_ROOT,
                'site', 'img', 'watermark.png'),
            )
        ],
```

```
        format="PNG"
    )
```

（5）使用一个图形程序创建一个半透明的 PNG 图像，其中，在透明背景上包含白色文本或一个 Logo，对应尺寸为 800 像素×400 像素，如图 7.4 所示。随后将该图像保存为 site_static/site/img/watermark.png。

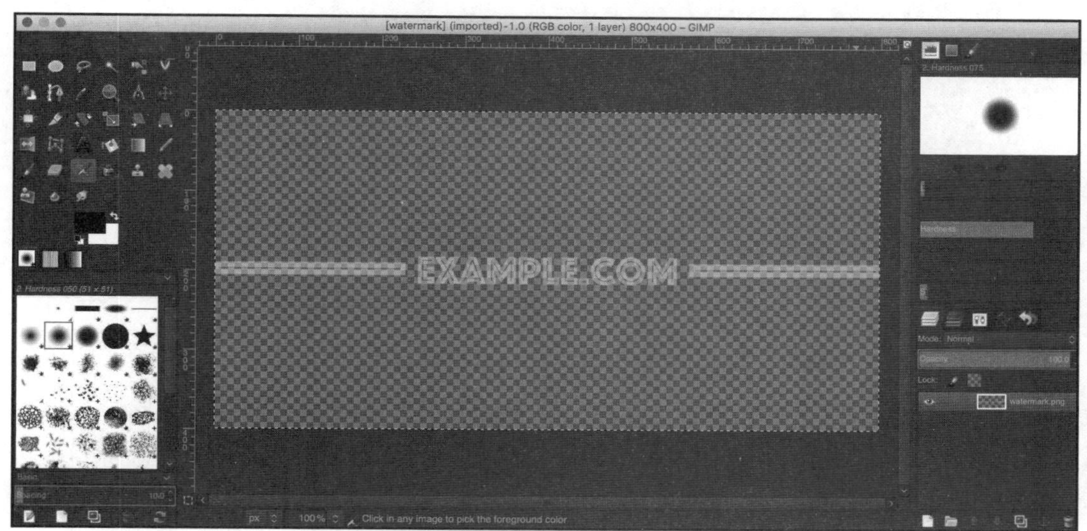

图 7.4

（6）运行 collectstatic 管理命令。

```
(env)$ export DJANGO_SETTINGS_MODULE=myproject.settings.dev
(env)$ python manage.py collectstatic
```

（7）编辑 idea 详细模板，并添加水印图像，如下所示。

```
{# ideas/idea_detail.html #}
{% extends "base.html" %}
{% load i18n %}

{% block content %}
    <a href="{% url "ideas:idea_list" %}">{% trans "List of ideas" %}</a>
    <h1>
        {% blocktrans trimmed with title=idea.translated_title %}
            Idea "{{ title }}"
        {% endblocktrans %}
    </h1>
```

```
    <img src="{{ idea.watermarked_picture_large.url }}" alt="" />
    {{ idea.translated_content|linebreaks|urlize }}
    <p>
        {% for category in idea.categories.all %}
            <span class="badge badge-pill badge-info">
             {{ category.translated_title }}</span>
        {% endfor %}
    </p>
    <a href="{% url 'ideas:download_idea_picture' pk=idea.pk %}"
     class="btn btn-primary">{% trans "Download picture" %}</a>
{% endblock %}
```

7.8.3 工作方式

当导航至 idea 详细页面时，即可看到一幅添加了水印后的图像，如图 7.5 所示。

图 7.5

在详细模板中，<img>标签的 src 属性使用了 idea 图像规范，即 watermarked_picture_large，并创建了一幅调整后的图像，随后存储于 media/CACHE/目录下。

django-imagekit 规范使用了预处理器调整图像，此处使用了两个预处理器。

（1）ResizeToFill 将图像重置为 800 像素×400 像素。

（2）自定义预处理器 WatermarkOverlay 应用了半透明重叠效果。

django-imagekit 预处理器需要包含一个 process()方法，该方法接收来自前一个预处理器的图像，并返回调整后的新图像。在当前示例中，我们从原始图像和半透明重叠中合成了最终的结果。

7.8.4　延伸阅读

读者还可参考“下载授权文件”示例。

7.9　基于 Auth0 的身份验证

随着人们每天接触的服务数量的增加，他们需要记住的用户名和密码的数量也随之增加。不仅如此，一旦出现安全漏洞，用户信息存储之处很可能是信息被窃取的地方。为了缓解这种情况，Auth0 等服务允许用户将身份验证服务集中在一个单一的、安全的平台上。

除了用户名和密码证书方面的支持，Auth0 还可通过社交平台对用户进行身份验证，如 Google、Facebook 或 Twitter。我们可通过 SMS 或电子邮件发送的单次代码使用无密码登录，甚至针对不同服务还存在企业级的支持。在当前示例中，我们将学习如何将 Auth0 应用程序连接至 Django 中，以及如何对其进行整合以处理用户身份验证问题。

7.9.1　准备工作

访问 https://auth0.com/并创建 Auth0 应用程序，随后遵循其中的指令配置 Auth0。其中，免费方案中提供了两个社交连接，因此我们将激活 Google 和 Twitter 并在此基础上登录。另外，我们还可尝试采用其他服务。注意，其中一些服务需要注册应用程序以获取 API 密钥或密码。

接下来需要在项目中安装 python-social-auth 和一些依赖项，在 pip 需求中包含这些依赖项。

```
# requirements/_base.txt
social-auth-app-django~=3.1
python-jose~=3.0
python-dotenv~=0.9
```

注意：

social-auth-app-django 是一个 python-social-auth 项目的 Django 包，并可通过多个社交连接之一对站点进行验证。

利用 pip 将依赖项安装至虚拟环境中。

7.9.2　实现方式

下列步骤将 Auth0 连接至 Django 项目。

（1）在设置项文件中，将社交身份验证应用程序添加至 INSTALLED_APPS 中，如下所示。

```
# myproject/settings/_base.py
INSTALLED_APPS = [
    # …
    "social_django",
]
```

（2）添加 social_django 应用程序所需的 Auth0 设置项，如下所示。

```
# myproject/settings/_base.py
SOCIAL_AUTH_AUTH0_DOMAIN = get_secret("AUTH0_DOMAIN")
SOCIAL_AUTH_AUTH0_KEY = get_secret("AUTH0_KEY")
SOCIAL_AUTH_AUTH0_SECRET = get_secret("AUTH0_SECRET")
SOCIAL_AUTH_AUTH0_SCOPE = ["openid", "profile", "email"]
SOCIAL_AUTH_TRAILING_SLASH = False
```

确保在密码或环境变量中定义了 AUTH0_DOMAIN、AUTH0_KEY 和 AUTH0_SECRET。这些变量值可在 Auth0 应用程序的设置项中找到。

（3）创建 Auth0 连接的后端，如下所示。

```
# myproject/apps/external_auth/backends.py
from urllib import request
from jose import jwt
from social_core.backends.oauth import BaseOAuth2

class Auth0(BaseOAuth2):
    """Auth0 OAuth authentication backend"""

    name = "auth0"
    SCOPE_SEPARATOR = " "
    ACCESS_TOKEN_METHOD = "POST"
    REDIRECT_STATE = False
```

```
    EXTRA_DATA = [("picture", "picture"), ("email", "email")]

    def authorization_url(self):
        return "https://" + self.setting("DOMAIN") + "/authorize"

    def access_token_url(self):
        return "https://" + self.setting("DOMAIN") + "/oauth/token"

    def get_user_id(self, details, response):
        """Return current user id."""
        return details["user_id"]

    def get_user_details(self, response):
        # Obtain JWT and the keys to validate the signature
        id_token = response.get("id_token")
        jwks = request.urlopen(
            "https://" + self.setting("DOMAIN") + "/.well-known/
              jwks.json"
        )
        issuer = "https://" + self.setting("DOMAIN") + "/"
        audience = self.setting("KEY") # CLIENT_ID
        payload = jwt.decode(
            id_token,
            jwks.read(),
            algorithms=["RS256"],
            audience=audience,
            issuer=issuer,
        )
        first_name, last_name = (payload.get("name") or " ").split(" ", 1)
        return {
            "username": payload.get("nickname") or "",
            "first_name": first_name,
            "last_name": last_name,
            "picture": payload.get("picture") or "",
            "user_id": payload.get("sub") or "",
            "email": payload.get("email") or "",
        }
```

（4）将新后端添加至 AUTHENTICATION_BACKENDS 设置项中，如下所示。

```
# myproject/settings/_base.py
AUTHENTICATION_BACKENDS = {
    "myproject.apps.external_auth.backends.Auth0",
    "django.contrib.auth.backends.ModelBackend",
}
```

（5）我们希望可以从任何模板访问社交身份验证用户，因而需要为其创建一个上下文预处理器。

```
# myproject/apps/external_auth/context_processors.py
def auth0(request):
    data = {}
    if request.user.is_authenticated:
        auth0_user = request.user.social_auth.filter(
            provider="auth0",
        ).first()
        data = {
            "auth0_user": auth0_user,
        }
    return data
```

（6）在设置项中进行注册。

```
# myproject/settings/_base.py
TEMPLATES = [
    {
        "BACKEND":
        "django.template.backends.django.DjangoTemplates",
        "DIRS": [os.path.join(BASE_DIR, "myproject", "templates")],
        "APP_DIRS": True,
        "OPTIONS": {
            "context_processors": [
                "django.template.context_processors.debug",
                "django.template.context_processors.request",
                "django.contrib.auth.context_processors.auth",
                "django.contrib.messages.context_processors.messages",
                "django.template.context_processors.media",
                "django.template.context_processors.static",
                "myproject.apps.core.context_processors.website_url",
                "myproject.apps.external_auth
                .context_processors.auth0",
            ]
        },
    }
]
```

（7）针对索引页面、仪表板和注销创建视图。

```
# myproject/apps/external_auth/views.py
from urllib.parse import urlencode
```

```
from django.shortcuts import render, redirect

from django.contrib.auth.decorators import login_required
from django.contrib.auth import logout as log_out
from django.conf import settings

def index(request):
    user = request.user
    if user.is_authenticated:
        return redirect(dashboard)
    else:
        return render(request, "index.html")

@login_required
def dashboard(request):
    return render(request, "dashboard.html")

def logout(request):
    log_out(request)
    return_to = urlencode({"returnTo":
     request.build_absolute_uri("/")})
    logout_url = "https://%s/v2/logout?client_id=%s&%s" % (
        settings.SOCIAL_AUTH_AUTH0_DOMAIN,
        settings.SOCIAL_AUTH_AUTH0_KEY,
        return_to,
    )
    return redirect(logout_url)
```

（8）创建索引模板，如下所示。

```
{# index.html #}
{% extends "base.html" %}
{% load i18n utility_tags %}

{% block content %}
<div class="login-box auth0-box before">
    <h3>{% trans "Please log in for the best user experience" %}</h3>
    <a class="btn btn-primary btn-lg" href="{% url "social:begin"
     backend="auth0" %}">{% trans "Log in" %}</a>
</div>
{% endblock %}
```

（9）创建仪表板模板。

```
{# dashboard.html #}
{% extends "base.html" %}
{% load i18n %}

{% block content %}
    <div class="logged-in-box auth0-box logged-in">
        <img alt="{% trans 'Avatar' %}" src="{{
         auth0_user.extra_data.picture }}"
         width="50" height="50" />
        <h2>{% blocktrans with name=request.user
         .first_name %}Welcome, {{ name }}
         {% endblocktrans %}!</h2>

        <a class="btn btn-primary btn-logout" href="{% url
         "auth0_logout" %}">{% trans "Log out" %}</a>
    </div>
{% endblock %}
```

（10）更新 URL 规则。

```
# myproject/urls.py
from django.conf.urls.i18n import i18n_patterns
from django.urls import path, include

from myproject.apps.external_auth import views as
external_auth_views

urlpatterns = i18n_patterns(
    path("", external_auth_views.index, name="index"),
    path("dashboard/", external_auth_views.dashboard, name="dashboard"),
    path("logout/", external_auth_views.logout, name="auth0_logout"),
    path("", include("social_django.urls")),
    # …
)
```

（11）添加登录 URL 设置项。

```
LOGIN_URL = "/login/auth0"
LOGIN_REDIRECT_URL = "dashboard"
```

7.9.3　工作方式

当浏览器访问项目的索引页面时，可以看到一个邀请用户登录的链接。单击该链接，用户将被重定向至 Auth0 身份验证系统，如图 7.6 所示。

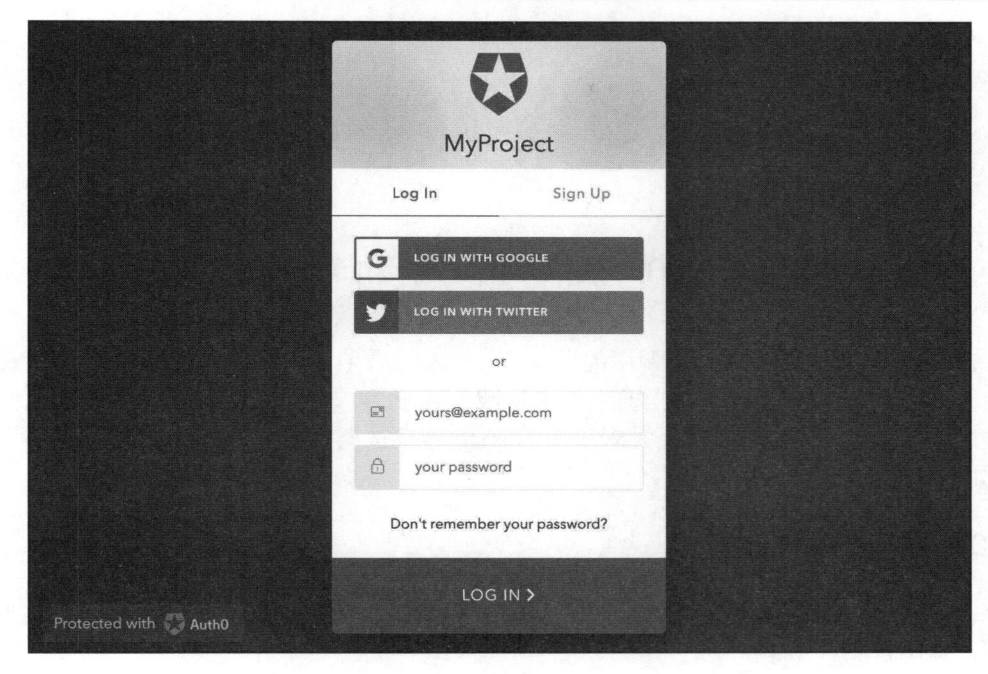

图 7.6

通过配置关联的 SOCIAL_AUTH_*设置项，这一功能可通过 python-social-auth（一个 Auth0 后端）予以启用。

一旦登录成功，Auth0 后端接收源自响应结果的数据并对其进行处理。所关联的数据绑定至与请求相关的用户对象上。在仪表板视图（对 LOGIN_REDIRECT_URL 进行身份验证的结果）中，用户详细信息被析取并添加至模板上下文中，随后渲染 dashboard.html，对应结果如图 7.7 所示。

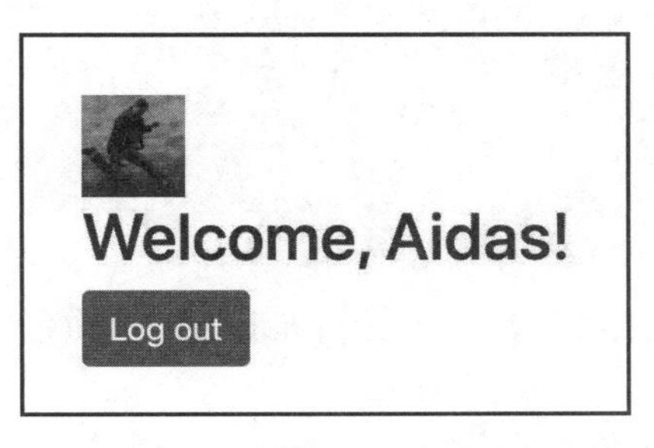

图 7.7

仪表板上的按钮负责处理用户的注销操作。

7.9.4　延伸阅读

- “实现密码验证”示例。
- “下载授权文件”示例。

7.10　缓存方法的返回值

在请求-响应循环中，如果调用的模型方法包含大量计算或多次的数据库查询，视图的性能将受到较大的影响。在当前示例中，我们将学习一种模式，以缓存方法的返回值以供后续重复使用。注意，此处并未采用 Django 缓存框架，这也是 Python 默认状态下提供的内容。

7.10.1　准备工作

选择一个应用程序，其中的模型包含了一个较为耗时的方法，该方法将在同一请求-响应循环中重复使用。

7.10.2　实现方式

具体实现方式如下列步骤所示。

（1）当前模式用于缓存模型的方法返回值，进而可在视图、表单或模板中重复使用，如下所示。

```
class SomeModel(models.Model):
    def some_expensive_function(self):
        if not hasattr(self, "_expensive_value_cached"):
            # do some heavy calculations...
            # ... and save the result to result variable
            self._expensive_value_cached = result
        return self._expensive_value_cached
```

（2）针对 ViralVideo 模型创建一个 get_thumbnail_url()方法（参见第 10 章“使用数据库查询表达式”示例）。

```
# myproject/apps/viral_videos/models.py
import re
from django.db import models
```

```
from django.utils.translation import ugettext_lazy as _

from myproject.apps.core.models import
CreationModificationDateBase, UrlBase

class ViralVideo(CreationModificationDateBase, UrlBase):
    embed_code = models.TextField(
        _("YouTube embed code"),
        blank=True)

    # …

    def get_thumbnail_url(self):
        if not hasattr(self, "_thumbnail_url_cached"):
            self._thumbnail_url_cached = ""
            url_pattern = re.compile(
                r'src="https://www.youtube.com/embed/([^"]+)"'
            )
            match = url_pattern.search(self.embed_code)
            if match:
                video_id = match.groups()[0]
                self._thumbnail_url_cached = (
                    f"https://img.youtube.com/vi/{video_id}/0.jpg"
                )
        return self._thumbnail_url_cached
```

7.10.3　工作方式

在当前示例中，对应方法检查模型实例是否存在_expensive_value_cached 属性。如果不存在，则执行较为耗时的计算，相应结果被赋予新属性中。在该方法结尾处，将返回缓存值。如果存在多个重量级的方法，则需要使用不同的属性名称保存每个计算值。

我们可在模板的标头和页脚中使用{{ object.some_expensive_function }}，且耗时的计算仅执行一次。

在某个模板中，我们还可在{% if %}条件和值的输出结果中使用函数，如下所示。

```
{% if object.some_expensive_function %}
    <span class="special">
        {{ object.some_expensive_function }}
    </span>
{% endif %}
```

在另一个示例中，我们通过解析视频嵌入代码的 URL、获取其 ID，然后组合缩略图图像的 URL 检查 YouTube 视频的缩略图，如下所示。

```
{% if video.get_thumbnail_url %}
   <figure>
       <img src="{{ video.get_thumbnail_url }}"
           alt="{{ video.title }}"
       />
       <figcaption>{{ video.title }}</figcaption>
   </figure>
{% endif %}
```

7.10.4　更多内容

我们前面描述的方法仅适用于无参方法调用，以便结果总是保持相同。如果输入发生变化，情况又当如何？自 Python 3.2 起，我们可使用一个装饰器，并根据参数哈希值提供方法调用的最近最少使用（Least Recently Used，LRU）缓存。

如某个函数接收两个值，并返回逻辑复杂的结果，如下所示。

```
def busy_bee(a, b):
    # expensive logic
    return result
```

如果我们持有这样一个函数，并希望提供一个缓存存储一些常用输入变量的结果，那么可以通过 functools 包中的@lru_cache 装饰器轻松做到这一点，如下所示。

```
from functools import lru_cache

@lru_cache(maxsize=100, typed=True)
def busy_bee(a, b):
    # expensive logic
    return result
```

当前，我们提供了一个缓存机制，可在输入哈希获得的键下存储多达 100 个结果。Python 3.3 中加入了 typed 选项，通过将该选项指定为 True，我们使 a=1 和 b=2 的调用与 a=1.0 和 b=2.0 的调用分开存储。根据逻辑操作方式和返回值结果的不同，这种变化可能是合适的，也可能是不合适的。

注意：

关于 functools 文档中的@lru_cache 装饰器，读者可访问 https://docs.python.org/3/library/functools.html#functools.lru_cache 以了解更多内容。

此外，还可针对之前示例使用该装饰器，以简化代码，如下所示。

```
# myproject/apps/viral_videos/models.py
from functools import lru_cache
# ...

class ViralVideo(CreationModificationDateMixin, UrlMixin):
    # ...
    @lru_cache
    def get_thumbnail_url(self):
        # ...
```

7.10.5 延伸阅读

- 第 4 章。
- “使用 Memcached 缓存 Django 视图”示例。
- “使用 Redis 缓存 Django 视图”示例。

7.11 使用 Memcached 缓存 Django 视图

通过缓存相对复杂的部分（如数据库查询或模板渲染），Django 可加速请求-响应周期。其中，Django 所支持的最快、最可靠的本地缓存机制是基于内存的缓存服务器 Memcached。在当前示例中，我们将学习如何使用 Memcached 缓存 viral_videos 应用程序的视图。关于缓存问题，我们将在第 10 章进一步讨论。

7.11.1 准备工作

下列步骤将准备 Django 项目的缓存机制。

（1）安装 memcached 服务。如在 macOS 上最简单的方式是使用 Homebrew。

```
$ brew install memcached
```

（2）下列命令用于启动、终止或重启 Memcached 服务。

```
$ brew services start memcached
$ brew services stop memcached
$ brew services restart memcached
```

注意：

在其他操作系统上，可通过 apt-get、yum 或其他默认的包管理实用程序安装 Memcached。此外，也可从源代码编译 Memcached，具体内容可参考 https://memcached.org/downloads。

（3）在虚拟环境中安装 Memcached Python 绑定，如下所示。

```
(env)$ pip install python-memcached==1.59
```

7.11.2　实现方式

执行下列步骤将集成特定视图的缓存。

（1）在项目设置项中设置 CACHES，如下所示。

```
# myproject/settings/_base.py
CACHES = {
    "memcached": {
        "BACKEND":
        "django.core.cache.backends.memcached.MemcachedCache",
        "LOCATION": get_secret("CACHE_LOCATION"),
        "TIMEOUT": 60, # 1 minute
        "KEY_PREFIX": "myproject",
    },
}
CACHES["default"] = CACHES["memcached"]
```

（2）确保 CACHE_LOCATION 在密码或环境变量中设置为"localhost:11211"。

（3）调整 viral_videos 应用程序的视图，如下所示。

```
# myproject/apps/viral_videos/views.py
from django.shortcuts import render
from django.views.decorators.cache import cache_page
from django.views.decorators.vary import vary_on_cookie

@vary_on_cookie
@cache_page(60)
def viral_video_detail(request, pk):
    # …
    return render(
        request,
        "viral_videos/viral_video_detail.html",
        {'video': video}
    )
```

注意：

如果读者查看下一个示例中遵循 Redis 的设置步骤，将会看到 views.py 文件中的内容没有发生任何变化。因此，无须调整使用缓存的代码即可修改底层的缓存机制。

7.11.3　工作方式

在第 10 章中将会看到，viral video 的详细视图显示了认证用户和匿名用户的观看次数。在访问一个 viral video 并在启用缓存的情况下多次刷新页面，我们将看到观看次数的变化仅一分钟一次。其原因在于，每次响应针对每个用户缓存 60 秒。相应地，我们通过 @cache_page 装饰器为视图设置缓存。

Memcached 是一个键-值存储，并在默认时使用完整的 URL 生成每个缓存页面的键。当两名访问者同时访问同一个页面，第一个访问者的请求将接收 Python 代码生成的页面，第二个访问者获取相同的 HTML 代码，但从 Memcached 服务器中获得。

在当前示例中，为了保证每位访问者被独立对待（即使他们访问了同一 URL），我们将使用@vary_on_cookie 装饰器。该装饰器检查 HTTP 请求中 Cookie 头的唯一性。

提示：

关于 Django 的缓存框架，读者可访问 https://docs.djangoproject.com/en/3.0/topics/cache/。类似地，读者可访问 https://memcached.org/以了解更多与 Memcached 相关的内容。

7.11.4　延伸阅读

- “缓存方法的返回值”示例。
- “使用 Redis 缓存 Django 视图”示例。
- 第 10 章中的“使用数据库查询表达式”示例。

7.12　使用 Redis 缓存 Django 视图

作为一种缓存机制，虽然 Memcached 在市场上反映良好，并得到了 Django 的较好地支持，但 Redis 提供了更为丰富的功能。此处将重新讨论上一个示例，并学习如何使用 Redis 实现相同的任务。

7.12.1　准备工作

下列步骤将准备 Django 项目的缓存机制。

（1）安装 Redis 服务。如在 macOS 环境下，最为简单的方式是使用 Homebrew。

```
$ brew install redis
```

（2）利用下列命令启动、终止或重启 Redis 服务。

```
$ brew services start redis
$ brew services stop redis
$ brew services restart redis
```

注意：

在其他操作系统上，可通过 apt-get、yum 或其他默认的包管理实用工具安装 Redis。此外，也可从源代码编译 Redis，具体内容可参考 https://redis.io/download。

（3）在虚拟环境中安装 Django 的 Redis 缓存后端及其依赖项，如下所示。

```
(env)$ pip install redis==3.3.11
(env)$ pip install hiredis==1.0.1
(env)$ pip install django-redis-cache==2.1.0
```

7.12.2　实现方式

下列步骤将集成特定视图的缓存。

（1）在项目设置项中设置 CACHES，如下所示。

```
# myproject/settings/_base.py
CACHES = {
    "redis": {
        "BACKEND": "redis_cache.RedisCache",
        "LOCATION": [get_secret("CACHE_LOCATION")],
        "TIMEOUT": 60, # 1 minute
        "KEY_PREFIX": "myproject",
    },
}
CACHES["default"] = CACHES["redis"]
```

（2）确保 CACHE_LOCATION 在密码或环境变量中设置为"localhost:6379"。

（3）调整 viral_videos 应用程序的视图。

```
# myproject/apps/viral_videos/views.py
from django.shortcuts import render
from django.views.decorators.cache import cache_page
from django.views.decorators.vary import vary_on_cookie

@vary_on_cookie
@cache_page(60)
def viral_video_detail(request, pk):
    # …
    return render(
        request,
        "viral_videos/viral_video_detail.html",
        {'video': video}
    )
```

注意：

如果遵循前述示例中的 Memcached 设置，则会看到此处的 views.py 没有任何变化。这表明，我们可修改底层缓存机制，且无须调整使用它的代码。

7.12.3　工作方式

类似于 Memcached，我们通过@cache_page 装饰器设置视图的缓存。因此，每个响应结果针对每个用户缓存 60 秒。Viral video 详细视图（如 http://127.0.0.1:8000/en/videos/1/）显示了验证用户和匿名用户的观看次数。当开启缓存时，如果多次刷新页面，将会看到观看次数一分钟仅变化一次。

Redis 也是一个键-值存储，当用于缓存时，Redis 将根据完整的 URL 针对每个缓存页面生成键。当两个访问者同时访问同一页面时，第一个访问者的请求将接收由 Python 代码生成的页面，而第二个访问者的请求则从 Redis 服务器上获取相同的 HTML 代码。

在当前示例中，为了确保每个访问者的请求被独立对待（即使他们访问的是相同的 URL），我们采用了@vary_on_cookie 装饰器。该装饰器检查 HTTP 请求中 Cookie 头的唯一性。

提示：

关于 Django 的缓存框架，读者可访问 https://docs.djangoproject.com/en/3.0/topics/cache/。类似地，读者可访问 https://redis.io/以了解更多与 Memcached 相关的内容。

7.12.4 更多内容

虽然 Redis 能够采用与 Memcached 相同的方式处理缓存，但对于系统中内建的缓存算法，Redis 还存在许多其他选项。除了缓存机制，Redis 还可用作数据库或消息存储，并支持不同的数据结构、事务、pub/sub 和自动故障转移等功能。

通过 django-redis-cache 后端，Redis 还可轻松地配置为会话后端，如下所示。

```
# myproject/settings/_base.py
SESSION_ENGINE = "django.contrib.sessions.backends.cache"
SESSION_CACHE_ALIAS = "default"
```

7.12.5 延伸阅读

- “缓存方法的返回值”示例。
- “使用 Memcached 缓存 Django 视图”示例。
- 第 10 章中的“使用数据库查询表达式”示例。

第8章 层次结构

本章主要涉及下列主题。

- ❑ 利用 django-mptt 创建层次分类。
- ❑ 利用 django-mptt-admin 创建分类管理界面。
- ❑ 利用 django-mptt 在模板中渲染分类。
- ❑ 利用 django-mptt 在表单中通过单选字段选择分类。
- ❑ 利用 django-mptt 在表单中通过复选框列表选择多个分类。
- ❑ 利用 django-treebeard 创建层次分类。
- ❑ 利用 django-treebeard 创建基本的分类管理界面。

8.1 简　　介

当构建自己的论坛、评论回复或分类系统时，都需要将层次结构保存至数据库中。虽然关系数据库（如 MySQL 和 PostgreSQL）中的表是平面的，但一种快速有效的方法可存储层次结构，即预排序遍历树（Modified Preorder Tree Traversal，MPTT）。MPTT 可读取树形结构，且无须递归调用数据库。

下面首先介绍一下树形结构中的术语。树形数据结构是一种嵌套的节点集合，从根节点开始并引用子节点。其间也存在一些限制条件，如节点不应向回引用进而形成一个循环，同时节点也不应重复引用。除此之外，其他注意事项还包括：

- ❑ 父节点是引用子节点的任何节点。
- ❑ 后代节点是通过递归地从父节点遍历其子节点可以到达的节点。因此，节点的后代将是它的子节点、子节点的子节点，以此类推。
- ❑ 祖先节点是可递归地从子节点遍历到父节点的节点。因此，一个节点的祖先将是它的父节点、父节点的父节点，以此类推，直到根节点。
- ❑ 兄弟节点是具有相同父节点的节点。
- ❑ 叶节点是不包含子节点的节点。

下面解释 MPTT 的工作方式。想象一下，若将树水平放置且根节点位于顶部。树中的每个节点都包含左值和右值，这里，可以将其想象为节点左右两侧的手柄。随后沿逆时针方向遍历树形结构。首先是根节点，同时用数字（1、2、3 等）标记每次遇到的左值和右值，以此类推，如图 8.1 所示。

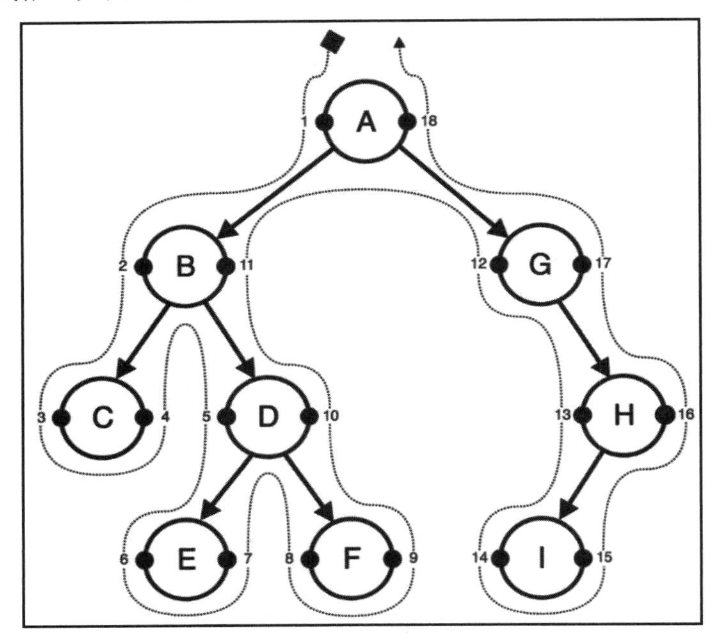

图 8.1

在这种层次结构的数据库表中，每个节点将持有一个标题、左值和右值。

现在，如果想要得到 B 节点左值为 2、右值为 11 的子树，则需要选择所有左值在 2 到 11 之间的节点，即 C、D、E 和 F。

相应地，如果需要获得 D 节点左值为 5、右值为 10 的所有祖先，则需要选择左值小于 5、右值大于 10 的所有节点，即 B 和 A。

当获取节点后代的数量时，可使用下列公式。

后代数量=(右值−左值−1)/2

因此，B 节点的后代数量可计算为：

(11−2−1)/2=4

如果打算将 E 节点绑定至 C 节点上，则需要更新二者第一个共同祖先（即 B 节点）的左右值。随后，C 节点的左值仍为 3，E 节点的左值为 4，而右值为 5；C 节点的右值变为 6，D 节点的左值变为 7；F 节点的左值仍然为 8，其他内容保持不变。

类似地，MPTT 中还存在其他树形节点操作，其过程可能较为复杂。对此，有一个名为 django-mptt 的 Django 应用程序负责处理这些算法，并提供了一个直观的 API 处理树形结构。此外，django-treebeard 作为一个强大的替代方案在 django CMS 3.1 中取代了

MPTT。在当前示例中，我们将学习如何使用这些辅助应用程序。

8.2 技术需求

读者需要在虚拟环境中安装最新稳定版本的 Python3、MySQL 或 PostgreSQL 和 Django 项目。

读者可访问 GitHub 储存库的 ch08 目录查看本章的所有代码，对应网址为 https://github.com/PacktPublishing/Django-3-Web-Development-Cookbook-Fourth-Edition。

8.3 利用 django-mptt 创建层次分类

为了进一步展示如何处理 MPTT，当前示例将在第 3 章 ideas 应用程序的基础上完成。其间，我们将利用一个 Category 层次模型替换分类，并更新 Idea 模型，以包含多对多的分类关系。另外，我们还将从头创建一个应用程序，并实现一个 Idea 模型的基础版本。

8.3.1 准备工作

具体准备工作如下列步骤所示。

（1）通过下列命令在虚拟环境中安装 django-mptt。

```
(env)$ pip install django-mptt==0.10.0
```

（2）创建 categories 和 ideas 应用程序，并连同 mptt 添加至设置项的 INSTALLED_APPS 中，如下所示。

```
# myproject/settings/_base.py
INSTALLED_APPS = [
    # …
    "mptt",
    # …
    "myproject.apps.categories",
    "myproject.apps.ideas",
]
```

8.3.2 实现方式

我们将创建一个 Category 层次模型，并将其连接至 Idea 模型上，这将形成多对多的分类关系，如下所示。

（1）打开 categories 应用程序中的 models.py 文件，并添加一个 Category 模型，该模型扩展了 mptt.models.MPTTModel 和 CreationModificationDateBase（参见第 2 章）。除了来自混入的字段，Category 模型还需要包含 TreeForeignKey 类型的 parent 字段和一个 title 字段。

```
# myproject/apps/ideas/models.py
from django.db import models
from django.utils.translation import ugettext_lazy as _
from mptt.models import MPTTModel
from mptt.fields import TreeForeignKey

from myproject.apps.core.models import CreationModificationDateBase

class Category(MPTTModel, CreationModificationDateBase):
    parent = TreeForeignKey(
        "self", on_delete=models.CASCADE,
        blank=True, null=True, related_name="children"
    )
    title = models.CharField(_("Title"), max_length=200)

    class Meta:
        ordering = ["tree_id", "lft"]
        verbose_name = _("Category")
        verbose_name_plural = _("Categories")

    class MPTTMeta:
        order_insertion_by = ["title"]

    def __str__(self):
        return self.title
```

（2）更新 Idea 模型并包含 TreeManyToManyField 类型的 categories 字段。

```
# myproject/apps/ideas/models.py
from django.utils.translation import gettext_lazy as _

from mptt.fields import TreeManyToManyField

from myproject.apps.core.models import
CreationModificationDateBase, UrlBase

class Idea(CreationModificationDateBase, UrlBase):
```

```
    # ...
    categories = TreeManyToManyField(
        "categories.Category",
        verbose_name=_("Categories"),
        related_name="category_ideas",
    )
```

（3）通过生成、运行迁移以更新数据库。

```
(env)$ python manage.py makemigrations
(env)$ python manage.py migrate
```

8.3.3　工作方式

MPTTModel 混入将把 tree_id、lft、rght 和 level 字段添加至 Category 模型中。

- 当在数据库表中包含多棵树时，使用 tree_id 字段。实际上，每个根分类被保存于独立的树中。
- lft 和 rght 字段存储 MPTT 算法使用的左右值。
- level 字段存储节点在树形结构中的深度。根节点的 level 为 0。

通过特定于 MPTT 的 order_insertion_by 元选项，我们确保在添加新类别时，它们能够按标题的字母顺序排列。

除了新字段，MPTTModel 混入还添加了一些方法遍历树形结构，这类似于利用 JavaScript 导航 DOM 元素。这些方法包括：

- 如果打算访问某个分类的祖先，可使用下列代码。这里，ascending 参数定义了读取节点的方向（默认为 False）；include_self 参数则定义了是否在 QuerySet 中包含分类自身（默认为 False）。

```
ancestor_categories = category.get_ancestors(
    ascending=False,
    include_self=False,
)
```

- 当只获取根分类时，可使用下列代码。

```
root = category.get_root()
```

- 如果打算获取某个分类的直接子节点，则可使用下列代码。

```
children = category.get_children()
```

- 当获取某个分类的全部后代时，可使用下列代码。这里，include_self 参数再次定义了是否在 QuerySet 中包含分类自身。

```
descendants = category.get_descendants(include_self=False)
```

- 如果打算获得后代计数且无须查询数据库，则可使用下列代码。

```
descendants_count = category.get_descendant_count()
```

- 当获取全部兄弟节点时，可调用下列方法。

```
siblings = category.get_siblings(include_self=False)
```

注意：

根分类被视为其他根分类的兄弟节点。

- 当获取上一个或下一个兄弟节点时，可调用下列方法。

```
previous_sibling = category.get_previous_sibling()
next_sibling = category.get_next_sibling()
```

- 下列方法用于检查分类是根节点、子节点或者是叶节点。

```
category.is_root_node()
category.is_child_node()
category.is_leaf_node()
```

所有这些方法均可用于视图、模板或管理命令中。如果打算操控树形结构，我们还可使用 insert_at()和 move_to()方法。在当前示例中，可访问 https://django-mptt.readthedocs.io/en/stable/models.html 以了解更多内容。

在前述模型中，我们使用了 TreeForeignKey 和 TreeManyToManyField，除了在管理界面中按层次结构缩进展示了选择样式，它们与 ForeignKey 和 ManyToManyField 十分类似。

注意，在 Category 模型的 Meta 类中，我们按照 tree_id 和 lft 值对分类进行排序，以便在树形结构中自然地显示类别。

8.3.4　延伸阅读

- 第 2 章中的“创建一个模型混入以处理日期的创建和修改”示例。
- “利用 django-mptt-admin 创建分类管理界面”示例。

8.4　利用 django-mptt-admin 创建分类管理界面

django-mptt 应用程序包含了一个简单的模型管理混入，允许我们创建一个树形结构并以缩进格式列出。当重新排序树形结构时，一般需要亲自实现或使用第三方方案。针对层次模型，django-mptt-admin 可帮助我们创建一个可拖曳的管理界面，当前示例将对此进行探讨。

8.4.1　准备工作

首先设置上一个示例中的 categories 应用程序，随后通过下列步骤安装 django-mptt-admin 应用程序。

（1）利用下列命令在虚拟环境中安装应用程序。

```
(env)$ pip install django-mptt-admin==0.7.2
```

（2）将应用程序置于设置项的 INSTALLED_APPS 中，如下所示。

```
# myproject/settings/_base.py
INSTALLED_APPS = [
    # …
    "mptt",
    "django_mptt_admin",
]
```

（3）确保 django-mptt-admin 的静态文件对项目有效。

```
(env)$ python manage.py collectstatic
```

8.4.2　实现方式

创建一个 admin.py 文件，并于其中定义 Category 模型的管理界面。这将扩展 DjangoMpttAdmin 而非 admin.ModelAdmin，如下所示。

```
# myproject/apps/categories/admin.py
from django.contrib import admin
from django_mptt_admin.admin import DjangoMpttAdmin

from .models import Category

@admin.register(Category)
class CategoryAdmin(DjangoMpttAdmin):
    list_display = ["title", "created", "modified"]
    list_filter = ["created"]
```

8.4.3　工作方式

分类的管理界面包含两个模型，即树形视图和网格视图。其中，树形视图如图 8.2 所示。

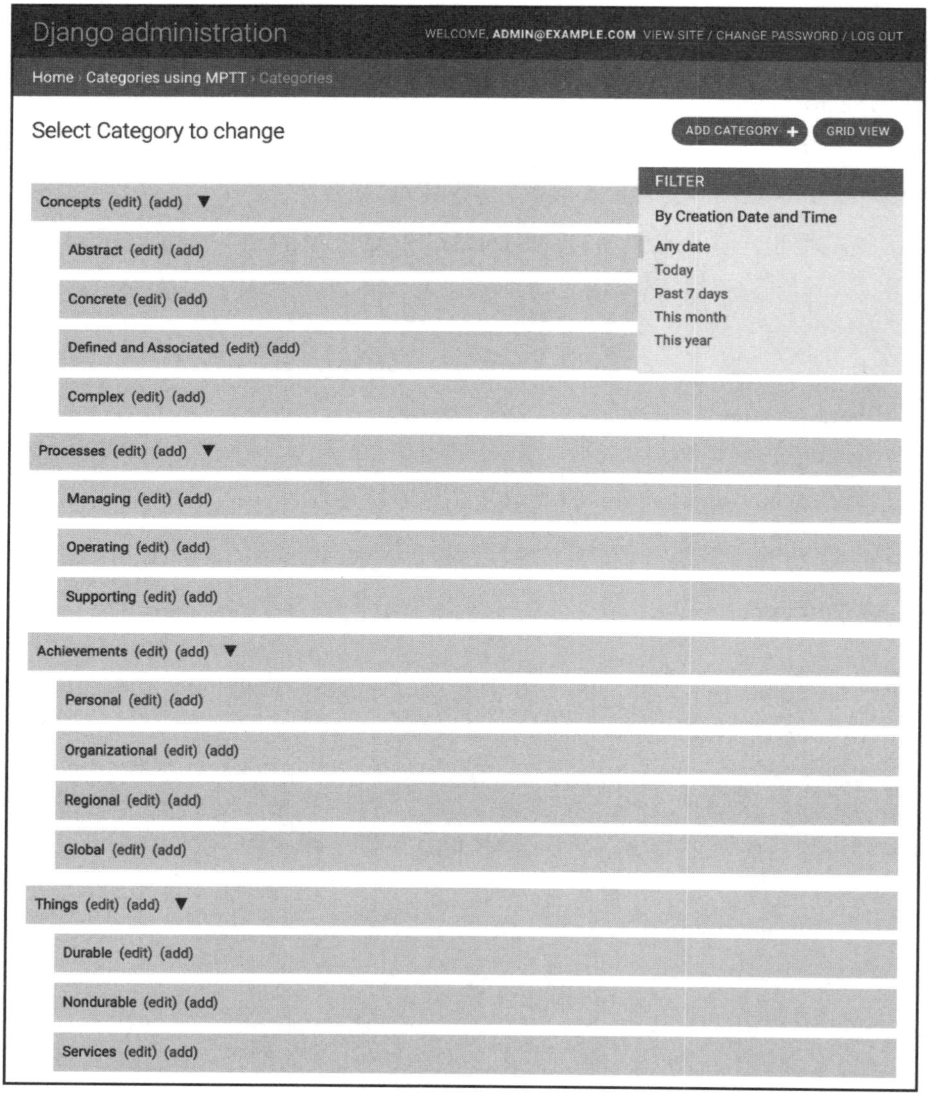

Django administration　WELCOME, ADMIN@EXAMPLE.COM VIEW SITE / CHANGE PASSWORD / LOG OUT

Home › Categories using MPTT › Categories

Select Category to change　ADD CATEGORY +　GRID VIEW

FILTER

By Creation Date and Time

Any date

Today

Past 7 days

This month

This year

Concepts (edit) (add) ▼

Abstract (edit) (add)

Concrete (edit) (add)

Defined and Associated (edit) (add)

Complex (edit) (add)

Processes (edit) (add) ▼

Managing (edit) (add)

Operating (edit) (add)

Supporting (edit) (add)

Achievements (edit) (add) ▼

Personal (edit) (add)

Organizational (edit) (add)

Regional (edit) (add)

Global (edit) (add)

Things (edit) (add) ▼

Durable (edit) (add)

Nondurable (edit) (add)

Services (edit) (add)

图 8.2

树形视图采用 jqTree jQuery 库管理节点。我们可以扩展或折叠分类以获取更好的概览效果。当对其重新排序或修改依赖项时，可以拖曳列表视图中的标题。在重新排序过程中，User Interface（UI）如图 8.3 所示。

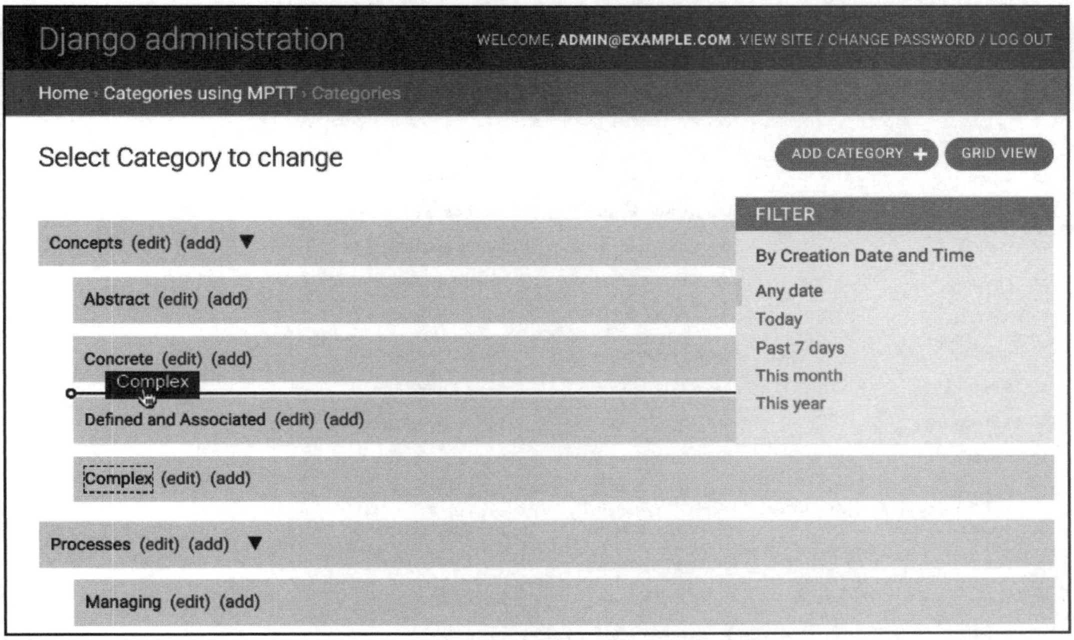

图 8.3

注意，任何与列表相关的设置项，如 list_display 或 list_filter，都将在树形视图中被忽略。另外，任何以 order_insertion_by 元属性驱动的排序都将被手动排序覆写。

如果打算过滤分类，可按照特定的字段对其进行排序，或者应用管理动作，我们可以切换至网格视图，这将显示默认的分类更改列表，如图 8.4 所示。

8.4.4　延伸阅读

- “利用 django-mptt 创建层次分类”示例。
- “利用 django-treebeard 创建分类管理界面”示例。

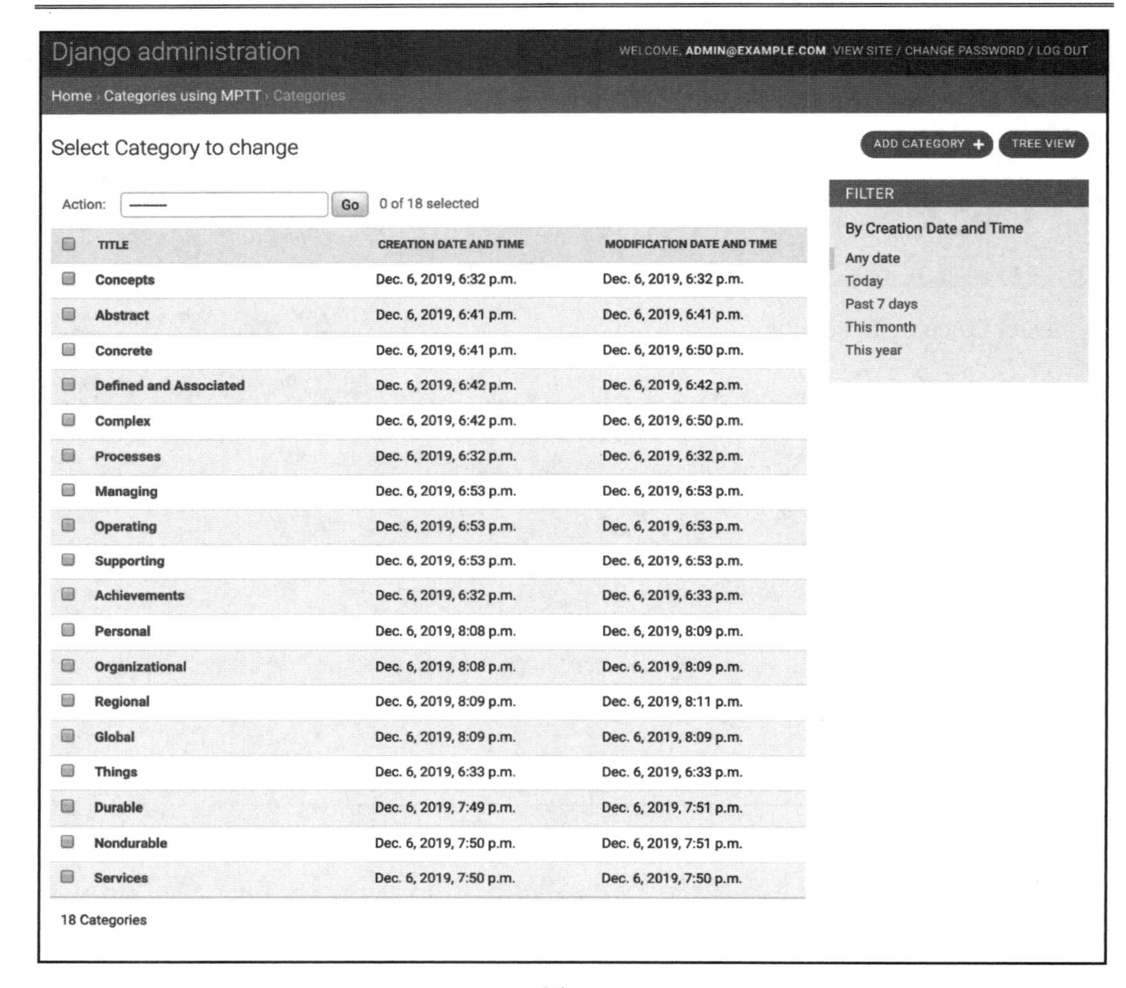

图 8.4

8.5　利用 django-mptt 在模板中渲染分类

当在应用程序中创建了分类后，需要在模板中以层次结构方式对其进行显示。对此，最简单的基于 MPTT 树的方式是使用来自 django-mptt 应用程序中的{% recursetree %}模板标签。当前示例将展示如何实现这一任务。

8.5.1　准备工作

确保已经创建了 categories 和 ideas 应用程序。其中，Idea 模型与 Category 模型之间具有多对多的关系。参见“利用 django-mptt 创建层次分类”示例并在数据库中输入某些分类。

8.5.2　实现方式

这里，将层次分类的 QuerySet 传递至模板，并于随后使用{% recursetree %}模板标签，如下所示。

（1）创建一个视图，加载所有分类，并将它们传递给模板。

```
# myproject/apps/categories/views.py
from django.views.generic import ListView

from .models import Category

class IdeaCategoryList(ListView):
    model = Category
    template_name = "categories/category_list.html"
    context_object_name = "categories"
```

（2）创建模板，输出分类的层次，如下所示。

```
{# categories/category_list.html #}
{% extends "base.html" %}
{% load mptt_tags %}

{% block content %}
    <ul class="root">
        {% recursetree categories %}
            <li>
                {{ node.title }}
                {% if not node.is_leaf_node %}
                    <ul class="children">
                        {{ children }}
                    </ul>
                {% endif %}
            </li>
        {% endrecursetree %}
```

```
    </ul>
{% endblock %}
```

（3）创建 URL 规则以显示视图。

```
# myproject/urls.py
from django.conf.urls.i18n import i18n_patterns
from django.urls import path

from myproject.apps.categories import views as categories_views

urlpatterns = i18n_patterns(
    # …
    path(
        "idea-categories/",
        categories_views.IdeaCategoryList.as_view(),
        name="idea_categories",
    ),
)
```

8.5.3 工作方式

当前模板将作为嵌套列表被渲染，如图 8.5 所示。

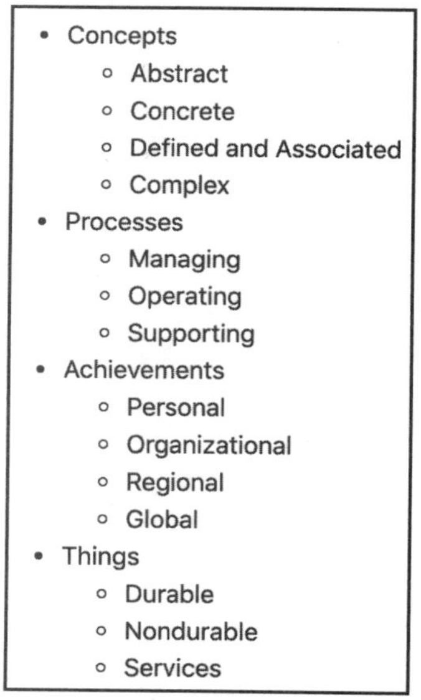

- Concepts
    - Abstract
    - Concrete
    - Defined and Associated
    - Complex
- Processes
    - Managing
    - Operating
    - Supporting
- Achievements
    - Personal
    - Organizational
    - Regional
    - Global
- Things
    - Durable
    - Nondurable
    - Services

图 8.5

{% recursetree %}块模板标签接收分类的 QuerySet，并通过嵌套在标签中的模板内容渲染列表，此处使用了两个特殊的变量：

（1）node 变量表示为 Category 模型的实例，其字段或方法可用于添加特定的 CSS 类，或 JavaScript 的 HTML 5 data-*属性，如{{ node.get_descendent_count }}、{{ node.level}} 或{{ node.is_root }}。

（2）我们持有一个 children 变量，用于定义当前分类的渲染的子节点将置于何处。

8.5.4 更多内容

如果层次结构十分复杂，如超过 20 个级别，则建议使用{% full_tree_for_model %}和{% drilldown_tree_for_node %}迭代标签，或非递归的 tree_info 模板过滤器。

注意：

关于具体实现的更多信息，读者可访问 https://django-mptt.readthedocs.io/en/latest/templates.html#iterative-tags。

8.5.5 延伸阅读

- 第 4 章中的“使用 HTML 5 数据属性”示例。
- “利用 django-mptt 创建层次分类”示例。
- “利用 django-treebeard 创建层次分类”示例。
- “利用 django-mptt 和单选字段在表单中选择分类”示例。

8.6 利用 django-mptt 和单选字段在表单中选择分类

如果需要在表单中显示分类选择结果，情况又当如何？层次结构将如何被表达？在 django-mptt 中，存在一个特殊的 TreeNodeChoiceField 表单字段，可用于在选取字段中显示层次结构。接下来让我们探讨如何实现这一任务。

8.6.1 准备工作

当前示例将在前述示例定义的 categories 和 ideas 应用程序的基础上完成。对于当前示例，我们还需使用 django-crispy-forms，读者可参考第 3 章查看其安装方式。

8.6.2 实现方式

接下来添加一个字段，并按照分类进行过滤，进而增强 ideas 的过滤器表单。该表单在第 3 章中的“过滤对象列表”示例中已创建。

（1）在 idesa 应用程序的 forms.py 文件中，利用一个分类字段创建一个表单，如下所示。

```
# myproject/apps/ideas/forms.py
from django import forms
from django.utils.safestring import mark_safe
from django.utils.translation import ugettext_lazy as _
from django.contrib.auth import get_user_model

from crispy_forms import bootstrap, helper, layout
from mptt.forms import TreeNodeChoiceField

from myproject.apps.categories.models import Category

from .models import Idea, RATING_CHOICES

User = get_user_model()

class IdeaFilterForm(forms.Form):
    author = forms.ModelChoiceField(
        label=_("Author"),
        required=False,
        queryset=User.objects.all(),
    )
    category = TreeNodeChoiceField(
        label=_("Category"),
        required=False,
        queryset=Category.objects.all(),
        level_indicator=mark_safe("    ")
    )
    rating = forms.ChoiceField(
        label=_("Rating"), required=False, choices=RATING_CHOICES
    )
    def __init__(self, *args, **kwargs):
        super().__init__(*args, **kwargs)

        author_field = layout.Field("author")
```

```
        category_field = layout.Field("category")
        rating_field = layout.Field("rating")
        submit_button = layout.Submit("filter", _("Filter"))
        actions = bootstrap.FormActions(submit_button)

        main_fieldset = layout.Fieldset(
            _("Filter"),
            author_field,
            category_field,
            rating_field,
            actions,
        )

        self.helper = helper.FormHelper()
        self.helper.form_method = "GET"
        self.helper.layout = layout.Layout(main_fieldset)
```

（2）我们已经创建了 IdeaListView、一个关联的 URL 规则，以及 idea_list.html 模板以显示表单。确保利用{% crispy %}模板标签在模板中渲染过滤器表单，如下所示。

```
{# ideas/idea_list.html #}
{% extends "base.html" %}
{% load i18n utility_tags crispy_forms_tags %}

{% block sidebar %}
    {% crispy form %}
{% endblock %}

{% block main %}
    {# … #}
{% endblock %}
```

8.6.3　工作方式

此时，分类选择下拉菜单如图 8.6 所示。

TreeNodeChoiceField 的行为类似于 ModelChoiceField，但以缩进模式显示层次选择结果。默认状态下，TreeNodeChoiceField 表示每个更深的层次（前缀为---）。在当前示例中，通过将 level_indicator 参数传递给字段，我们将层次指示器修改为 4 个不间断的空格（ HTML 实体）。为了确保不间断的空格不会被转义，我们使用了 mark_safe()函数。

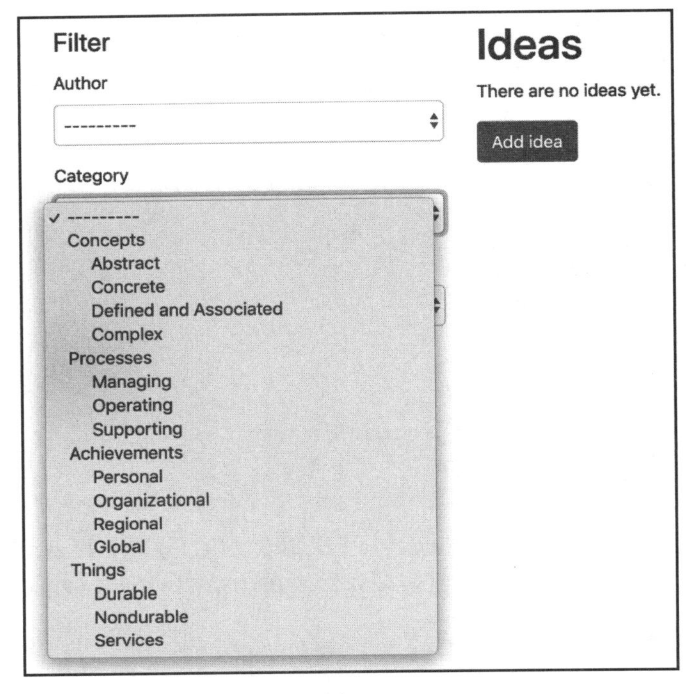

图 8.6

8.6.4　延伸阅读

- “利用 django-mptt 在模板中渲染分类”示例。
- “利用 django-mptt 在表单中通过复选框列表选择多个分类”示例。

8.7　利用 django-mptt 在表单中通过复选框列表选择多个分类

当一个或多个分类需要在表单中一次性地被选取，我们可以使用 django-mptt 提供的 TreeNodeMultipleChoiceField 多选字段。然而，从界面的角度来看，多选字段（如<select multiple>）缺乏一定的友好性，因为用户需要滚动并按住控制键或命令键，同时单击以做出多个选择。特别地，若存在多个选取条目，用户需要一次性地选择多项内容，这对于障碍人士来说可能会导致非常糟糕的用户体验。对此，一种较好的解决方案是使用复选框列表，并以此选择分类。在当前示例中，我们将创建一个字段，并在表单中以缩进的复选框形式显示层次树形结构。

8.7.1　准备工作

当前示例将在前述示例定义的 categories、ideas 和 core 应用程序的基础上完成。

8.7.2　实现方式

当渲染一个包含复选框的缩进分类列表时，我们将创建和使用一个新的 MultipleChoiceTreeField 表单字段，同时为该字段生成一个 HTML 模板。

这一特殊的模板将被传递至 crispy_forms 布局中。对此，需要执行下列步骤。

（1）在 core 应用程序中，添加一个 form_fields.py 文件，并创建一个扩展了 ModelMultipleChoiceField 的 MultipleChoiceTreeField 表单字段，如下所示。

```
# myproject/apps/core/form_fields.py
from django import forms

class MultipleChoiceTreeField(forms.ModelMultipleChoiceField):
    widget = forms.CheckboxSelectMultiple

    def label_from_instance(self, obj):
        return obj
```

（2）使用基于分类的新字段，在新的表单中进行选择以创建 idea。另外，在表单布局中，将一个自定义模板传递至 categories 字段，如下所示。

```
# myproject/apps/ideas/forms.py
from django import forms
from django.utils.translation import ugettext_lazy as _
from django.contrib.auth import get_user_model

from crispy_forms import bootstrap, helper, layout

from myproject.apps.categories.models import Category
from myproject.apps.core.form_fields import MultipleChoiceTreeField

from .models import Idea, RATING_CHOICES

User = get_user_model()

class IdeaForm(forms.ModelForm):
    categories = MultipleChoiceTreeField(
```

```
        label=_("Categories"),
        required=False,
        queryset=Category.objects.all(),
    )

    class Meta:
        model = Idea
        exclude = ["author"]

    def __init__(self, request, *args, **kwargs):
        self.request = request
        super().__init__(*args, **kwargs)

        title_field = layout.Field("title")
        content_field = layout.Field("content", rows="3")
        main_fieldset = layout.Fieldset(_("Main data"),
         title_field, content_field)

        picture_field = layout.Field("picture")
        format_html = layout.HTML(
            """{% include "ideas/includes/picture_guidelines.html" %}"""
        )

        picture_fieldset = layout.Fieldset(
            _("Picture"),
            picture_field,
            format_html,
            title=_("Image upload"),
            css_id="picture_fieldset",
        )

        categories_field = layout.Field(
            "categories",
            template="core/includes
            /checkboxselectmultiple_tree.html"
        )
        categories_fieldset = layout.Fieldset(
            _("Categories"), categories_field,
            css_id="categories_fieldset"
        )

        submit_button = layout.Submit("save", _("Save"))
        actions = bootstrap.FormActions(submit_button, css_class="my-4")
```

```
        self.helper = helper.FormHelper()
        self.helper.form_action = self.request.path
        self.helper.form_method = "POST"
        self.helper.layout = layout.Layout(
            main_fieldset,
            picture_fieldset,
            categories_fieldset,
            actions,
        )

    def save(self, commit=True):
        instance = super().save(commit=False)
        instance.author = self.request.user
        if commit:
            instance.save()
            self.save_m2m()
        return instance
```

（3）根据 crispy 表单创建一个 Bootstrap 样式复选框列表的模板，即 bootstrap4/layout/checkboxselectmultiple.html，如下所示。

```
{# core/include/checkboxselectmultiple_tree.html #}
{% load crispy_forms_filters l10n %}

<div class="{% if field_class %} {{ field_class }}{% endif %}"{% if
flat_attrs %} {{ flat_attrs|safe }}{% endif %}>

    {% for choice_value, choice_instance in field.field.choices %}
    <div class="{%if use_custom_control%}custom-control Custom-
     checkbox{% if inline_class %} custom-control-inline{% endif
     %}{% else %}form-check{% if inline_class %} form-checkinline{%
     endif %}{% endif %}">
        <input type="checkbox" class="{%if use_custom_control%}
         custom-control-input{% else %}form-check-input
         {% endif %}{% if field.errors %} is-invalid{% endif %}"
         {% if choice_value in field.value or choice_
         value|stringformat:"s" in field.value or
         choice_value|stringformat:"s" == field.value
         |default_if_none:""|stringformat:"s" %} checked=
         "checked"{% endif %} name="{{ field.html_name }}"
          id="id_{{ field.html_name }}_{{ forloop.counter }}"
          value="{{ choice_value|unlocalize }}" {{ field.field
           .widget.attrs|flatatt }}>
        <label class="{%if use_custom_control%}custom-controllabel{%
```

```
        else %}form-check-label{% endif %} level-{{
        choice_instance.level }}" for="id_{{ field.html_name
        }}_{{ forloop.counter }}">
            {{ choice_instance|unlocalize }}
        </label>
        {% if field.errors and forloop.last and not inline_class %}
            {% include 'bootstrap4/layout/field_errors_block.html' %}
        {% endif %}
    </div>
    {% endfor %}
    {% if field.errors and inline_class %}
    <div class="w-100 {%if use_custom_control%}custom-control
    custom-checkbox{% if inline_class %} custom-control-inline
    {% endif %}{% else %}form-check{% if inline_class %} form-
    check-inline{% endif %}{% endif %}">
        <input type="checkbox" class="custom-control-input {% if
        field.errors %}is-invalid{%endif%}">
        {% include 'bootstrap4/layout/field_errors_block.html' %}
    </div>
    {% endif %}

    {% include 'bootstrap4/layout/help_text.html' %}
</div>
```

（4）利用刚刚生成的表单创建一个新的视图用于添加一个 idea。

```
# myproject/apps/ideas/views.py
from django.contrib.auth.decorators import login_required
from django.shortcuts import render, redirect, get_object_or_404

from .forms import IdeaForm
from .models import Idea

@login_required
def add_or_change_idea(request, pk=None):
    idea = None
    if pk:
        idea = get_object_or_404(Idea, pk=pk)
    if request.method == "POST":
        form = IdeaForm(request, data=request.POST,
         files=request.FILES, instance=idea)
        if form.is_valid():
            idea = form.save()
            return redirect("ideas:idea_detail", pk=idea.pk)
    else:
```

```
        form = IdeaForm(request, instance=idea)

    context = {"idea": idea, "form": form}
    return render(request, "ideas/idea_form.html", context)
```

（5）添加关联的模板并利用{% crispy %}模板标签显示表单（关于该应用的更多内容，可参考第 3 章）。

```
{# ideas/idea_form.html #}
{% extends "base.html" %}
{% load i18n crispy_forms_tags static %}

{% block content %}
    <a href="{% url "ideas:idea_list" %}">{% trans "List of ideas" %}</a>
    <h1>
        {% if idea %}
            {% blocktrans trimmed with title=idea.translated_title %}
                Change Idea "{{ title }}"
            {% endblocktrans %}
        {% else %}
            {% trans "Add Idea" %}
        {% endif %}
    </h1>
  {% crispy form %}
{% endblock %}
```

（6）此外还需要一个指向新视图的 URL 规则，如下所示。

```
# myproject/apps/ideas/urls.py
from django.urls import path

from .views import add_or_change_idea

urlpatterns =
    # …
    path("add/", add_or_change_idea, name="add_idea"),
    path("<uuid:pk>/change/", add_or_change_idea, name="change_idea"),
]
```

（7）通过设置 margin-left 参数，向 CSS 文件添加规则，使用复选框树字段模板中生成的类来缩进标签，如.level-0、.level-1 和.level-2。对于上下文中所期望树的最大深度，确保持有合理数量的 CSS 类，如下所示。

```
/* myproject/site_static/site/css/style.css */
```

```
.level-0 {margin-left: 0;}
.level-1 {margin-left: 20px;}
.level-2 {margin-left: 40px;}
```

8.7.3　工作方式

最终，我们得到如图 8.7 所示的表单。

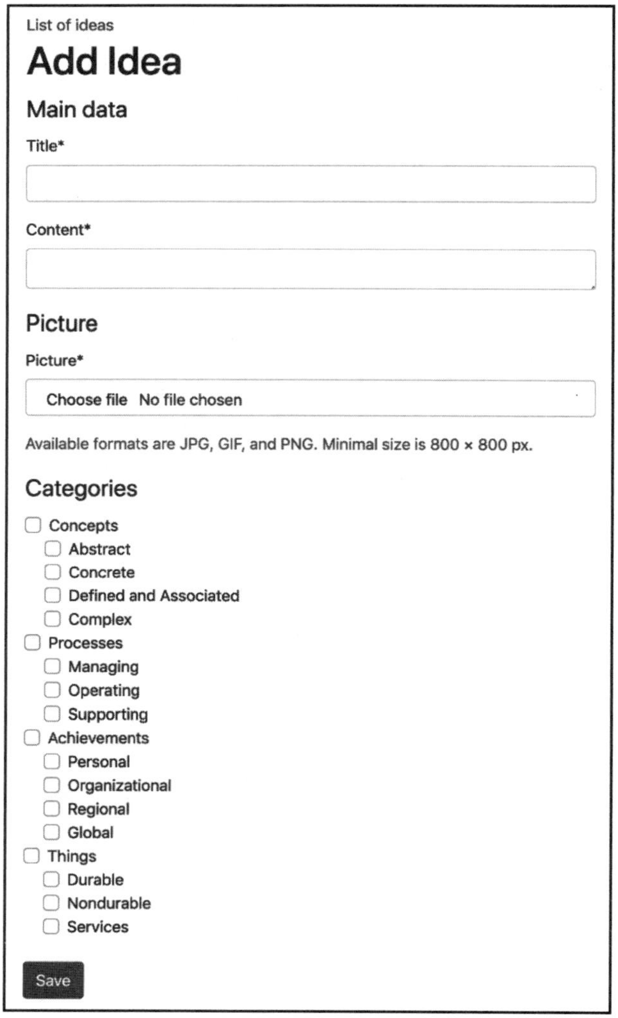

图 8.7

与 Django 在 Python 代码中硬编码字段生成的默认行为相反，django-crispy-forms 应用程序使用模板渲染字段。我们可以在 crispy_forms/templates/bootstrap4 下浏览它们，并将其中一些复制至项目模板目录中的类似的路径中，以便在必要时覆写它们。

在 idea 的创建和表单编辑过程中，我们传递了 categories 字段的自定义模板，该模板将.level-* CSS 类添加到<label>标签，并封装了复选框。普通 CheckboxSelectMultiple 微件的一个问题在于当渲染时该微件仅使用选择值和选择文本；而我们还需要分类的其他属性，如深度层次。为了解决这一问题，我们还创建了一个自定义 MultipleChoiceTreeField 表单字段，该表单字段扩展了 ModelMultipleChoiceField，并覆写了 label_from_instance()方法以返回分类实例自身，而非其 Unicode 表示。该字段的模板看起来较为复杂，然而，它主要是一个重构的多复选框字段模板（crispy_forms/templates/bootstrap4/layout/checkboxselectmultiple.html），并带有所有必要的 Bootstrap 标记。我们仅仅做了少量的调整并添加了.level-* CSS 类。

8.7.4 延伸阅读

- 第 3 章中的“利用 django-crispy-forms 创建一个表单布局”示例。
- “利用 django-mptt 在模板中渲染分类”示例。
- “利用 django-mptt 和单选字段在表单中选择分类”示例。

8.8 利用 django-treebeard 创建层次分类

树形结构包含多种算法且每种算法都有各自的优点。django-treebeard 应用程序（Django CMS 所采用的 django-mptt 的替代方案）提供了 3 种树形表单。

（1）邻接列表（Adjacency List）树是一种较为简单的结构，其中，每个节点包含一个父属性。虽然邻接列表树的读取操作很快，但这是以慢速写入为代价的。

（2）嵌套集（Nested Set）树和 MPTT 树相同，并将节点构建为嵌套在父节点下的集合。这种结构提供了较快的读取访问速度，但写入和删除代价高昂，特别是写入操作需要某些特定的顺序时。

（3）物化路径（Materialized Path）树基于树中的每个节点加以构造，这些节点包含关联的路径属性。该属性是一个字符串，表示从根到节点的完整路径——这很像在网站上查找特定页面的 URL 路径。

在当前示例中，我们将使用 django-treebeard 及与其一致的 API，并扩展第 3 章中的 categories 应用程序。其间，我们还通过树算法增强 Category 模型，以使该模型支持层次结构。

8.8.1 准备工作

准备工作如下列步骤所示。

（1）使用下列命令在虚拟环境中安装 django-treebeard。

```
(env)$ pip install django-treebeard==4.3
```

（2）创建 categories 和 ideas 应用程序。向设置项中的 INSTALLED_APPS 添加 categories 应用程序和 treebeard，如下所示。

```
# myproject/settings/_base.py
INSTALLED_APPS = [
    # …
    "treebeard",
    # …
    "myproject.apps.categories",
    "myproject.apps.ideas",
]
```

8.8.2 实现方式

我们将利用物化路径算法增强 Category 模型，如下所示。

（1）打开 models.py 文件，更新 Category 模型并扩展 treebeard.mp_tree.MP_Node，而非标准的 Django 模型。此外，该模型还继承自第 2 章定义的 CreationModificationDateMixin。除了来自混入的字段，Category 模型还需要一个 title 字段。

```
# myproject/apps/categories/models.py
from django.db import models
from django.utils.translation import ugettext_lazy as _
from treebeard.mp_tree import MP_Node

from myproject.apps.core.models import CreationModificationDateBase

class Category(MP_Node, CreationModificationDateBase):
    title = models.CharField(_("Title"), max_length=200)

    class Meta:
        verbose_name = _("Category")
        verbose_name_plural = _("Categories")
```

```
    def __str__(self):
        return self.title
```

（2）更新数据库。接下来迁移 categories 应用程序。

```
(env)$ python manage.py makemigrations
(env)$ python manage.py migrate
```

（3）当采用抽象模型继承时，treebeard 树节点可通过标准关系与其他模型关联。因此，Idea 模型可与 Category 模型包含一个简单的 ManyToManyField 关系。

```
# myproject/apps/ideas/models.py
from django.db import models
from django.utils.translation import gettext_lazy as _

from myproject.apps.core.models import
CreationModificationDateBase, UrlBase

class Idea(CreationModificationDateBase, UrlBase):
    # …
    categories = models.ManyToManyField(
        "categories.Category",
        verbose_name=_("Categories"),
        related_name="category_ideas",
    )
```

8.8.3　工作方式

MP_Node 抽象模型向 Category 模型提供了 path、depth 和 numchild 字段，以及 steplen、alphabet 和 node_order_by 属性，用以构建树形结构。

- depth 和 numchild 字段提供了与节点位置和后代相关的元数据。
- path 字段被索引化，并通过 LIKE 启用了快速的数据库查询。
- path 字段由固定长度的编码段构成，每段的大小由 steplen 属性值（默认值为 4）确定。编码使用了 alphabet 属性值中的字符（默认为拉丁字符）。

path、depth 和 numchild 字段应被视为具有只读属性。另外，在将第 1 个对象保存至树中后，steplen、alphabet 和 node_order_by 值不可更改，否则数据将被损坏。

除了新的字段和属性，MP_Node 抽象类通过树形结构添加了导航方法。这些方法中较为重要的示例如下所示。

- 如果需要获得某个分类的祖先，那么最终结果将作为祖先的 QuerySet 返回（从根至当前节点的父节点）。

```
ancestor_categories = category.get_ancestors()
```

- 若仅获取 root 分类，该结果可通过深度为 1 加以识别，如下所示。

```
root = category.get_root()
```

- 获取分类的直接 children，如下所示。

```
children = category.get_children()
```

- 获取某个分类的全部 descendants，最终结果将以所有子节点（及其子节点）的 QuerySet 返回，但不包括当前节点自身，如下所示。

```
descendants = category.get_descendants()
```

- 获取 descendant 计数，如下所示。

```
descendants_count = category.get_descendant_count()
```

- 获取所有的 siblings（包括引用节点），如下所示。

```
siblings = category.get_siblings()
```

注意：

Root 分类可视为其他根分类的兄弟节点。

- 仅获取上一个或下一个 siblings，如下所示（对于最左侧的兄弟节点，get_prev_sibling()返回 None；对于最右侧的兄弟节点，get_next_sibling()返回 None）。

```
previous_sibling = category.get_prev_sibling()
next_sibling = category.get_next_sibling()
```

- 此外，还存在一些方法可检查分类是否为 root、leaf 或与另一个节点关联的其他节点。

```
category.is_root()
category.is_leaf()
category.is_child_of(another_category)
category.is_descendant_of(another_category)
category.is_sibling_of(another_category)
```

8.8.4 更多内容

当前示例仅描述了 django-treebeard 的基本内容及其物化路径。此外还存在许多其他方法可用于导航和树形结构的构建。除此之外，Materialized Path 树形结构的 API 很大程度上等同于 Nested Sets 树形结构和 Adjacency List 树形结构的 API，后者分别通过 NS_Node 或 AL_Node 抽象类实现了模型，而非使用 MP_Node。

注意：

针对每种树形结构的实现，读者可阅读 django-treebeard API 文档以查看完整的属性和方法列表，对应网址为 https://django-treebeard.readthedocs.io/en/latest/api.html。

8.8.5　延伸阅读

- 第 3 章。
- “利用 django-mptt 创建层次分类”示例。
- “利用 django-treebeard 创建分类管理界面”示例。

8.9　利用 django-treebeard 创建分类管理界面

django-treebeard 应用程序提供了自己的 TreeAdmin，并扩展自标准的 ModelAdmin，进而可在管理界面中以层次结构方式查看树节点，且界面特性依赖于所采用的树算法。当前示例将对此加以探讨。

8.9.1　准备工作

首先设置前述 categories 应用程序和 django-treebeard。此外还应确保项目的 django-treebeard 静态文件有效。

```
(env)$ python manage.py collectstatic
```

8.9.2　实现方式

从 categories 应用程序中为 Category 模型创建一个管理界面，它扩展了 treebeard.adminTree.Admin 而不是 admin.ModelAdmin，并使用一个自定义的表单工厂，如下所示。

```
# myproject/apps/categories/admin.py
from django.contrib import admin
from treebeard.admin import TreeAdmin
from treebeard.forms import movenodeform_factory

from .models import Category

@admin.register(Category)
class CategoryAdmin(TreeAdmin):
```

```
    form = movenodeform_factory(Category)
    list_display = ["title", "created", "modified"]
    list_filter = ["created"]
```

8.9.3 工作方式

根据所用的树形结构实现，分类的管理界面将包含两个模式之一。对于 Materialized Path 和 Nested Sets 树形结构，将显示如图 8.8 所示的高级 UI。

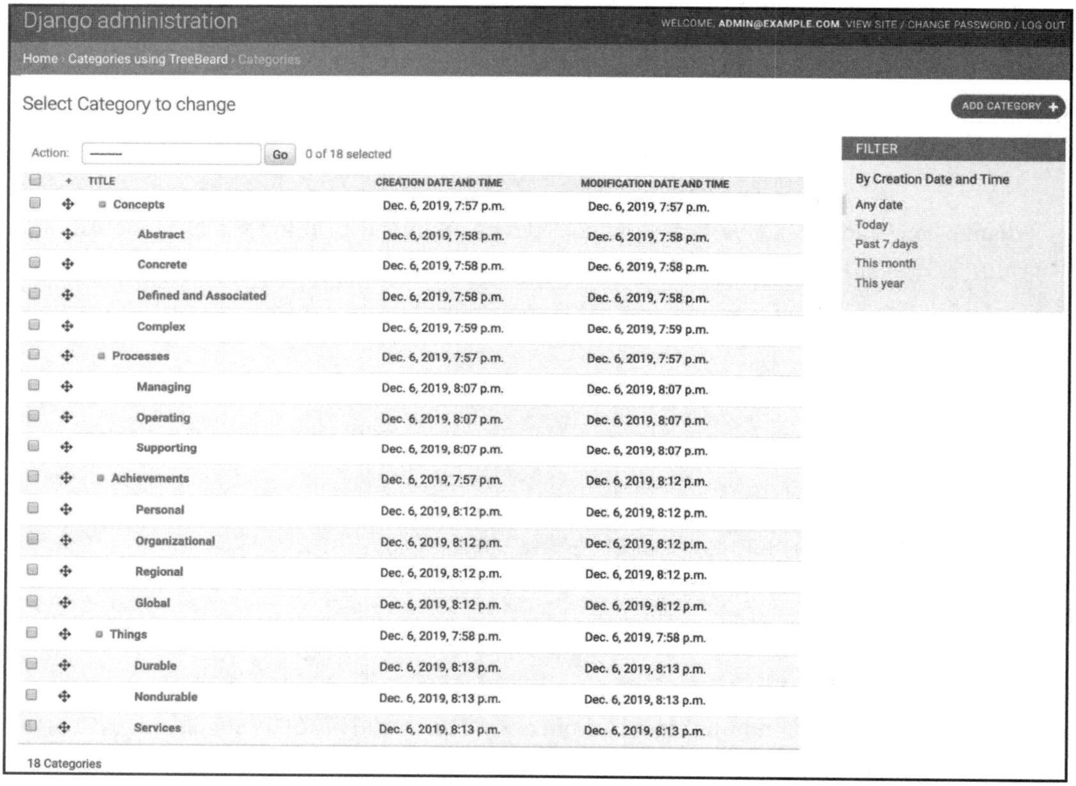

图 8.8

高级视图可扩展或折叠分类，以获得较好的概览效果。当对其重新排序或修改依赖项时，拖曳标题即可。在重新排序过程中，用户界面如图 8.9 所示。

如果按照特定字段应用分类的过滤机制或排序，高级功能将被禁用，但会保持高级界面较好的观感。这里，我们可以查看中间视图，其中仅显示 Past 7 days 中创建的分类，如图 8.10 所示。

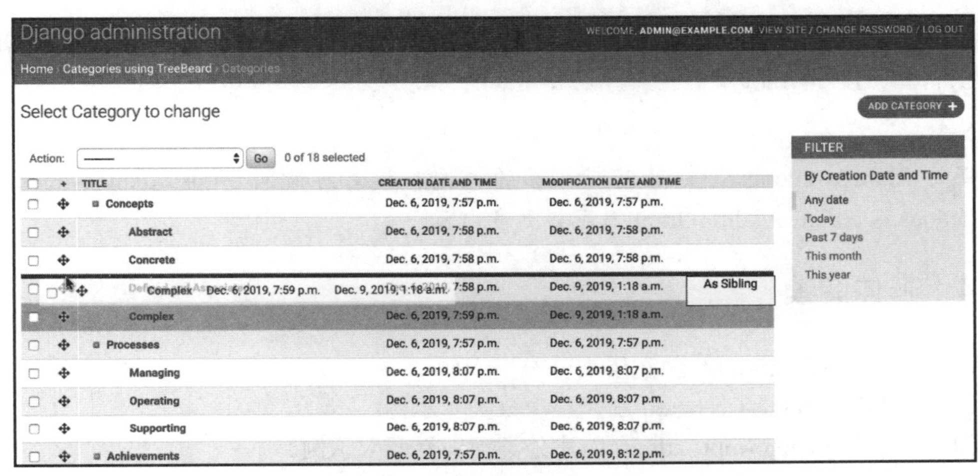
Django administration
WELCOME, ADMIN@EXAMPLE.COM VIEW SITE / CHANGE PASSWORD / LOG OUT
Home › Categories using TreeBeard › Categories
Select Category to change
ADD CATEGORY +
Action:
Go
0 of 18 selected
FILTER
By Creation Date and Time
Any date
Today
Past 7 days
This month
This year
TITLE
CREATION DATE AND TIME
MODIFICATION DATE AND TIME
Concepts
Abstract
Concrete
Complex
Processes
Managing
Operating
Supporting
Achievements
As Sibling

图 8.9

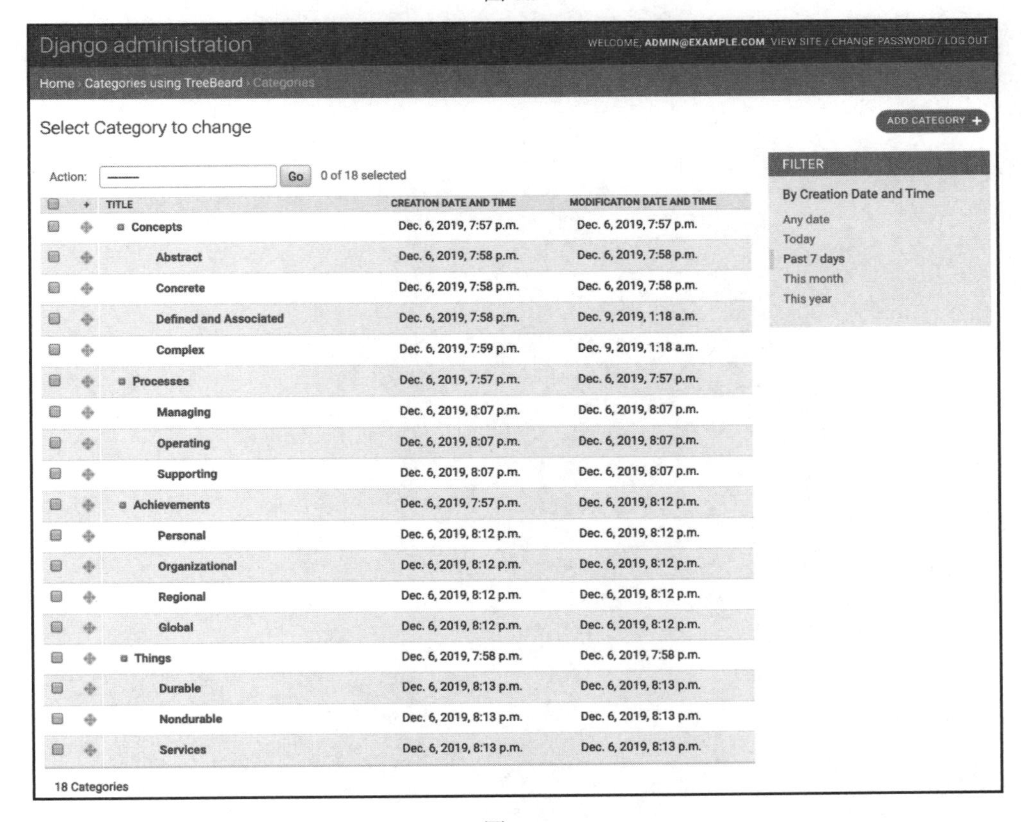
Django administration
WELCOME, ADMIN@EXAMPLE.COM VIEW SITE / CHANGE PASSWORD / LOG OUT
Home › Categories using TreeBeard › Categories
Select Category to change
ADD CATEGORY +
Action:
Go
0 of 18 selected
FILTER
By Creation Date and Time
Any date
Today
Past 7 days
This month
This year
TITLE
CREATION DATE AND TIME
MODIFICATION DATE AND TIME
Concepts
Abstract
Concrete
Defined and Associated
Complex
Processes
Managing
Operating
Supporting
Achievements
Personal
Organizational
Regional
Global
Things
Durable
Nondurable
Services
18 Categories

图 8.10

然而，若树采用 Adjacency List 算法则仅会提供较为基础的 UI，且不包含高级 UI 中的切换和重新排序功能。

注意：

关于 django-treebeard 以及基础界面的截图，读者可访问 https://django-treebeard.readthedocs.io/en/latest/admin.html 以了解更多内容。

8.9.4 延伸阅读

- “利用 django-mptt 创建层次分类”示例。
- “利用 django-treebeard 创建层次分类”示例。
- “利用 django-mptt-admin 创建分类管理界面”示例。

第 9 章　导入和导出数据

本章主要涉及下列内容。

- 从本地 CSV 文件中导入数据。
- 从本地 Excel 文件中导入数据。
- 从外部 JSON 文件中导入数据。
- 从外部 XML 文件中导入数据。
- 准备搜索引擎的分页站点图。
- 创建可过滤的 RSS 订阅。
- 使用 Django REST 框架创建 API。

9.1　简　　介

某些时候，数据需要以本地格式传输至数据库、从外部资源导入或者提供至第三方。在本章中，我们将探讨如何编写管理命令和 API 实现此类任务。

9.2　技 术 需 求

当与本章代码协同工作时，需要使用最新稳定版本的 Python、MySQL 或 PostgreSQL 数据库，以及基于虚拟环境的 Django 项目。另外，还应确保已将 Django、Pillow 和数据库绑定安装至虚拟环境中。

读者可访问 GitHub 储存库的 ch09 目录查看本章代码，对应网址为 https://github.com/PacktPublishing/Django-3-Web-Development-Cookbook-Fourth-Edition。

9.3　从本地 CSV 文件中导入数据

逗号分隔值（Comma-Separated Values，CSV）格式可能是将列表数据存储至文本文件中的最简单的方法。在当前示例中，我们将创建一个管理命令，并将数据从 CSV 文件导入至 Django 数据库中。其间，我们将使用一个 CSV 歌曲列表。我们可通过 Excel、Calc 或其他电子表格应用程序创建此类文件。

9.3.1　准备工作

下面将创建本章使用的 music 应用程序。

（1）创建 music 应用程序并将其置于设置项的 INSTALLED_APPS 下。

```
# myproject/settings/_base.py
INSTALLED_APPS = [
    # …
    "myproject.apps.core",
    "myproject.apps.music",
]
```

（2）Song 模型应包含 uuid、artist、title、url 和 image 字段。另外，我们将扩展 CreationModificationDateBase 并加入时间戳的创建和修改功能，以及 UrlBase 添加方法进而处理模型的详细 URL。

```
# myproject/apps/music/models.py
import os
import uuid
from django.urls import reverse
from django.utils.translation import ugettext_lazy as _
from django.db import models
from django.utils.text import slugify
from myproject.apps.core.models import
CreationModificationDateBase, UrlBase

def upload_to(instance, filename):
    filename_base, filename_ext = os.path.splitext(filename)
    artist = slugify(instance.artist)
    title = slugify(instance.title)
    return f"music/{artist}--{title}{filename_ext.lower()}"

class Song(CreationModificationDateBase, UrlBase):
    uuid = models.UUIDField(primary_key=True, default=None,
     editable=False)
    artist = models.CharField(_("Artist"), max_length=250)
    title = models.CharField(_("Title"), max_length=250)
    url = models.URLField(_("URL"), blank=True)
```

```
    image = models.ImageField(_("Image"), upload_to=upload_to,
    blank=True, null=True)

    class Meta:
        verbose_name = _("Song")
        verbose_name_plural = _("Songs")
        unique_together = ["artist", "title"]

    def __str__(self):
        return f"{self.artist} - {self.title}"

    def get_url_path(self):
        return reverse("music:song_detail", kwargs={"pk": self.pk})

    def save(self, *args, **kwargs):
        if self.pk is None:
            self.pk = uuid.uuid4()
        super().save(*args, **kwargs)
```

（3）创建并运行迁移，如下所示。

```
(env)$ python manage.py makemigrations
(env)$ python manage.py migrate
```

（4）针对 Song 模型添加简单的管理。

```
# myproject/apps/music/admin.py
from django.contrib import admin
from .models import Song

@admin.register(Song)
class SongAdmin(admin.ModelAdmin):
    list_display = ["title", "artist", "url"]
    list_filter = ["artist"]
    search_fields = ["title", "artist"]
```

（5）此外还需要一个表单，并在导入脚本中验证和创建 Song 模型。这也是最直观的模型表单，如下所示。

```
# myproject/apps/music/forms.py
from django import forms
from django.utils.translation import ugettext_lazy as _
from .models import Song
```

```
class SongForm(forms.ModelForm):
    class Meta:
        model = Song
        fields = "__all__"
```

9.3.2 实现方式

下列步骤将创建和使用管理命令，进而从本地 CSV 文件中导入歌曲。

（1）创建一个 CSV 文件，将列名、artist、title 和 url 放在第一行。在下一个匹配列的行中添加一些歌曲数据。如这可能是一个 data/music.csv 文件，其内容如下所示。

```
artist,title,url
Capital Cities,Safe And
Sound,https://open.spotify.com/track/40Fs0YrUGuwLNQSaHGVfqT?si=2OUa
wusIT-evyZKonT5GgQ
Milky Chance,Stolen
Dance,https://open.spotify.com/track/3miMZ2IlJiaeSWo1DohXlN?si=g-xM
M4m9S_yScOm02C2MLQ
Lana Del Rey,Video Games -
Remastered,https://open.spotify.com/track/5UOo694cVvjcPFqLFiNWGU?si=
maZ7JCJ7Rb6WzESLXg1Gdw
Men I
Trust,Tailwhip,https://open.spotify.com/track/2DoO0sn4SbUrz7Uay9ACTM?
si=SC_MixNKSnuxNvQMf3yBBg
```

（2）在 music 应用程序中，创建 management 目录，随后在新的 management 目录中创建一个 commands 目录。将__init__.py 空文件置于这两个新目录中，使其成为 Python 包。

（3）添加 import_music_from_csv.py 文件，如下所示。

```
# yproject/apps/music/management/commands/import_music_from_csv.py
from django.core.management.base import BaseCommand

class Command(BaseCommand):
    help = (
        "Imports music from a local CSV file. "
        "Expects columns: artist, title, url"
    )
    SILENT, NORMAL, VERBOSE, VERY_VERBOSE = 0, 1, 2, 3

    def add_arguments(self, parser):
        # Positional arguments
        parser.add_argument("file_path", nargs=1, type=str)
```

```
    def handle(self, *args, **options):
        self.verbosity = options.get("verbosity", self.NORMAL)
        self.file_path = options["file_path"][0]
        self.prepare()
        self.main()
        self.finalize()
```

（4）在同一文件中，针对 Command 类创建 prepare()方法。

```
def prepare(self):
    self.imported_counter = 0
    self.skipped_counter = 0
```

（5）创建 main()方法。

```
def main(self):
    import csv
    from ...forms import SongForm

    if self.verbosity >= self.NORMAL:
        self.stdout.write("=== Importing music ===")

    with open(self.file_path, mode="r") as f:
        reader = csv.DictReader(f)
        for index, row_dict in enumerate(reader):
            form = SongForm(data=row_dict)
            if form.is_valid():
                song = form.save()
                if self.verbosity >= self.NORMAL:
                    self.stdout.write(f" - {song}\n")
                self.imported_counter += 1
            else:
                if self.verbosity >= self.NORMAL:
                    self.stderr.write(
                        f"Errors importing song "
                        f"{row_dict['artist']} -
                        {row_dict['title']}:\n"
                    )
                    self.stderr.write(f"{form.errors.as_json()}\n")
                self.skipped_counter += 1
```

（6）最后是 finalize()方法。

```
def finalize(self)
```

```
    if self.verbosity >= self.NORMAL:
        self.stdout.write(f"-------------------------\n")
        self.stdout.write(f"Songs imported:
         {self.imported_counter}\n")
        self.stdout.write(f"Songs skipped:
         {self.skipped_counter}\n\n")
```

（7）在命令行中调用下列内容运行导入操作。

```
(env)$ python manage.py import_music_from_csv data/music.csv
```

9.3.3　工作方式

Django 管理命令可视为脚本，其中 Command 类继承自 BaseCommand，并覆写了 add_arguments()和 handle()方法。help 属性定义了管理命令的帮助文本，并在命令行中输入下列内容时可见。

```
(env)$ python manage.py help import_music_from_csv
```

Django 管理命令使用了内建的 argparse 模块解析所传递的参数。add_arguments()方法定义了应传递至管理命令的位置或命名参数。在当前示例中，我们添加了一个 Unicode 类型的 file_path 位置参数。通过将 nargs 变量设置为 1，因而仅支持一个值。

注意：

关于其他参数的定义和操作方式，读者可参考 argparse 官方文档，对应网址为 https://docs.python.org/3/library/argparse.html#adding-arguments。

handle()方法于开始处检查 verbosity 参数，verbosity 参数定义了命令应该提供多少 Terminal 输出，从 0（不提供任何日志记录）到 3（提供大量日志记录）。我们按照下列方式将这一命名参数传递至命令中。

```
(env)$ python manage.py import_music_from_csv data/music.csv --verbosity=0
```

此外，我们期望文件名为第一个位置参数。options["file_path"]返回一个长度在 nargs 中定义的值列表。在当前示例中，nargs 等于 1，因此，options["file_path"]等于一个元素的列表。

较好的做法是将管理命令的逻辑划分至多个较小的方法中。例如，在当前示例中，我们在脚本中使用了 prepare()、main()和 finalize()。

- prepare()方法将导入计数器设置为 0，此外还用于脚本所需的其他设置。
- main()方法执行管理命令的主要逻辑。首先打开给定的文件进行读取，并将其指针传递至 csv.DictReader 中。假定文件中的第一行包含每个列的标题，DictReader 将此用作每行中字典的键。当遍历行时，我们将字典传递至模型表单中并尝试对其进行验证。如果验证成功，歌曲将被保存且 imported_counter 递增。如果验证失败，如过长的值、缺失所需值、错误类型或其他验证错误，skipped_counter 将递增。如果 verbosity 等于或大于 MORMAL（值 1），每首导入或被略过的歌曲将连同可能的验证错误一同被输出。

- ❑ finalize()方法输出导入歌曲的数量，以及由于验证错误被略过的歌曲数量。

如果调试管理命令错误时，可传递--traceback 参数至管理命令。当出现错误时，我们将会看到该问题的全栈跟踪。

假设我们利用--verbosity=1 或更大值调用命令两次，对应的输出结果如图 9.1 所示。

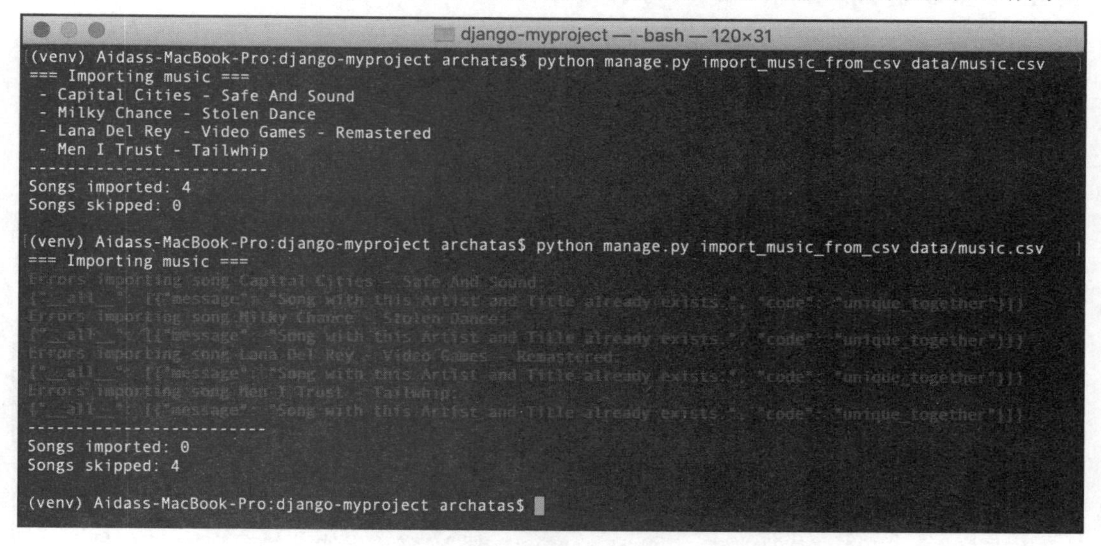

图 9.1

可以看到，当一首歌曲第二次被导入时，并未传递 unique_together 约束，因而将被略过。

9.3.4 延伸阅读

- ❑ “从本地 Excel 文件中导入数据”示例。
- ❑ “从外部 JSON 文件中导入数据”示例。
- ❑ “从外部 XML 文件中导入数据”示例。

9.4　从本地 Excel 文件中导入数据

另一种较为流行的表格数据存储格式是 Excel 电子表格。在当前示例中，我们将从 Excel 格式的文件中导入歌曲。

9.4.1　准备工作

当前示例将在前述示例创建的 music 应用程序的基础上完成。当读取 Excel 文件时，我们需要安装 openpyxl 包，如下所示。

```
(env)$ pip install openpyxl==3.0.2
```

9.4.2　实现方式

下列步骤将创建和使用管理命令并从本地 XLSX 文件中导入歌曲。

（1）创建一个 XLSX 文件，其中，第一行包含列名 Artist、Title 和 URL。在与列匹配的下一行中，向文件中添加一些歌曲数据。我们可以在电子表格应用程序中完成该操作，并将之前的 CSV 文件保存为 XLSX 文件 data/music.xlsx，如图 9.2 所示。

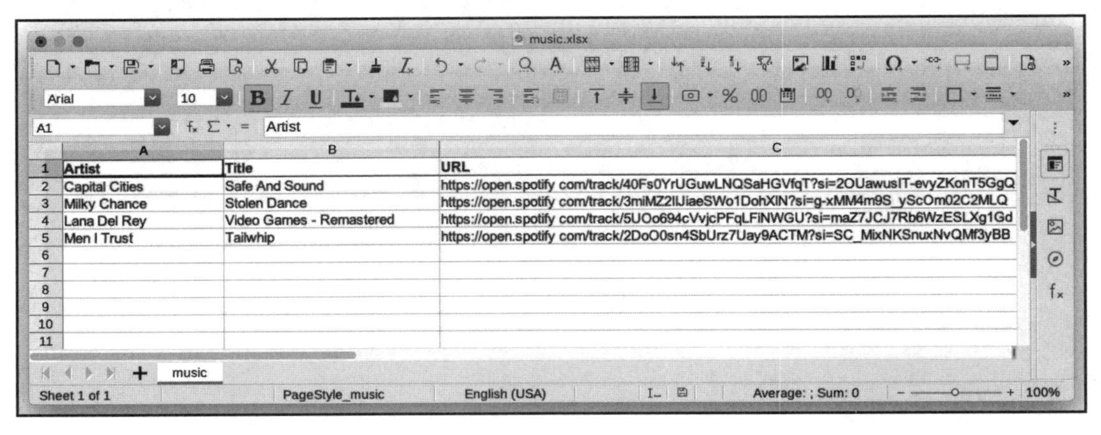

图 9.2

（2）在 music 应用程序中，创建一个 management 目录，随后在该目录中创建一个子目录 commands。将__init__.py 空文件添加至两个新目录中，以使其成为 Python 包。

（3）添加 import_music_from_xlsx.py 文件，如下所示。

```
# myproject/apps/music/management/commands
# /import_music_from_xlsx.py
from django.core.management.base import BaseCommand

class Command(BaseCommand):

    help = (
        "Imports music from a local XLSX file. "
        "Expects columns: Artist, Title, URL"
    )
    SILENT, NORMAL, VERBOSE, VERY_VERBOSE = 0, 1, 2, 3

    def add_arguments(self, parser):
        # Positional arguments
        parser.add_argument("file_path",
                            nargs=1,
                            type=str)

    def handle(self, *args, **options):
        self.verbosity = options.get("verbosity", self.NORMAL)
        self.file_path = options["file_path"][0]
        self.prepare()
        self.main()
        self.finalize()
```

（4）在同一个文件中，针对 Command 类，创建一个 prepare()方法。

```
def prepare(self):
    self.imported_counter = 0
    self.skipped_counter = 0
```

（5）创建 main()方法。

```
def main(self):
    from openpyxl import load_workbook
    from ...forms import SongForm

    wb = load_workbook(filename=self.file_path)
    ws = wb.worksheets[0]

    if self.verbosity >= self.NORMAL:
        self.stdout.write("=== Importing music ===")

    columns = ["artist", "title", "url"]
    rows = ws.iter_rows(min_row=2) # skip the column captions
```

```
    for index, row in enumerate(rows, start=1):
        row_values = [cell.value for cell in row]
        row_dict = dict(zip(columns, row_values))
        form = SongForm(data=row_dict)
        if form.is_valid():
            song = form.save()
            if self.verbosity >= self.NORMAL:
                self.stdout.write(f" - {song}\n")
            self.imported_counter += 1
        else:
            if self.verbosity >= self.NORMAL:
                self.stderr.write(
                    f"Errors importing song "
                    f"{row_dict['artist']} -
                     {row_dict['title']}:\n"
                )
                self.stderr.write(f"{form.errors.as_json()}\n")
            self.skipped_counter += 1
```

（6）添加 finalize()方法。

```
def finalize(self):
    if self.verbosity >= self.NORMAL:
        self.stdout.write(f"-------------------------\n")
        self.stdout.write(f"Songs imported:
         {self.imported_counter}\n")
        self.stdout.write(f"Songs skipped:
         {self.skipped_counter}\n\n")
```

（7）当运行导入操作时，可在命令行中调用下列内容。

```
(env)$ python manage.py import_music_from_xlsx data/music.xlsx
```

9.4.3　工作方式

XLSX 文件的导入原则等同于 CSV 文件。我们打开相关文件、逐行读取、表单数据字典、通过模型表单进行验证，并从提供的数据中创建 Song 对象。

再次说明，我们采用 prerare()、main()和 finalize()方法将逻辑划分为原子部分。

考虑到 main()方法是管理命令中唯一的不同部分，其解释如下所示。

- Excel 文件是包含电子表格作为不同标签页的工作簿。
- 我们采用 openyxl 库打开一个文件，该文件作为位置参数被传递至命令中。随后从工作簿中读取第一个电子表格。

- ❑ 第一行包含列名，并略过该行内容。
- ❑ 随后作为值列表逐行读取、使用 zip()函数创建字典、将字典传递至模型表单、验证并创建 Song 对象。
- ❑ 如果存在验证错误且 verbosity 大于或等于 NORMAL，则输出错误消息。
- ❑ 再次强调，管理命令将导入的歌曲输出至控制台，除非--verbosity=0。

如果--verbosity=1 或更大值，运行命令两次后输出结果如图 9.3 所示。

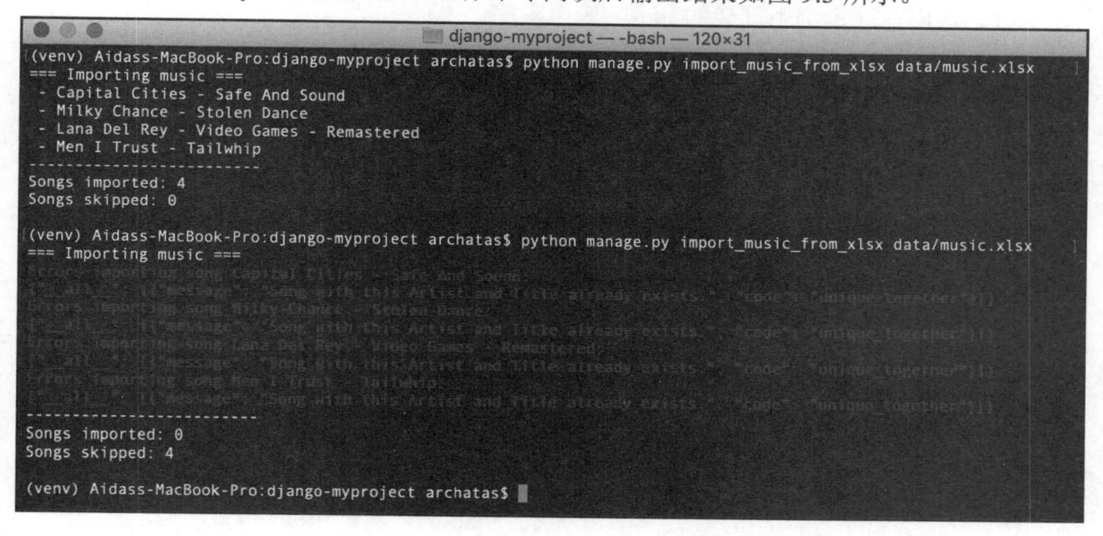

图 9.3

注意：

关于 Excel 文件的处理方式，读者可访问 http://www.python-excel.org/。

9.4.4　延伸阅读

- ❑ “从本地 CSV 文件中导入数据”示例。
- ❑ “从外部 JSON 文件中导入数据”示例。
- ❑ “从外部 XML文件中导入数据”示例。

9.5　从外部 JSON 文件中导入数据

Last.fm 音乐网站在 https://ws.audioscrobbler.com/域下包含一个 API，可用于读取专

辑、艺术家、曲目、事件等。该 API 可采用 JSON 或 XML 格式。在当前示例中，我们将利用 JSON 格式导入标记为 indie 的曲目。

9.5.1　准备工作

下列步骤将从 Last.fm 中导入 JSON 格式的数据。

（1）启用前一示例中创建的 music 应用程序。

（2）当使用 Last.fm 时，需要注册并获取一个 API 密钥。读者可访问 https://www.ast.fm/api/account/create 创建该 API 密钥。

（3）在设置项中，该 API 密钥需要设置为 LAST_FM_API_KEY。我们建议从密码文件或环境变量中提供该 API 密钥，并将其置入设置项中，如下所示。

```
# myproject/settings/_base.py
LAST_FM_API_KEY = get_secret("LAST_FM_API_KEY")
```

（4）利用下列命令在虚拟环境中安装 requests 库。

```
(env)$ pip install requests==2.22.0
```

（5）针对 indie 曲目，检查 JSON 端点结构（https://ws.audioscrobbler.com/2.0/? method=tag.gettoptracks&tag=indie&api_key=YOUR_API_KEY&format=json），如下所示。

```
{
   "tracks": {
      "track": [
         {
           "name": "Mr. Brightside",
           "duration": "224",
           "mbid": "37d516ab-d61f-4bcb-9316-7a0b3eb845a8",
           "url": "https://www.last.fm/music
            /The+Killers/_/Mr.+Brightside",
           "streamable": {
              "#text": "0",
              "fulltrack": "0"
           },
           "artist": {
              "name": "The Killers",
              "mbid": "95e1ead9-4d31-4808-a7ac-32c3614c116b",
              "url": "https://www.last.fm/music/The+Killers"
           },
           "image": [
```

```
              {
                  "#text":
                  "https://lastfm.freetls.fastly.net/i/u/34s
                   /2a96cbd8b46e442fc41c2b86b821562f.png",
                  "size": "small"
              },
              {
                  "#text":
                  "https://lastfm.freetls.fastly.net/i/u/64s
                   /2a96cbd8b46e442fc41c2b86b821562f.png",
                  "size": "medium"
              },
              {
                  "#text":
                  "https://lastfm.freetls.fastly.net/i/u/174s
                   /2a96cbd8b46e442fc41c2b86b821562f.png",
                  "size": "large"
              },
              {
                  "#text":
                  "https://lastfm.freetls.fastly.net/i/u/300x300
                   /2a96cbd8b46e442fc41c2b86b821562f.png",
                  "size": "extralarge"
              }
          ],
          "@attr": {
              "rank": "1"
          }
        },
        ...
      ],
      "@attr": {
        "tag": "indie",
        "page": "1",
        "perPage": "50",
        "totalPages": "4475",
        "total": "223728"
      }
    }
}
```

我们打算读取曲目的 name、artist、URL 和中等尺寸的图像。另外，我们还将关注页面的总数量，这是作为元信息在 JSON 文件结尾提供的。

9.5.2　实现方式

下列步骤创建 Song 模型和一个管理命令，并以 JSON 格式将 Last.fm 中的曲目导入至数据库中。

（1）在 music 应用程序中，创建 management 目录，并于其中创建一个子目录 commands。在两个新创建的目录中添加__init__.py 空文件，并使其成为 Python 包。

（2）添加 import_music_from_lastfm_json.py 文件，如下所示。

```
# myproject/apps/music/management/commands
# /import_music_from_lastfm_json.py
from django.core.management.base import BaseCommand

class Command(BaseCommand):
    help = "Imports top songs from last.fm as JSON."
    SILENT, NORMAL, VERBOSE, VERY_VERBOSE = 0, 1, 2, 3
    API_URL = "https://ws.audioscrobbler.com/2.0/"

    def add_arguments(self, parser):
        # Named (optional) arguments
        parser.add_argument("--max_pages", type=int, default=0)

    def handle(self, *args, **options):
        self.verbosity = options.get("verbosity", self.NORMAL)
        self.max_pages = options["max_pages"]
        self.prepare()
        self.main()
        self.finalize()
```

（3）在同一文件中，针对 Command 类，创建 prepare()方法。

```
def prepare(self):
    from django.conf import settings

    self.imported_counter = 0
    self.skipped_counter = 0
    self.params = {
        "method": "tag.gettoptracks",
        "tag": "indie",
        "api_key": settings.LAST_FM_API_KEY,
```

```
        "format": "json",
        "page": 1,
    }
```

（4）创建 main()方法。

```
def main(self):
    import requests

    response = requests.get(self.API_URL, params=self.params)
    if response.status_code != requests.codes.ok:
        self.stderr.write(f"Error connecting to
         {response.url}")
        return
    response_dict = response.json()
    pages = int(
        response_dict.get("tracks", {})
        .get("@attr", {}).get("totalPages", 1)
    )

    if self.max_pages > 0:
        pages = min(pages, self.max_pages)

    if self.verbosity >= self.NORMAL:
        self.stdout.write(f"=== Importing {pages} page(s) of tracks ===")

    self.save_page(response_dict)

    for page_number in range(2, pages + 1):
        self.params["page"] = page_number
        response = requests.get(self.API_URL, params=self.params)
        if response.status_code != requests.codes.ok:
            self.stderr.write(f"Error connecting to
             {response.url}")
            return
        response_dict = response.json()
        self.save_page(response_dict)
```

（5）分页订阅中的每个页面将通过 save_page()方法保存，如下所示。

```
def save_page(self, data):
    import os
    import requests
    from io import BytesIO
```

```
from django.core.files import File

from ...forms import SongForm

for track_dict in data.get("tracks", {}).get("track"):
    if not track_dict:
        continue

    song_dict = {
        "artist": track_dict.get("artist", {}).get("name", ""),
        "title": track_dict.get("name", ""),
        "url": track_dict.get("url", ""),
    }
    form = SongForm(data=song_dict)
    if form.is_valid():
        song = form.save()

        image_dict = track_dict.get("image", None)
        if image_dict:
            image_url = image_dict[1]["#text"]
            image_response = requests.get(image_url)
            song.image.save(
                os.path.basename(image_url),
                File(BytesIO(image_response.content)),
            )

        if self.verbosity >= self.NORMAL:
            self.stdout.write(f" - {song}\n")
        self.imported_counter += 1
    else:
        if self.verbosity >= self.NORMAL:
            self.stderr.write(
                f"Errors importing song "
                f"{song_dict['artist']} -
                 {song_dict['title']}:\n"
            )
            self.stderr.write(f"{form.errors.as_json()}\n")
        self.skipped_counter += 1
```

（6）创建 finalize()方法。

```
def finalize(self):
    if self.verbosity >= self.NORMAL:
```

```
self.stdout.write(f"------------------------\n")
self.stdout.write(f"Songs imported:
 {self.imported_counter}\n")
self.stdout.write(f"Songs skipped:
 {self.skipped_counter}\n\n")
```

（7）在命令行中调用下列内容运行导入操作。

```
(env)$ python manage.py import_music_from_lastfm_json --max_pages=3
```

9.5.3 工作方式

如前所述，如果仅是一个字符串列表，脚本参数定义为位置参数；如果始于--或一个变量名，那么脚本参数表示为命名参数。其中，命名的--max_pages 参数将导入数据限制为 3 页。如果希望下载全部曲目，则可对此予以忽略或显式地传递 0 值。

我们可以看到，totalPages 值中包含了大约 4500 个页面，这将占用较长时间并包含大量的处理工作。

当前的脚本结果类似于之前的导入脚本。

- prepare()方法用于设置操作。
- mian()方法处理请求和响应。
- save_page()方法保存单一分页页面中的歌曲。
- finalize()输出导入统计信息。

在 main()方法中，我们使用了 requests.get()方法，通过传递 params 查询参数读取 Last.fm 中的数据。响应对象包含了一个名为 json()的内建方法，该方法将 JSON 字符串转换为解析后的字典对象。从第一次请求起，我们获得了全部页面数量，随后读取每个页面并调用 save_page()方法解析信息和保存歌曲。

在 save_page()方法中，我们读取歌曲中的值，并对模型表单构建一个字段，验证该表单。如果数据有效，则创建 Song 对象。

导入过程中一个值得关注的问题是下载和保存图像。这里，我们还使用了 requests.get()方法检索图像数据，并于随后通过 BytesIO 将其传递至 File 中。相应地，BytesIO 则用于 image.save()方法中。image.save()方法的第一个参数是一个文件名，它将被 upload_to 函数的值覆写，并且只对文件扩展名有必要。

如果命令通过--verbosity=1 或更大值被调用，我们将看到与导入相关的详细信息。

注意：

关于 Last.fm 处理方式的更多信息，读者可访问 https://www.last.fm/api/。

9.5.4　延伸阅读

- “从本地 CSV 文件中导入数据”示例。
- “从本地 Excel 文件中导入数据”示例。
- “从外部 XML 文件中导入数据”示例。

9.6　从外部 XML 文件中导入数据

类似于 JSON 文件，Last.fm 文件还可以使用其 XML 格式的服务数据。在当前示例中，我们将讨论如何实现这一任务。

9.6.1　准备工作

下列步骤将从 Last.fm 中导入 XML 格式的数据。

（1）启用之前创建的 music 应用程序。

（2）当使用 Last.fm 时，需要注册和获取 API 密钥。这里，API 密钥可在 https://www.last.fm/api/account/create 中生成。

（3）API 密钥需要在设置项中被设置为 LAST_FM_API_KEY。我们建议将其保存至密码或环境变量中，并将其拖到设置项中。

```
# myproject/settings/_base.py
LAST_FM_API_KEY = get_secret("LAST_FM_API_KEY")
```

（4）利用下列命令在虚拟环境中安装 requests 和 defusedxml 库。

```
(env)$ pip install requests==2.22.0
(env)$ pip install defusedxml==0.6.0
```

（5）检查 indie 曲目的 JSON 端点结构（https://ws.audioscrobbler.com/2.0/?method=tag.gettoptracks&tag=indie&api_key=YOUR_API_KEY&format=xml），如下所示。

```
<?xml version="1.0" encoding="UTF-8" ?>
<lfm status="ok">
   <tracks tag="indie" page="1" perPage="50"
    totalPages="4475" total="223728">
      <track rank="1">
         <name>Mr. Brightside</name>
```

```
            <duration>224</duration>
        <mbid>37d516ab-d61f-4bcb-9316-7a0b3eb845a8</mbid>
        <url>https://www.last.fm/music
        /The+Killers/_/Mr.+Brightside</url>
        <streamable fulltrack="0">0</streamable>
        <artist>
            <name>The Killers</name>
            <mbid>95e1ead9-4d31-4808-a7ac-32c3614c116b</mbid>
            <url>https://www.last.fm/music/The+Killers</url>
        </artist>
        <image size="small">https://lastfm.freetls.fastly.net/i
        /u/34s/2a96cbd8b46e442fc41c2b86b821562f.png</image>
        <image size="medium">
        https://lastfm.freetls.fastly.net/i
        /u/64s/2a96cbd8b46e442fc41c2b86b821562f.png</image>
        <image size="large">https://lastfm.freetls.fastly.net/i
        /u/174s/2a96cbd8b46e442fc41c2b86b821562f.png</image>
        <image size="extralarge">
            https://lastfm.freetls.fastly.net/i/u/300x300
            /2a96cbd8b46e442fc41c2b86b821562f.png
        </image>
    </track>
    ...
  </tracks>
</lfm>
```

9.6.2 实现方式

下列步骤将创建一个 Song 模型和一个管理命令，并以 XML 格式将曲目从 Last.fm 导入至数据库中。

（1）在 music 应用程序中，创建一个 management 目录和其下的一个 commands 子目录。将__init__.py 空文件置于这两个目录下，并生成 Python 包。

（2）添加 import_music_from_lastfm_xml.py 文件，如下所示。

```
# myproject/apps/music/management/commands
# /import_music_from_lastfm_xml.py
from django.core.management.base import BaseCommand

class Command(BaseCommand):
    help = "Imports top songs from last.fm as XML."
    SILENT, NORMAL, VERBOSE, VERY_VERBOSE = 0, 1, 2, 3
```

```
    API_URL = "https://ws.audioscrobbler.com/2.0/"

    def add_arguments(self, parser):
        # Named (optional) arguments
        parser.add_argument("--max_pages", type=int, default=0)

    def handle(self, *args, **options):
        self.verbosity = options.get("verbosity", self.NORMAL)
        self.max_pages = options["max_pages"]
        self.prepare()
        self.main()
        self.finalize()
```

（3）在同一文件中，针对 Command 类，创建 prepare()方法。

```
def prepare(self):
    from django.conf import settings

    self.imported_counter = 0
    self.skipped_counter = 0
    self.params = {
        "method": "tag.gettoptracks",
        "tag": "indie",
        "api_key": settings.LAST_FM_API_KEY,
        "format": "xml",
        "page": 1,
    }
```

（4）创建 main()方法。

```
def main(self):
    import requests
    from defusedxml import ElementTree

    response = requests.get(self.API_URL, params=self.params)
    if response.status_code != requests.codes.ok:
        self.stderr.write(f"Error connecting to {response.url}")
        return
    root = ElementTree.fromstring(response.content)

    pages = int(root.find("tracks").attrib.get("totalPages", 1))
    if self.max_pages > 0:
        pages = min(pages, self.max_pages)
```

```
    if self.verbosity >= self.NORMAL:
        self.stdout.write(f"=== Importing {pages} page(s)
         of songs ===")

    self.save_page(root)

    for page_number in range(2, pages + 1):
        self.params["page"] = page_number
        response = requests.get(self.API_URL, params=self.params)
        if response.status_code != requests.codes.ok:
            self.stderr.write(f"Error connecting to {response.url}")
            return
        root = ElementTree.fromstring(response.content)
        self.save_page(root)
```

（5）每个分页订阅中的页面将通过 save_page()方法保存，如下所示。

```
def save_page(self, root):
    import os
    import requests
    from io import BytesIO
    from django.core.files import File
    from ...forms import SongForm

    for track_node in root.findall("tracks/track"):
        if not track_node:
            continue

        song_dict = {
            "artist": track_node.find("artist/name").text,
            "title": track_node.find("name").text,
            "url": track_node.find("url").text,
        }
        form = SongForm(data=song_dict)
        if form.is_valid():
            song = form.save()

            image_node = track_node.find("image[@size='medium']")
            if image_node is not None:
                image_url = image_node.text
                image_response = requests.get(image_url)
                song.image.save(
                    os.path.basename(image_url),
```

```
                File(BytesIO(image_response.content)),
            )

        if self.verbosity >= self.NORMAL:
            self.stdout.write(f" - {song}\n")
        self.imported_counter += 1
    else:
        if self.verbosity >= self.NORMAL:
            self.stderr.write(
                f"Errors importing song "
                f"{song_dict['artist']} - {song_dict['title']}:\n"
            )
            self.stderr.write(f"{form.errors.as_json()}\n")
        self.skipped_counter += 1
```

（6）创建 finalize()方法。

```
def finalize(self):
    if self.verbosity >= self.NORMAL:
        self.stdout.write(f"------------------------\n")
        self.stdout.write(f"Songs imported: {self.imported_counter}\n")
        self.stdout.write(f"Songs skipped: {self.skipped_counter}\n\n")
```

（7）当运行导入操作时，可在命令行中运行下列内容。

```
(env)$ python manage.py import_music_from_lastfm_xml --max_pages=3
```

9.6.3　工作方式

全部过程与 JSON 方案类似。当采用 requests.get()方法时，我们从 Last.fm 中读取数据，并以 params 形式传递查询参数。响应结果的 XML 内容被传递至 defusedxml 模块中的 ElementTree 解析器中，并返回 root 节点。

注意：

defusedxml 模块是 xml 模块更安全的替代品，可防止 XML 炸弹——即攻击者使用几百个字节的 XML 数据占用吉字节级的内存漏洞。

ElementTree 节点包含 find()和 findall()方法，其中，可传递 Xpath 查询以过滤特定的子节点。

表 9.1 显示了 ElementTree 所支持的有效 XPath 语法。

表 9.1

XPath 语法成分	含　义
tag	选择具有给定标记的所有子元素
*	选择全部子元素
.	选择当前节点
//	选择当前元素下全部层次上的所有子元素
..	选择父元素
[@attrib]	选择包含给定属性的全部元素
[@attrib='value']	选择给定属性具有给定值的所有元素
[tag]	选择具有子命名标签的所有元素。仅支持直接子节点
[position]	选择给定位置处的所有元素。对应位置可以是一个整数（1 表示首位置）、last()表达式（最后一个位置），或者是相对于最后一个位置的位置（如 last()-1）

因此，在 main()方法中，当使用 root.find("tracks").attrib.get("totalPages", 1)时，我们将读取全部页面数量（如果数据缺失，则默认值为 1）。这里，我们将保存第一个页面，并逐一访问其他页面同时予以保存。

在 save_page()方法中，通过<tracks>节点下的<track>节点，root.findall("tracks/track")返回一个迭代器。基于 track_node.find("image[@size='medium']")，我们将得到中等尺寸的图像。再次说明，Song 的创建通过模型表单（用于验证传入数据）进行。

如果利用--verbosity=1 或更大值调用命令，我们将会看到与导入歌曲相关的详细信息。

9.6.4 更多内容

下列链接提供了更多信息。

- 如何处理 Last.fm：https://www.last.fm/api/。
- XPath：https://en.wikipedia.org/wiki/XPath。
- ElementTree 的完整文档：https://docs.python.org/3/library/xml.etree.elementtree. html。

9.6.5 延伸阅读

- “从本地 CSV 文件中导入数据”示例。
- “从本地 Excel 文件中导入数据”示例。
- “从外部 XML 文件中导入数据”示例。

9.7 针对搜索引擎准备分页网站地图

网站地图协议通知搜索引擎网站上所有不同页面的信息。通常情况下，这是一个单一的 sitemap.xml 文件，并通知索引内容和索引频率。如果网站上包含大量不同的页面，我们还可以划分和分页 XML 文件，以更快地渲染每个资源列表。

在当前示例中，我们将讨论如何创建一个分页网站地图，以及如何在 Django 网站中使用站点地图。

9.7.1 准备工作

当前示例需要扩展 music 应用程序，并添加列表和详细视图。

（1）创建 views.py 文件，如下所示。

```
# myproject/apps/music/views.py
from django.views.generic import ListView, DetailView
from django.utils.translation import ugettext_lazy as _
from .models import Song

class SongList(ListView):
    model = Song

class SongDetail(DetailView):
    model = Song
```

（2）创建 urls.py 文件，如下所示。

```
# myproject/apps/music/urls.py
from django.urls import path
from .views import SongList, SongDetail

app_name = "music"

urlpatterns = [
    path("", SongList.as_view(), name="song_list"),
    path("<uuid:pk>/", SongDetail.as_view(), name="song_detail"),
]
```

（3）将 URL 配置包含至项目的 URL 配置中。

```
# myproject/urls.py
from django.conf.urls.i18n import i18n_patterns
from django.urls import include, path

urlpatterns = i18n_patterns(
    # …
    path("songs/", include("myproject.apps.music.urls",
     namespace="music")),
)
```

（4）创建歌曲列表视图的模板。

```
{# music/song_list.html #}
{% extends "base.html" %}
{% load i18n %}

{% block main %}
    <ul>
        {% for song in object_list %}
            <li><a href="{{ song.get_url_path }}">
             {{ song }}</a></li>
        {% endfor %}
    </ul>
{% endblock %}
```

（5）创建歌曲详细视图的模板。

```
{# music/song_detail.html #}
{% extends "base.html" %}
{% load i18n %}

{% block content %}
    {% with song=object %}
        <h1>{{ song }}</h1>
        {% if song.image %}
            <img src="{{ song.image.url }}" alt="{{ song }}" />
        {% endif %}
        {% if song.url %}
            <a href="{{ song.url }}" target="_blank"
             rel="noreferrer noopener">
               {% trans "Check this song" %}
            </a>
        {% endif %}
    {% endwith %}
{% endblock %}
```

9.7.2 实现方式

下列步骤将添加分页站点地图。

（1）在设置项的 INSTALLED_APPS 中包含 django.contrib.sitemaps。

```
# myproject/settings/_base.py
INSTALLED_APPS = [
    # …
    "django.contrib.sitemaps",
    # …
]
```

（2）调整项目的 urls.py 文件，如下所示。

```
# myproject/urls.py
from django.conf.urls.i18n import i18n_patterns
from django.urls import include, path
from django.contrib.sitemaps import views as sitemaps_views
from django.contrib.sitemaps import GenericSitemap
from myproject.apps.music.models import Song

class MySitemap(GenericSitemap):
    limit = 50

    def location(self, obj):
        return obj.get_url_path()

song_info_dict = {
    "queryset": Song.objects.all(),
    "date_field": "modified",
}
sitemaps = {"music": MySitemap(song_info_dict, priority=1.0)}

urlpatterns = [
    path("sitemap.xml", sitemaps_views.index,
     {"sitemaps": sitemaps}),
    path("sitemap-<str:section>.xml", sitemaps_views.sitemap,
     {"sitemaps": sitemaps},
        name="django.contrib.sitemaps.views.sitemap"
    ),
]
```

```
urlpatterns += i18n_patterns(
    # …
    path("songs/", include("myproject.apps.music.urls",
     namespace="music")),
)
```

9.7.3 工作方式

当访问 http://127.0.0.1:8000/sitemap.xml 时，可以看到基于分页网站地图的索引，如下所示。

```
<?xml version="1.0" encoding="UTF-8"?>
<sitemapindex xmlns="http://www.sitemaps.org/schemas/sitemap/0.9">
    <sitemap>
        <loc>http://127.0.0.1:8000/sitemap-music.xml</loc>
    </sitemap>
    <sitemap>
        <loc>http://127.0.0.1:8000/sitemap-music.xml?p=2</loc>
    </sitemap>
    <sitemap>
        <loc>http://127.0.0.1:8000/sitemap-music.xml?p=3</loc>
    </sitemap>
</sitemapindex>
```

这里，每个页面将显示多达 50 个条目，包括 URL、最后一次修改和优先级。

```
<?xml version="1.0" encoding="UTF-8"?>
<urlset xmlns="http://www.sitemaps.org/schemas/sitemap/0.9">
    <url>
        <loc>http://127.0.0.1:8000/en/songs/b2d3627b-dbc7
         -4c11-a13e-03d86f32a719/</loc>
        <lastmod>2019-12-15</lastmod>
        <priority>1.0</priority>
    </url>
    <url>
        <loc>http://127.0.0.1:8000/en/songs/f5c386fd-1952
         -4ace-9848-717d27186fa9/</loc>
        <lastmod>2019-12-15</lastmod>
        <priority>1.0</priority>
    </url>
    <url>
        <loc>http://127.0.0.1:8000/en/songs/a59cbb5a-16e8
         -46dd-9498-d86e24e277a5/</loc>
```

```
        <lastmod>2019-12-15</lastmod>
        <priority>1.0</priority>
    </url>
    ...
</urlset>
```

当网站准备发布至产品阶段时，可利用网站框架提供的 ping_google 管理命令通知 Google Search Engine 与页面相关的信息。对此，可在产品服务器上执行下列命令。

```
(env)$ python manage.py ping_google --
settings=myproject.settings.production
```

9.7.4 更多内容

读者可访问下列链接以了解更多内容。

- 网站地图协议：https://www.sitemaps.org/。
- Django 网站地图框架：https://docs.djangoproject.com/en/3.0/ref/contrib/sitemaps/。

9.7.5 延伸阅读

此外，读者还可参考“创建可过滤的 RSS 订阅”示例。

9.8 创建可过滤的 RSS 订阅

Django 包含一个聚合订阅（feed）框架，并可创建 RSS（Really Simple Syndication）和 Atom 订阅。这里，RSS 和 Atom 订阅可视为包含特殊语义的 XML 文档，并可订阅至 RSS 阅读器，如 Feedly；此外也可聚合至其他网站、移动应用程序或桌面应用程序中。在当前示例中，我们将创建一个 RSS 订阅，进而提供与歌曲相关的信息。而且，最终结果还可根据 URL 查询参数过滤。

9.8.1 准备工作

首先启用之前创建的 music 应用程序。特别地，我们需要先设置模型、表单、视图、URL 配置和模板。

当查看列表歌曲时，需要添加基于艺术家的过滤机制，稍后将被 RSS 订阅使用。

（1）向 forms.py 文件添加一个过滤表单，其中包含 artist 选择字段，所有艺术家的名字按照字母顺序排序且忽略大小写。

```
# myproject/apps/music/forms.py
from django import forms
from django.utils.translation import ugettext_lazy as _
from .models import Song

# ...

class SongFilterForm(forms.Form):
    def __init__(self, *args, **kwargs):
        super().__init__(*args, **kwargs)
        artist_choices = [
            (artist, artist)
            for artist in sorted(
                Song.objects.values_list("artist", flat=True).distinct(),
                key=str.casefold
            )
        ]
        self.fields["artist"] = forms.ChoiceField(
            label=_("Artist"),
            choices=artist_choices,
            required=False,
        )
```

（2）利用相关方法增强 SongList 视图以管理过滤机制。具体来说，get()方法将处理过滤机制并显示结果；get_form_kwargs()方法将准备过滤器表单的关键字参数；get_queryset()方法将根据艺术家过滤歌曲。

```
# myproject/apps/music/views.py
from django.http import Http404
from django.views.generic import ListView, DetailView, FormView
from django.utils.translation import ugettext_lazy as _
from .models import Song
from .forms import SongFilterForm

class SongList(ListView, FormView):
    form_class = SongFilterForm
    model = Song

    def get(self, request, *args, **kwargs):
        form_class = self.get_form_class()
```

```
        self.form = self.get_form(form_class)

        self.object_list = self.get_queryset()
        allow_empty = self.get_allow_empty()
        if not allow_empty and len(self.object_list) == 0:
            raise Http404(_(u"Empty list and '%(class_name)s
             .allow_empty' is False.")
                          % {'class_name':
                          self.__class__.__name__})

        context = self.get_context_data(object_list=
         self.object_list, form=self.form)
        return self.render_to_response(context)

    def get_form_kwargs(self):
        kwargs = {
            'initial': self.get_initial(),
            'prefix': self.get_prefix(),
        }
        if self.request.method == 'GET':
            kwargs.update({
                'data': self.request.GET,
            })
        return kwargs

    def get_queryset(self):
        queryset = super().get_queryset()
        if self.form.is_valid():
            artist = self.form.cleaned_data.get("artist")
            if artist:
                queryset = queryset.filter(artist=artist)
        return queryset
```

（3）调整歌曲列表模板，并添加过滤表单。

```
{# music/song_list.html #}
{% extends "base.html" %}
{% load i18n %}

{% block sidebar %}
    <form action="" method="get">
        {{ form.errors }}
        {{ form.as_p }}
```

```
        <button type="submit" class="btn btn-primary">
         {% trans "Filter" %}</button>
    </form>
{% endblock %}

{% block main %}
    <ul>
        {% for song in object_list %}
            <li><a href="{{ song.get_url_path }}">
             {{ song }}</a></li>
        {% endfor %}
    </ul>
{% endblock %}
```

当前，如果在浏览器中检查歌曲列表视图，并按照 Lana Del Rey 过滤歌曲，最终结果如图 9.4 所示。

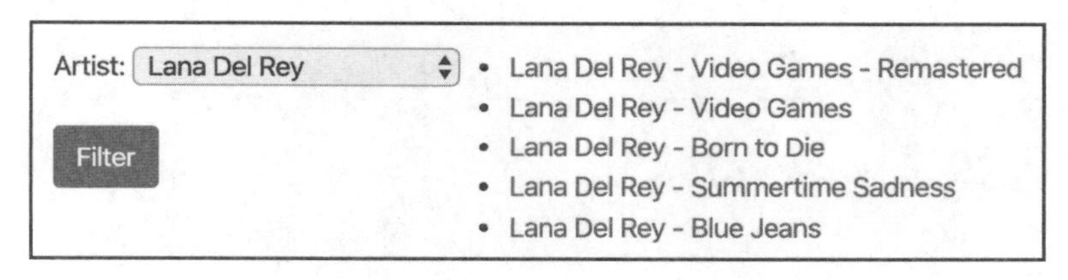

图 9.4

过滤歌曲列表的 URL 位于 http://127.0.0.1:8000/en/songs/?artist=Lana+Del+Rey 中。

9.8.2　实现方式

下列步骤将向 music 应用程序中添加 RSS 订阅。

（1）在 music 应用程序中，创建 feeds.py 文件并添加如下所示的内容。

```
# myproject/apps/music/feeds.py
from django.contrib.syndication.views import Feed
from django.urls import reverse

from .models import Song
from .forms import SongFilterForm

class SongFeed(Feed):
    description_template = "music/feeds/song_description.html"
```

```
    def get_object(self, request, *args, **kwargs):
        form = SongFilterForm(data=request.GET)
        obj = {}
        if form.is_valid():
            obj = {"query_string": request.META["QUERY_STRING"]}
            for field in ["artist"]:
                value = form.cleaned_data[field]
                obj[field] = value
        return obj

    def title(self, obj):
        the_title = "Music"
        artist = obj.get("artist")
        if artist:
            the_title = f"Music by {artist}"
        return the_title

    def link(self, obj):
        return self.get_named_url("music:song_list", obj)

    def feed_url(self, obj):
        return self.get_named_url("music:song_rss", obj)

    @staticmethod
    def get_named_url(name, obj):
        url = reverse(name)
        qs = obj.get("query_string", False)
        if qs:
            url = f"{url}?{qs}"
        return url

    def items(self, obj):
        queryset = Song.objects.order_by("-created")

        artist = obj.get("artist")
        if artist:
            queryset = queryset.filter(artist=artist)

        return queryset[:30]

    def item_pubdate(self, item):
        return item.created
```

（2）在 RSS 订阅中创建歌曲描述模板。

```
{# music/feeds/song_description.html #}
{% load i18n %}
{% with song=obj %}
    {% if song.image %}
        <img src="{{ song.image.url }}" alt="{{ song }}" />
    {% endif %}
    {% if song.url %}
        <a href="{{ song.url }}" target="_blank"
         rel="noreferrer noopener">
            {% trans "Check this song" %}
        </a>
    {% endif %}
{% endwith %}
```

（3）在应用程序的 URL 配置中插入 RSS 订阅。

```
# myproject/apps/music/urls.py
from django.urls import path

from .feeds import SongFeed
from .views import SongList, SongDetail

app_name = "music"

urlpatterns = [
    path("", SongList.as_view(), name="song_list"),
    path("<uuid:pk>/", SongDetail.as_view(), name="song_detail"),
    path("rss/", SongFeed(), name="song_rss"),
]
```

（4）在歌曲列表视图的模板中，添加指向 RSS 订阅的链接。

```
{# music/song_list.html #}

{% url "music:songs_rss" as songs_rss_url %}
<p>
    <a href="{{ songs_rss_url }}?{{ request.META.QUERY_STRING }}">
        {% trans "Subscribe to RSS feed" %}
    </a>
</p>
```

9.8.3 工作方式

当刷新 http://127.0.0.1:8000/en/songs/?artist=Lana+Del+Rey 处的过滤列表视图时，可以看到指向 http://127.0.0.1:8000/en/songs/rss/?artist=Lana+Del+Rey 的 Subscribe to RSS feed 链接。这将是由艺术家过滤的多达 30 首歌曲的 RSS 订阅内容。

SongFeed 类关注 RSS 订阅的自动 XML 标记的生成。此处指定了下列方法。

- get_object()方法定义了 Feed 类的上下文字典，以供其他方法使用。
- title()方法根据对应结果是否过滤定义了订阅的标题。
- link()方法返回列表视图的 URL，而 feed_url()方法则返回订阅的 URL。这两个方法均使用了一个帮助方法 get_named_url()，该方法通过路径名和查询参数形成了一个 URL。
- items()方法返回歌曲的 queryset，并可以根据艺术家进行过滤（可选）。
- item_pubdate()方法返回歌曲的生成日期。

注意：

关于扩展的 Feed 类的全部方法和属性，读者可访问 https://docs.djangoproject.com/en/3.0/ref/contrib/syndication/#feed-class-reference。

9.8.4 延伸阅读

- “从本地 CSV 文件中导入数据”示例。
- “针对搜索引擎准备分页网站地图”示例。

9.9 使用 Django REST 框架创建一个 API

当为模型创建一个 RESTful API 并在第三方之间传输数据时，Django REST 框架是一种可以使用的最佳工具。该框架包含大量的文档和以 Django 为中心的实现，同时兼具可维护性。在当前示例中，我们将学习如何使用 Django REST 框架以使项目合作者、移动客户端或基于 Ajax 的站点可访问站点上的数据，进而创建、读取、更新和删除相关内容。

9.9.1 准备工作

首先利用下列命令在虚拟环境中安装 Django REST 框架。

```
(env)$ pip install djangorestframework==3.11.0
```

在设置项中将"rest_framework"添加至 INSTALLED_APPS 中。

随后增强之前定义的 music 应用程序。此外，还需要收集 Django REST 框架为页面提供的静态文件，以提供良好的样式。

```
(env)$ python manage.py collectstatic
```

9.9.2 实现方式

执行下列步骤将在 music 应用程序中集成新的 RESTful API。

（1）将 Django REST 框架的配置添加至当前设置项中，如下所示。

```
# myproject/settings/_base.py
REST_FRAMEWORK = {
    "DEFAULT_PERMISSION_CLASSES": [ "rest_framework.permissions
          .DjangoModelPermissionsOrAnonReadOnly"
    ],
    "DEFAULT_PAGINATION_CLASS":
    "rest_framework.pagination.LimitOffsetPagination",
    "PAGE_SIZE": 50,
}
```

（2）在 music 应用程序中，创建 serializers.py 文件，如下所示。

```
from rest_framework import serializers
from .models import Song

class SongSerializer(serializers.ModelSerializer):
    class Meta:
        model = Song
        fields = ["uuid", "artist", "title", "url", "image"]
```

（3）在 music 应用程序中，向 views.py 文件中添加两个基于类的新视图。

```
from rest_framework import generics

from .serializers import SongSerializer
from .models import Song

# ...
```

```
class RESTSongList(generics.ListCreateAPIView):
    queryset = Song.objects.all()
    serializer_class = SongSerializer

    def get_view_name(self):
        return "Song List"

class RESTSongDetail(generics.RetrieveUpdateDestroyAPIView):
    queryset = Song.objects.all()
    serializer_class = SongSerializer

    def get_view_name(self):
        return "Song Detail"
```

（4）向项目的 URL 配置中置入新的视图。

```
# myproject/urls.py
from django.urls import include, path
from myproject.apps.music.views import RESTSongList, RESTSongDetail

urlpatterns = [
    path("api-auth/", include("rest_framework.urls",
     namespace="rest_framework")),
    path("rest-api/songs/", RESTSongList.as_view(),
     name="rest_song_list"),
    path(
        "rest-api/songs/<uuid:pk>/", RESTSongDetail.as_view(),
         name="rest_song_detail"
    ),
    # …
]
```

9.9.3　工作方式

此处创建的是一个音乐 API，我们可以读取一个分页的歌曲列表、创建新的歌曲，并按照 ID 读取、修改或删除一首歌曲。其间，未经身份验证的用户具有读取权限，但需要持有一个授权的用户账户，以添加、修改或删除歌曲。Django REST 框架提供了基于 Web 的 API 文档，当通过 GET 在浏览器中访问 API 端点时，将会显示这些文档。在未登录的前提下，框架将显示如图 9.5 所示的内容。

表 9.2 展示了所创建的 API 的具体解释。

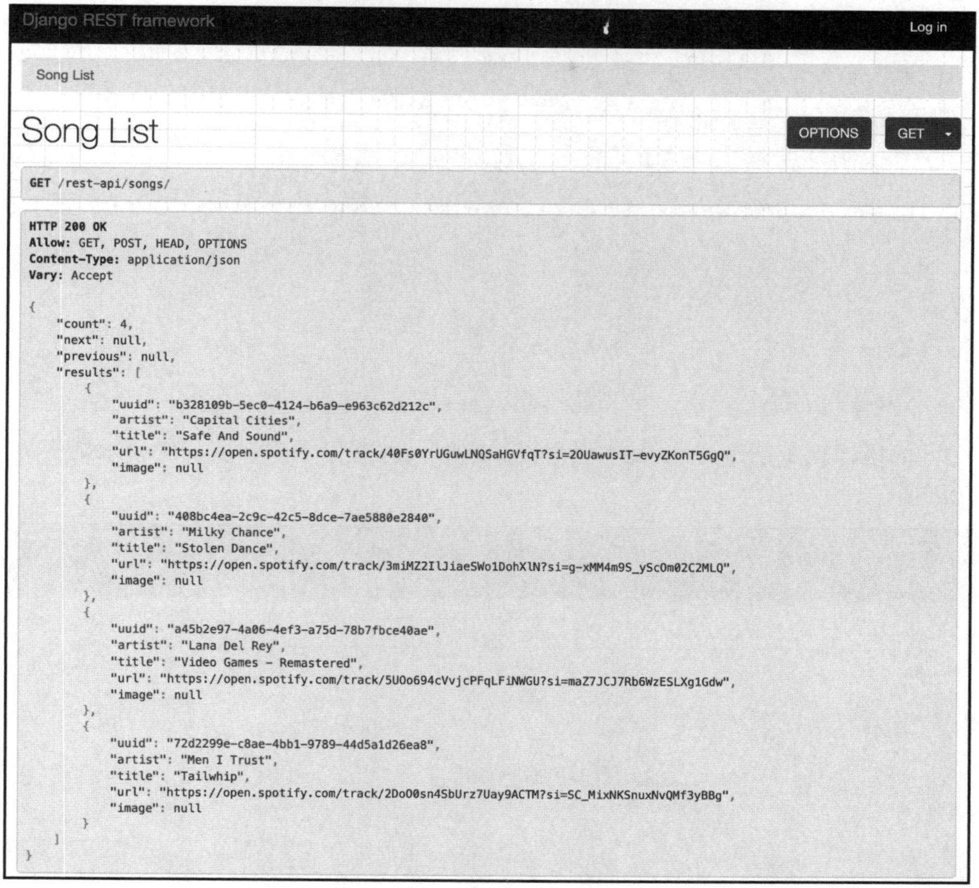

Django REST framework　　Log in

Song List

Song List

OPTIONS　GET

GET /rest-api/songs/

```
HTTP 200 OK
Allow: GET, POST, HEAD, OPTIONS
Content-Type: application/json
Vary: Accept
```


```
{
    "count": 4,
    "next": null,
    "previous": null,
    "results": [
        {
            "uuid": "b328109b-5ec0-4124-b6a9-e963c62d212c",
            "artist": "Capital Cities",
            "title": "Safe And Sound",
            "url": "https://open.spotify.com/track/40Fs0YrUGuwLNQSaHGVfqT?si=2OUawusIT-evyZKonT5GgQ",
            "image": null
        },
        {
            "uuid": "408bc4ea-2c9c-42c5-8dce-7ae5880e2840",
            "artist": "Milky Chance",
            "title": "Stolen Dance",
            "url": "https://open.spotify.com/track/3miMZ2IlJiaeSWo1DohXlN?si=g-xMM4m9S_yScOm02C2MLQ",
            "image": null
        },
        {
            "uuid": "a45b2e97-4a06-4ef3-a75d-78b7fbce40ae",
            "artist": "Lana Del Rey",
            "title": "Video Games - Remastered",
            "url": "https://open.spotify.com/track/5UOo694cVvjcPFqLFiNWGU?si=maZ7JCJ7Rb6WzESLXg1Gdw",
            "image": null
        },
        {
            "uuid": "72d2299e-c8ae-4bb1-9789-44d5a1d26ea8",
            "artist": "Men I Trust",
            "title": "Tailwhip",
            "url": "https://open.spotify.com/track/2DoO0sn4SbUrz7Uay9ACTM?si=SC_MixNKSnuxNvQMf3yBBg",
            "image": null
        }
    ]
}
```


图 9.5

表 9.2

URL	HTTP 方法	描　　述
/rest-api/songs/	GET	每页包含 50 个条目
/rest-api/songs/	POST	如果请求用户通过身份验证并授权创建歌曲，则创建一首新歌曲
/rest-api/songs/b328109b-5ec0-4124-b6a9-e963c62d212c/	GET	获取 ID 为 b328109b-5ec0-4124-b6a9-e963c62d212c 的一首歌曲
/rest-api/songs/b328109b-5ec0-4124-b6a9-e963c62d212c/	PUT	如果请求用户通过身份验证并授权修改歌曲，则更新 ID 为 b328109b-5ec0-4124-b6a9-e963c62d212c 的一首歌曲
/rest-api/songs/b328109b-5ec0-4124-b6a9-e963c62d212c/	DELETE	如果请求用户通过身份验证并授权删除歌曲，则删除 ID 为 b328109b-5ec0-4124-b6a9-e963c62d212c 的一首歌曲

这里，读者可能会询问 API 的实际操作方式。例如，我们可能使用 requests 库从 Python 脚本中创建一首新的歌曲，如下所示。

```
import requests

response = requests.post(
    url="http://127.0.0.1:8000/rest-api/songs/",
    data={
        "artist": "Luwten",
        "title": "Go Honey",
    },
    auth=("admin", "<YOUR_ADMIN_PASSWORD>"),
)
assert(response.status_code == requests.codes.CREATED)
```

相同的操作也可通过 Postman 应用程序完成，并提供了用户友好的请求提交界面，如图 9.6 所示。

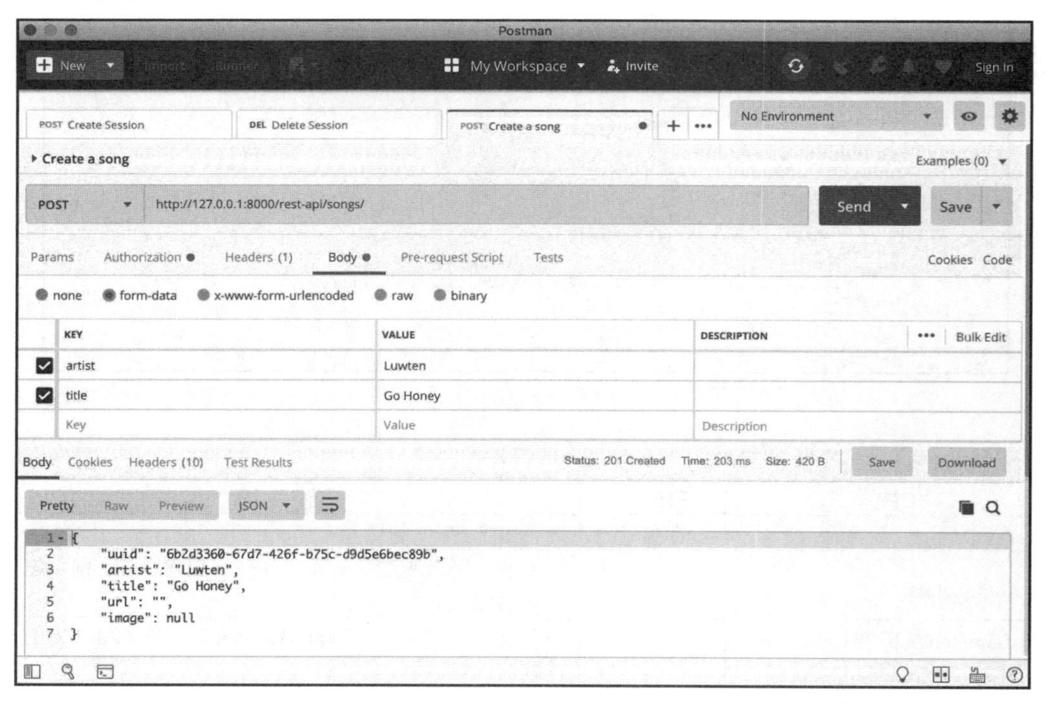

图 9.6

在登录后，在框架生成的 API 文档下，还可通过集成表单尝试使用 API，如图 9.7 所示。

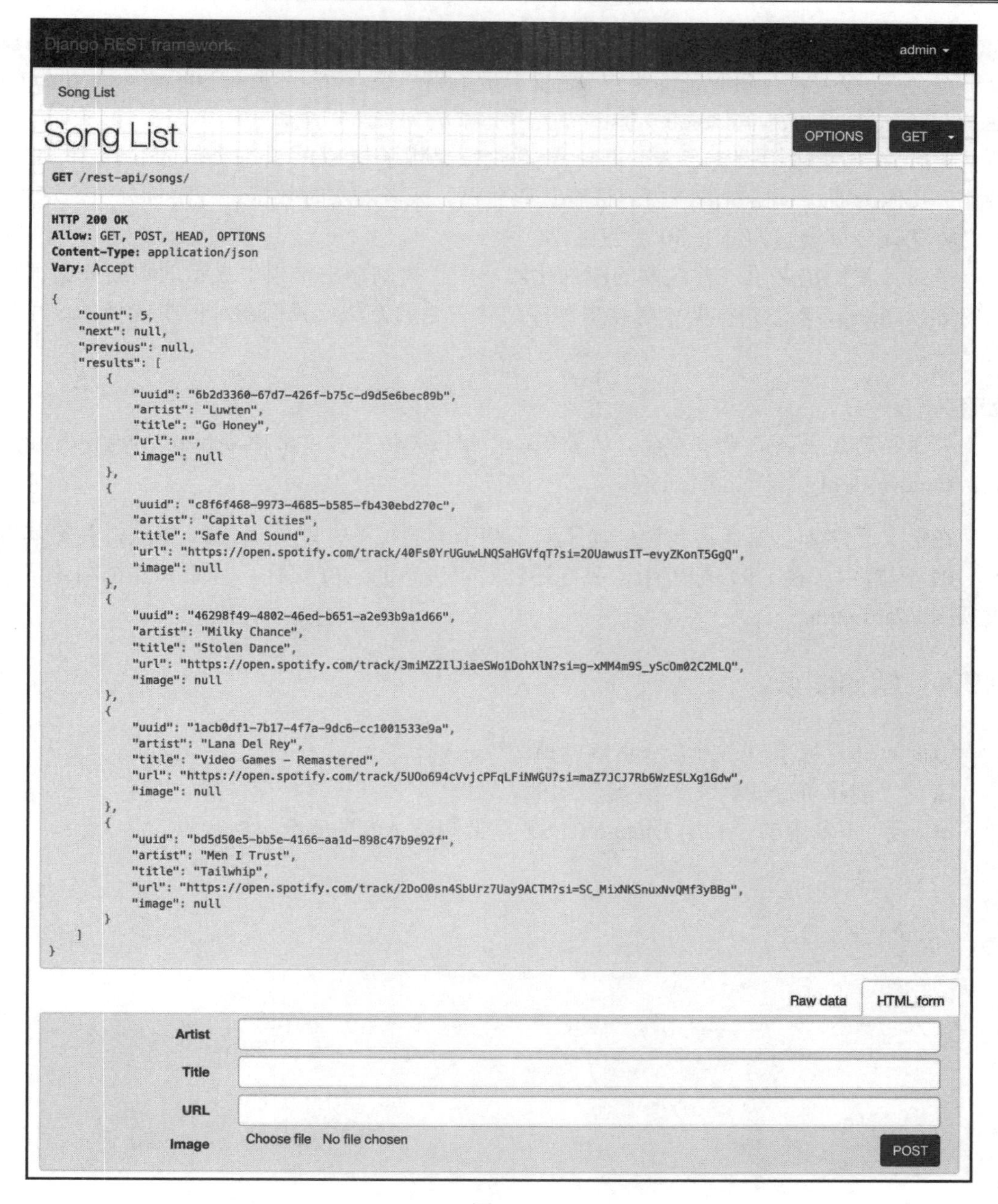

图 9.7

下面快速浏览一下代码的工作方式。在当前设置项中，我们设置了访问权限根据

Django 系统的授权。对于匿名请求，仅支持读取行为。相应地，其他访问选项还包括授权于每个人、仅授权身份验证的用户、授权于员工用户等。读者可访问 https://www.django-rest-framework.org/api-guide/permissions/查看完整的列表。

随后，在设置项中设置了分页机制。当前选项采用了与 SQL 查询类似的 limit 和 offset 参数。其他选项还包括对静态内容进行页码分页，或者对实时数据进行游标分页。这里，我们将默认分页设置为每页 50 个条目。

最后，我们定义了一首歌曲的序列化器，用于控制输出结果中显示的数据并验证输入内容。Django REST 框架中包含多种序列化关系的方法，我们在示例中选择了最冗长的一种。

提示：

关于序列化关系的更多内容，读者可参考相关文档，对应网址为 https://www.django-rest-framework.org/api-guide/relations/。

在定义了序列化器后，我们创建了两个基于类的视图，以处理 API 端点，并将其置入 URL 配置中。在 URL 配置中，我们还针对可浏览的 API 页面、登录和注销制定了一条规则（/api-auth/）。

9.9.4 延伸阅读

- "针对搜索引擎准备分页网站地图"示例。
- "创建可过滤的 RSS 订阅"示例。
- 第 11 章中的"利用 Django REST 框架测试 API"示例。

第 10 章　其 他 内 容

本章主要涉及下列内容。

- 使用 Django shell。
- 使用数据库查询表达式。
- slugify()函数的猴子补丁以获得更好的国际化支持。
- 切换 Debug 工具栏。
- 使用 ThreadLocalMiddleware。
- 使用信号通知管理员有关新的条目。
- 检查缺失的设置项。

10.1　简　　介

本章将介绍一些重点知识，以帮助读者更好地理解和使用 Django。在正式编写代码之前，我们将学习如何使用 Django shell 以实现各种尝试性的代码操作在文件中。另外，本章还将讨论猴子补丁（游击补丁）这一概念，这也是动态语言（如 Python 和 Ruby）的一个强大特性。接下来我们将介绍全文本搜索功能，读者将学习如何调试代码并检测其性能。随后，本章将介绍如何从任意模块中访问当前登录用户（或其他请求参数）。最后，我们还将阐述如何处理信号并创建系统检测。

10.2　技 术 需 求

当与本章代码协同工作时，需要安装最新稳定版本的 Python、MySQL 或 PostgreSQL 数据库，以及虚拟环境中的 Django 项目。

读者可访问 GitHub 储存库中的 ch10 查看本章的代码，对应网址为 https://github.com/PacktPublishing/Django-3-Web-Development-Cookbook-Fourth-Edition。

10.3　使用 Django shell

当虚拟环境处于激活状态，且项目目录选取为当前目录后，在命令行工具中输入下列命令。

```
(env)$ python manage.py shell
```

通过执行上述命令，用户将进入一个交互式 Python shell 中（针对 Django 项目配置）。其中，我们可以尝试操作代码、查看类、尝试各种方法或执行脚本。在当前示例中，我们将介绍一些重要的函数，进而更好地与 Django shell 协同工作。

10.3.1　准备工作

我们可以安装 IPython 或 bpython 进而为 Python shell 提供额外的界面选项。如果读者需要选择也可以安装这两种选项。对于 Django shell 的输出结果，它们能够高亮显示相关语法，同时还添加了其他一些帮助内容。我们可在虚拟环境中通过下列命令安装 IPython 或 bpython。

```
(env)$ pip install ipython
(env)$ pip install bpython
```

10.3.2　实现方式

下列步骤将展示一些 Django shell 的基本操作知识。

（1）输入下列命令运行 Django shell。

```
(env)$ python manage.py shell
```

当 IPython 或 bpython 安装完毕，进入 shell 后，IPython 或 bpython 会自动变为默认的界面。此外，还可向上述命令中添加-i <interface>选项并使用特定的界面。相应地，提示符根据所使用的界面而有所变化。图 10.1 显示了一个 IPython shell，且提示符以 In [1]:开始。

当采用 bpython 时，shell 的提示符为>>>。在输入过程中，还可实现代码高亮显示和文本自动完成功能，如图 10.2 所示。

默认的 Python 界面如图 10.3 所示，同样使用了>>>提示符，但提供了一些与系统相关的信息。

```
django-myproject — IPython: src/django-myproject — python manage.py shell -i ipython — 123×20
(venv) archatas@Aidass-MacBook-Pro django-myproject % python manage.py shell -i ipython
Python 3.7.4 (default, Jul  9 2019, 18:13:23)
Type 'copyright', 'credits' or 'license' for more information
IPython 7.10.2 -- An enhanced Interactive Python. Type '?' for help.

In [1]:
```

图 10.1

```
django-myproject — IPython: src/django-myproject — python manage.py shell -i bpython — 123×20
(venv) archatas@Aidass-MacBook-Pro django-myproject % python manage.py shell -i bpython
>>>
Welcome to bpython! Press <F1> for help.
```

图 10.2

```
django-myproject — IPython: src/django-myproject — python manage.py shell -i python — 123×20
(venv) archatas@Aidass-MacBook-Pro django-myproject % python manage.py shell -i python
Python 3.7.4 (default, Jul  9 2019, 18:13:23)
[Clang 10.0.1 (clang-1001.0.46.4)] on darwin
Type "help", "copyright", "credits" or "license" for more information.
(InteractiveConsole)
>>>
```

图 10.3

当前，我们可以导入类、函数或变量，并尝试对其进行各种操作。例如，当查看安装模块的版本时，我们可导入该模块并尝试读取__version__、VERSION 或 version 属性（此处采用 bpython 显示，同时还展示了高亮显示和自动完成功能），如图 10.4 所示。

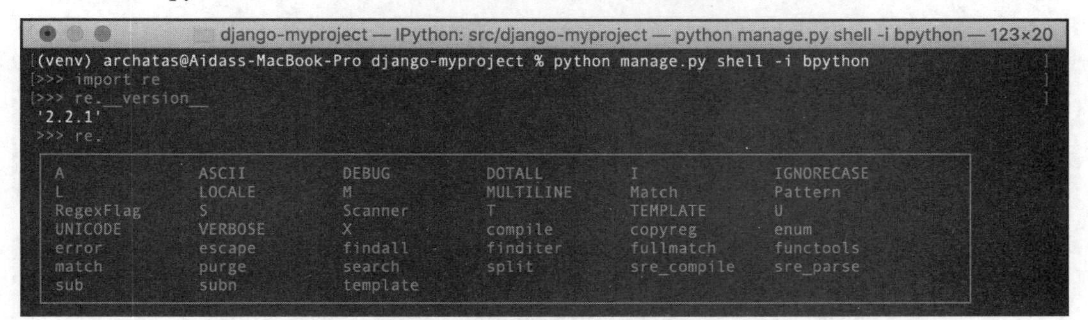

图 10.4

（2）当获取模块、类、函数、方法、关键字或文档主题的详细描述时，可使用 help()函数。我们可以向某一个特定的实体或实体自身中传递一个带有路径的字符串，如下所示。

```
>>> help("django.forms")
```

这将打开一个 django.forms 模块的帮助页面。我们可通过箭头键上下滚动页面。按 Q 键将返回至 shell。如果以无参数形式运行 help()函数，这将打开交互式帮助页面。其中，我们可输入模块的路径、类、函数等，并获得功能与使用方式相关的信息。若打算退出交互式帮助页面，可以按 Ctrl+D 快捷键。

（3）图 10.5 显示了使用 IPython 时如何将一个实体传递至 help()函数。

```
django-myproject — IPython: src/django-myproject — python manage.py shell -i ipython — 123×20
(venv) archatas@Aidass-MacBook-Pro django-myproject % python manage.py shell -i ipython
Python 3.7.4 (default, Jul  9 2019, 18:13:23)
Type 'copyright', 'credits' or 'license' for more information
IPython 7.10.2 -- An enhanced Interactive Python. Type '?' for help.

In [1]: from django.forms import ModelForm

In [2]: help(ModelForm)
```

图 10.5

这将打开 ModelForm 类的帮助页面，如图 10.6 所示。

```
django-myproject — IPython: src/django-myproject — less ◂ python manage.py shell -i ipython — 123×35
Help on class ModelForm in module django.forms.models:

class ModelForm(BaseModelForm)
 |  ModelForm(data=None, files=None, auto_id='id_%s', prefix=None, initial=None, error_class=<class 'django.forms.utils.Err
orList'>, label_suffix=None, empty_permitted=False, instance=None, use_required_attribute=None, renderer=None)
 |
 |  The main implementation of all the Form logic. Note that this class is
 |  different than Form. See the comments by the Form class for more info. Any
 |  improvements to the form API should be made to this class, not to the Form
 |  class.
 |
 |  Method resolution order:
 |      ModelForm
 |      BaseModelForm
 |      django.forms.forms.BaseForm
 |      builtins.object
 |
 |  Data descriptors defined here:
 |
 |  media
 |
 |  ----------------------------------------------------------------------
 |  Data and other attributes defined here:
 |
 |  base_fields = {}
 |
 |  declared_fields = {}
 |
 |  ----------------------------------------------------------------------
 |  Methods inherited from BaseModelForm:
 |
 |  __init__(self, data=None, files=None, auto_id='id_%s', prefix=None, initial=None, error_class=<class 'django.forms.util
s.ErrorList'>, label_suffix=None, empty_permitted=False, instance=None, use_required_attribute=None, renderer=None)
 |      Initialize self.  See help(type(self)) for accurate signature.
:
```

图 10.6

对于某个模型实例，当快速查看哪个字段和值有效时，可使用__dict__属性。我们可使用 pprint()函数并以更具可读的格式（而不是仅一长行内容）输出字典，如图 10.7 所示。注意，当使用__dict__时，我们无法获得多对多的关系，但这对于字段和值的快速浏览已然足够。

（4）当获取某个对象的全部有效属性和方法时，可使用 dir()函数，如图 10.8 所示。

```
django-myproject — IPython: src/django-myproject — python manage.py shell -i bpython — 123×20
(venv) archatas@Aidass-MacBook-Pro django-myproject % python manage.py shell -i bpython
>>> from django.contrib.contenttypes.models import ContentType
>>> content_type = ContentType.objects.first()
>>> content_type.__dict__
{'_state': <django.db.models.base.ModelState object at 0x10aee8490>, 'id': 1, 'app_label': 'admin', 'model': 'logentry'}
>>> from pprint import pprint
>>> pprint(content_type.__dict__)
{'_state': <django.db.models.base.ModelState object at 0x10aee8490>,
 'app_label': 'admin',
 'id': 1,
 'model': 'logentry'}
>>>
```

图 10.7

```
django-myproject — IPython: src/django-myproject — python manage.py shell -i bpython — 123×20
>>> pprint(content_type.__dict__)
{'_state': <django.db.models.base.ModelState object at 0x10aee8490>,
 'app_label': 'admin',
 'id': 1,
 'model': 'logentry'}
>>> dir(ContentType)
['DoesNotExist', 'MultipleObjectsReturned', '__class__', '__delattr__', '__dict__', '__dir__', '__doc__', '__eq__', '__form
at__', '__ge__', '__getattribute__', '__getstate__', '__gt__', '__hash__', '__init__', '__init_subclass__', '__le__', '__lt
__', '__module__', '__ne__', '__new__', '__reduce__', '__reduce_ex__', '__repr__', '__setattr__', '__setstate__', '__sizeof
__', '__str__', '__subclasshook__', '__weakref__', '_check_column_name_clashes', '_check_constraints', '_check_field_name_c
lashes', '_check_fields', '_check_id_field', '_check_index_together', '_check_indexes', '_check_local_fields', '_check_long
_column_names', '_check_m2m_through_same_relationship', '_check_managers', '_check_model', '_check_model_name_db_lookup_cla
shes', '_check_ordering', '_check_property_name_related_field_accessor_clashes', '_check_single_primary_key', '_check_swapp
able', '_check_unique_together', '_do_insert', '_do_update', '_get_FIELD_display', '_get_next_or_previous_by_FIELD', '_get_
next_or_previous_in_order', '_get_pk_val', '_get_unique_checks', '_meta', '_perform_date_checks', '_perform_unique_checks',
 '_save_parents', '_save_table', '_set_pk_val', 'app_label', 'app_labeled_name', 'check', 'clean', 'clean_fields', 'date_er
ror_message', 'delete', 'from_db', 'full_clean', 'get_all_objects_for_this_type', 'get_deferred_fields', 'get_object_for_th
is_type', 'id', 'logentry_set', 'model', 'model_class', 'name', 'natural_key', 'objects', 'permission_set', 'pk', 'prepare_
database_save', 'refresh_from_db', 'save', 'save_base', 'serializable_value', 'unique_error_message', 'validate_unique']
>>>
```

图 10.8

（5）若打算每行输出一个属性，则可使用如图 10.9 所示的代码。

```
django-myproject — IPython: src/django-myproject — python manage.py shell -i bpython — 123×20
>>> pprint(dir(ContentType))
['DoesNotExist',
 'MultipleObjectsReturned',
 '__class__',
 '__delattr__',
 '__dict__',
 '__dir__',
 '__doc__',
 '__eq__',
 '__format__',
 '__ge__',
 '__getattribute__',
 '__getstate__',
 '__gt__',
 '__hash__',
 '__init__',
 '__init_subclass__',
 '__le__',
 '__lt__',
 '__module__',
```

图 10.9

（6）在尝试对 QuerySet 或正则表达式进行各种操作时，Django shell 十分有用，随后可将其置于模型方法、视图或管理命令中。例如，当检查电子邮件验证正则表达式时，可在 Django shell 中输入下列内容。

```
>>> import re
>>> email_pattern = re.compile(r"[^@]+@[^@]+\.[^@]+")
>>> email_pattern.match("aidas@bendoraitis.lt")
<_sre.SRE_Match object at 0x1075681d0>
```

（7）当尝试对不同的 QuerySet 进行操作时，可使用下列代码。

```
>>> from django.contrib.auth.models import User
>>> User.objects.filter(groups__name="Editors")
[<User: admin>]
```

（8）当退出 Django shell 时，可按 Ctrl+D 快捷键或输入下列命令。

```
>>> exit()
```

10.3.3　工作方式

普通 Python shell 和 Django shell 之间的差别在于，当运行 Django shell 时，manage.py 将设置 DJANGO_SETTINGS_MODULE 环境变量，以便指向项目的 settings.py 路径，且 Django shell 中的全部代码将在项目的上下文中被处理。当采用第三方 IPython 或 bpython 界面时，我们可进一步增强 Python shell。包括语法高亮显示、自动完成等功能。

10.3.4　延伸阅读

- “使用数据库查询表达式”示例。
- “slugify()函数的猴子补丁以获得更好的国际支持”示例。

10.4　使用数据库查询表达式

Django 对象-关系映射（Object-Relational Mapping，ORM）包含了特定的抽象结构，可用于构建复杂的数据库查询。这被称作查询表达式，可用于过滤数据、排序数据、注解新的列和聚合关系。在当前示例中，我们将探讨查询表达式的应用方式。其间，我们将创建一个应用程序，显示 viral videos 并计算每部视频被匿名用户或登录用户的观看次数。

10.4.1　准备工作

利用 ViralVideo 模型创建一个 viral_videos 应用程序，并设置系统以便系统默认登录到日志文件。

创建 viral_videos 应用程序并将其添加至设置项的 INSTALLED_APPS 下。

```
# myproject/settings/_base.py
INSTALLED_APPS = [
    # …
    "myproject.apps.core",
    "myproject.apps.viral_videos",
]
```

接下来创建一个以 UUID 为主键的 viral videos 模型，以及创建和修改的时间戳、标题、嵌入代码、匿名用户观看次数和登录用户观看次数。

```
# myproject/apps/viral_videos/models.py
import uuid
from django.db import models
from django.utils.translation import ugettext_lazy as _

from myproject.apps.core.models import (
    CreationModificationDateBase,
    UrlBase,
)

class ViralVideo(CreationModificationDateBase, UrlBase):
    uuid = models.UUIDField(primary_key=True,default=None,editable=False)
    title = models.CharField(_("Title"), max_length=200, blank=True)
    embed_code = models.TextField(_("YouTube embed code"), blank=True)
    anonymous_views = models.PositiveIntegerField(_("Anonymous
     impressions"), default=0)
    authenticated_views = models.PositiveIntegerField(
        _("Authenticated impressions"), default=0
    )

    class Meta:
        verbose_name = _("Viral video")
        verbose_name_plural = _("Viral videos")

    def __str__(self):
        return self.title

    def get_url_path(self):
        from django.urls import reverse

        return reverse("viral_videos:viral_video_detail",
```

```
        kwargs={"pk": self.pk})

    def save(self, *args, **kwargs):
        if self.pk is None:
            self.pk = uuid.uuid4()
        super().save(*args, **kwargs)
```

为新的应用程序生成并运行迁移，以便数据库处于就绪状态。

```
(env)$ python manage.py makemigrations
(env)$ python manage.py migrate
```

向设置项添加登录配置。

```
LOGGING = {
    "version": 1,
    "disable_existing_loggers": False,
    "handlers": {
        "file": {
            "level": "DEBUG",
            "class": "logging.FileHandler",
            "filename": os.path.join(BASE_DIR, "tmp", "debug.log"),
        }
    },
    "loggers": {"django": {"handlers": ["file"], "level": "DEBUG",
     "propagate": True}},
}
```

这将把调试信息记录至名为 tmp/debug.log 的临时文件中。

10.4.2　实现方式

当展示查询表达式时，下面创建一个 viral video 详细视图，并将其置于 URL 配置中。

（1）在 views.py 文件中创建 viral video 列表和详细视图。

```
# myproject/apps/viral_videos/views.py
import logging

from django.conf import settings
from django.db import models
from django.utils.timezone import now, timedelta
from django.shortcuts import render, get_object_or_404
from django.views.generic import ListView

from .models import ViralVideo
```

```
POPULAR_FROM = getattr(settings, "VIRAL_VIDEOS_POPULAR_FROM", 500)

logger = logging.getLogger(__name__)

class ViralVideoList(ListView):
    template_name = "viral_videos/viral_video_list.html"
    model = ViralVideo

def viral_video_detail(request, pk):
    yesterday = now() - timedelta(days=1)

    qs = ViralVideo.objects.annotate(
        total_views=models.F("authenticated_views") +
         models.F("anonymous_views"),
        label=models.Case(
            models.When(total_views__gt=POPULAR_FROM,
             then=models.Value("popular")),
            models.When(created__gt=yesterday,
             then=models.Value("new")),
            default=models.Value("cool"),
            output_field=models.CharField(),
        ),
    )

    # DEBUG: check the SQL query that Django ORM generates
    logger.debug(f"Query: {qs.query}")

    qs = qs.filter(pk=pk)
    if request.user.is_authenticated:
        qs.update(authenticated_views=models.F("authenticated_views") + 1)
    else:
        qs.update(anonymous_views=models.F("anonymous_views") + 1)

    video = get_object_or_404(qs)

    return render(request, "viral_videos/viral_video_detail.html",
     {"video": video})
```

（2）定义应用程序的URL配置，如下所示。

```
# myproject/apps/viral_videos/urls.py
from django.urls import path
```

```
from .views import ViralVideoList, viral_video_detail

app_name = "viral_videos"

urlpatterns = [
    path("", ViralVideoList.as_view(), name="viral_video_list"),
    path("<uuid:pk>/", viral_video_detail,
     name="viral_video_detail"),
]
```

（3）将应用程序的 URL 包含至项目的根 URL 配置中，如下所示。

```
# myproject/urls.py
from django.conf.urls.i18n import i18n_patterns
from django.urls import include, path

urlpatterns = i18n_patterns(
path("viral-videos/", include("myproject.apps.viral_videos.urls",
namespace="viral_videos")),
)
```

（4）创建 viral video 列表视图的模板，如下所示。

```
{# viral_videos/viral_video_list.html #}
{% extends "base.html" %}
{% load i18n %}

{% block content %}
    <h1>{% trans "Viral Videos" %}</h1>
    <ul>
        {% for video in object_list %}
            <li><a href="{{ video.get_url_path }}">
             {{ video.title }}</a></li>
        {% endfor %}
    </ul>
{% endblock %}
```

（5）针对视频详细视图创建模板，如下所示。

```
{# viral_videos/viral_video_detail.html #}
{% extends "base.html" %}
{% load i18n %}

{% block content %}
    <h1>{{ video.title }}
```

```
        <span class="badge">{{ video.label }}</span>
    </h1>
    <div>{{ video.embed_code|safe }}</div>
    <div>
        <h2>{% trans "Impressions" %}</h2>
        <ul>
            <li>{% trans "Authenticated views" %}:
                {{ video.authenticated_views }}
            </li>
            <li>{% trans "Anonymous views" %}:
                {{ video.anonymous_views }}
            </li>
            <li>{% trans "Total views" %}:
                {{ video.total_views }}
            </li>
        </ul>
    </div>
{% endblock %}
```

（6）设置 viral_videos 应用程序的管理，如下所示，并在结束时向数据库中添加一些视频。

```
# myproject/apps/viral_videos/admin.py
from django.contrib import admin
from .models import ViralVideo

@admin.register(ViralVideo)
class ViralVideoAdmin(admin.ModelAdmin):
    list_display = ["title", "created", "modified"]
```

10.4.3 工作方式

查看视图中的 logger.debug()语句。当在 DEBUG 模式下运行服务器，并在浏览器中访问一个视频（如本地开发中的 http://127.0.0.1:8000/en/viral-videos/2b14ffd3-d1f1-4699-a07b-1328421d8312/），将会在日志（tmp/debug.log）中输出下列 SQL 查询。

```
SELECT "viral_videos_viralvideo"."created",
"viral_videos_viralvideo"."modified", "viral_videos_viralvideo"."uuid",
"viral_videos_viralvideo"."title",
"viral_videos_viralvideo"."embed_code",
"viral_videos_viralvideo"."anonymous_views",
"viral_videos_viralvideo"."authenticated_views",
("viral_videos_viralvideo"."authenticated_views" +
```

```
"viral_videos_viralvideo"."anonymous_views") AS "total_views", CASE WHEN
("viral_videos_viralvideo"."authenticated_views" +
"viral_videos_viralvideo"."anonymous_views") > 500 THEN 'popular' WHEN
"viral_videos_viralvideo"."created" >
'2019-12-21T05:01:58.775441+00:00'::timestamptz THEN 'new' ELSE 'cool' END
AS "label" FROM "viral_videos_viralvideo" WHERE
"viral_videos_viralvideo"."uuid" = '2b14ffd3-d1f1-4699-
a07b-1328421d8312'::uuid LIMIT 21; args=(500, 'popular',
datetime.datetime(2019, 12, 21, 5, 1, 58, 775441, tzinfo=<UTC>), 'new',
'cool', UUID('2b14ffd3-d1f1-4699-a07b-1328421d8312'))
```

接下来，在浏览器中，我们将会看到下列内容的一个简单的页面。

- 视频的标题。
- 视频的标记。
- 嵌入的视频。
- 来自匿名用户和登录用户的视图数量，以及视图的总计数量。

对应结果如图 10.10 所示。

图 10.10

Django QuerySet 中的 annotate()方法可在 SELECT SQL 语句中添加额外的列，以及从 QuerySet 中检索到的对象动态创建的属性。当采用 models.F()时，我们可从所选的数据库的表中引用不同的字段值。在当前示例中，我们创建了 total_views 属性，该属性为来自登录用户和匿名用户的视图总量。

通过 models.Case()和 models.When()，我们可根据不同的条件返回值。当标记值时，这里采用了 models.Value()。在当前示例中，我们创建了 SQL 查询的 label 列，以及 QuerySet

返回对象的属性。如果超出 500 次观看，则设置为 popular；如果在 24 小时内创建，则设置为 new；否则设置为 cool。

在视图结尾处，我们调用了 qs.update()方法，并根据观看视频的用户是否登录递增当前视频的 authenticated_views 或 anonymous_views。这里，递增不会出现于 Python 级别，而是出现于 SQL 级别。这解决了所谓的竞争条件问题，其中，两名或多名访问者同时访问视图，并尝试同步递增视图计数。

10.4.4　延伸阅读

- “使用 Django shell”示例。
- 第 2 章中的“利用与 URL 相关的方法创建一个模型混入”示例。
- 第 2 章中的“创建一个模型混入以处理日期的创建和修改”示例。

10.5　slugify()函数的猴子补丁以获得更好的国际支持

猴子补丁（或游击补丁）是一段代码，并在运行期间扩展或调整另一段代码。这里，不建议经常使用猴子补丁。但某些时候，在复杂的第三方模块中，猴子补丁是修复 bug 的唯一可能方式，且无须创建模块的独立分支。另外，猴子补丁可用于准备功能或单元测试，而无须采用复杂和耗时的数据库或文件操作。

在当前示例中，我们将学习如何将默认的 slugify()函数与第三方 transliterate 包中的函数进行交换，这将以更加智能的方式处理 Unicode 字符与 ASCII 字符之间的转换，并包含许多根据需要提供特定转换的语言包。作为提示，我们使用 slugfy()实用程序创建对象标题或上传文件名的 URL 友好版本。在处理过程中，该函数将去除开始和结尾的空格，将文本转换为小写格式，删除非字母数字字符，并将空格转换为连字符。

10.5.1　准备工作

下列步骤实现了相应的准备工作。

（1）在虚拟环境中安装 transliterate，如下所示。

```
(env)$ pip install transliterate==1.10.2
```

（2）在项目中创建 guerrilla_patches 应用程序，并将其置于设置项的 INSTALLED_APPS 下。

10.5.2　实现方式

在 guerrilla_patches 应用程序的 models.py 文件中，利用 transliterate 包中的函数覆写 django.utils.text 中的 slugify 函数。

```
# myproject/apps/guerrilla_patches/models.py
from django.utils import text
from transliterate import slugify

text.slugify = slugify
```

Django 默认的 slugfy()函数会错误地处理德语中的注音符号。对此，可尝试将一个较长的德语单词与德语注音符号混合。首先在缺少猴子补丁的情况下运行下列代码。

```
(env)$ python manage.py shell
>>> from django.utils.text import slugify
>>> slugify("Heizörückstoßbdäpfung")
'heizolruckstoabdampfung'
```

这在德语中是错误的。其中，字母 ß 被完全去掉，而不是被替换为 ss，字母 ä、ö 和 ü 被更改为 a、o 和 u，它们应该被替换为 ae、oe 和 ue。

我们创建的猴子补丁在初始化阶段加载 django.utils.text 模块，并重新分配 transliteration.slugify 以替代核心 slugify()函数。

当在 Django shell 中运行相同代码时，将会得到正确的结果，如下所示。

```
(env)$ python manage.py shell
>>> from django.utils.text import slugify
>>> slugify("Heizölrückstoßabdämpfung")
'heizoelrueckstossabdaempfung'
```

注意：

关于 transliterate 模块的应用方式，读者可访问 https://pypi.org/project/transliterate。

10.5.3　更多内容

在创建猴子补丁之前，我们应完全理解所修改代码的工作方式。这可通过分析现有代码并查看不同变量值这种方式来实现。对此，存在一个有用的内建 Python 调试器模块，即 pdb。pdb 可临时添加至 Django 代码（或第三方模块）中，并在任何断点处终止开发服务器的执行。相应地，使用以下代码调试 Python 模块中不清晰的部分。

```
breakpoint()
```

这将启用交互式 shell，并于其中输入变量以查看变量值。如果输入 c 或 continue，代码将持续执行直至遇到下一个断点。如果输入 q 或 quit，管理命令将退出。

注意：

关于 Python 调试器和如何查看代码的回溯，读者可访问 https://docs.python.org/3/library/pdb.html 以了解更多内容。

另一种查看开发服务器中变量值的方法是使用变量作为消息并发出警告，如下所示。

```
raise Warning, some_variable
```

在 DEBUG 模式下，Django 日志记录器将提供回溯和其他本地变量信息。

提示：

在将全部工作提交至储存库之前，不要忘记移除调试代码。

当采用 PyCharm 交互式开发环境时，则可通过可视化方式设置断点和调试变量，且不需要修改源代码。

10.5.4　延伸阅读

读者还可参考“使用 Django shell”示例。

10.6　切换调试工具栏

当使用 Django 进行开发时，我们可能希望查看请求头和参数，检查当前模板上下文环境，或者衡量 SQL 查询的性能。所有这些都可以在 Django 调试工具栏的帮助下完成。Django 调试工具栏是一个面板配置集，并显示与当前请求和响应相关的各种调试信息。在当前示例中，我们将讨论如何切换调试工具栏的可见性，这取决于 cookie（其值可由书签小工具设置）。这里，书签小工具是包含一段 JavaScript 代码的书签，可以在浏览器的任何页面上运行。

10.6.1　准备工作

当切换调试工具栏的可见性时，可按下列步骤执行。

（1）在虚拟环境中安装 Django 调试工具栏。

```
(env)$ pip install django-debug-toolbar==2.1
```

（2）将"debug_toolbar"添加至设置项的 INSTALLED_APPS 下。

```
# myproject/settings/_base.py
INSTALLED_APPS = [
    # …
    "debug_toolbar",
]
```

10.6.2 实现方式

下列步骤将设置 Django 调试工具栏，并通过浏览器中的书签小工具开启或关闭。

（1）添加下列项目设置项。

```
# myproject/settings/_base.py
DEBUG_TOOLBAR_CONFIG = {
    "DISABLE_PANELS": [],
    "SHOW_TOOLBAR_CALLBACK":
    "myproject.apps.core.misc.custom_show_toolbar",
    "SHOW_TEMPLATE_CONTEXT": True,
}

DEBUG_TOOLBAR_PANELS = [
    "debug_toolbar.panels.versions.VersionsPanel",
    "debug_toolbar.panels.timer.TimerPanel",
    "debug_toolbar.panels.settings.SettingsPanel",
    "debug_toolbar.panels.headers.HeadersPanel",
    "debug_toolbar.panels.request.RequestPanel",
    "debug_toolbar.panels.sql.SQLPanel",
    "debug_toolbar.panels.templates.TemplatesPanel",
    "debug_toolbar.panels.staticfiles.StaticFilesPanel",
    "debug_toolbar.panels.cache.CachePanel",
    "debug_toolbar.panels.signals.SignalsPanel",
    "debug_toolbar.panels.logging.LoggingPanel",
    "debug_toolbar.panels.redirects.RedirectsPanel",
]
```

（2）在 core 应用程序中，创建包含 custom_show_toolbar()函数的 misc.py 文件，如下所示。

```
# myproject/apps/core/misc.py
def custom_show_toolbar(request):
    return "1" == request.COOKIES.get("DebugToolbar", False)
```

（3）在项目的 urls.py 文件中，添加以下配置规则。

```
# myproject/urls.py
from django.conf.urls.i18n import i18n_patterns
from django.urls import include, path
from django.conf import settings
import debug_toolbar

urlpatterns = i18n_patterns(
    # …
)

urlpatterns = [
    path('__debug__/', include(debug_toolbar.urls)),
] + urlpatterns
```

（4）打开 Chrome 或 Firefox 浏览器访问书签管理器。随后创建两个包含 JavaScript 的新书签。其中，第一个链接显示了工具栏，如图 10.11 所示。

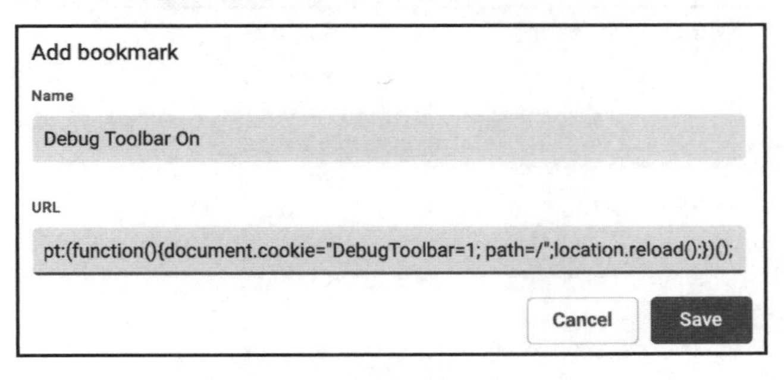

图 10.11

JavaScript 代码如下所示。

```
javascript:(function(){document.cookie="DebugToolbar=1;
path=/";location.reload();})();
```

（5）第二个 JavaScript 链接将隐藏工具栏，如图 10.12 所示。

图 10.12

完整的 JavaScript 代码如下所示。

```
javascript:(function(){document.cookie="DebugToolbar=0;
path=/";location.reload();})();
```

10.6.3　工作方式

DEBUG_TOOLBAR_PANELS 设置项定义了在工具栏中显示的面板。DEBUG_TOOLBAR_CONFIG 字典定义了工具栏的配置，包括一个函数的路径，该函数用于检测是否显示工具栏。

默认状态下，当浏览项目时，并不会显示 Django 调试工具栏。然而，当单击书签小工具 Debug Toolbar On 时，DebugToolbar cookie 将被设置为 1，页面经刷新后将看到包含调试面板的工具栏。如我们将能够查看已优化的 SQL 语句的性能，如图 10.13 所示。

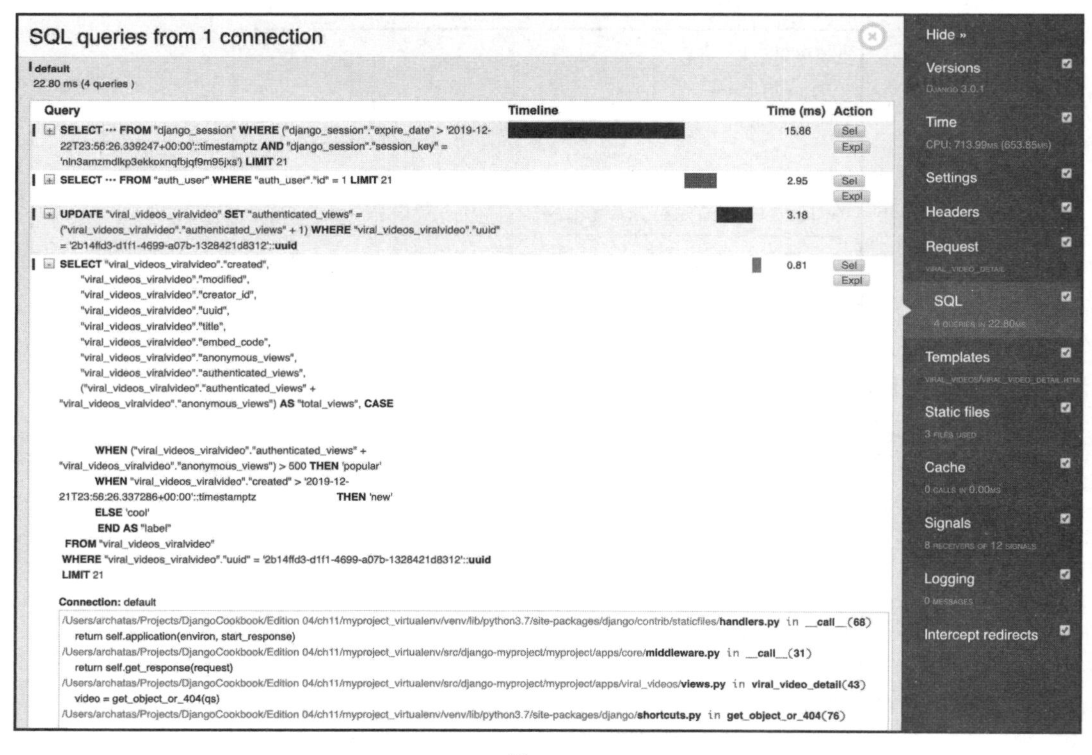

图 10.13

除此之外，我们还可检查当前视图的模板上下文变量，如图 10.14 所示。

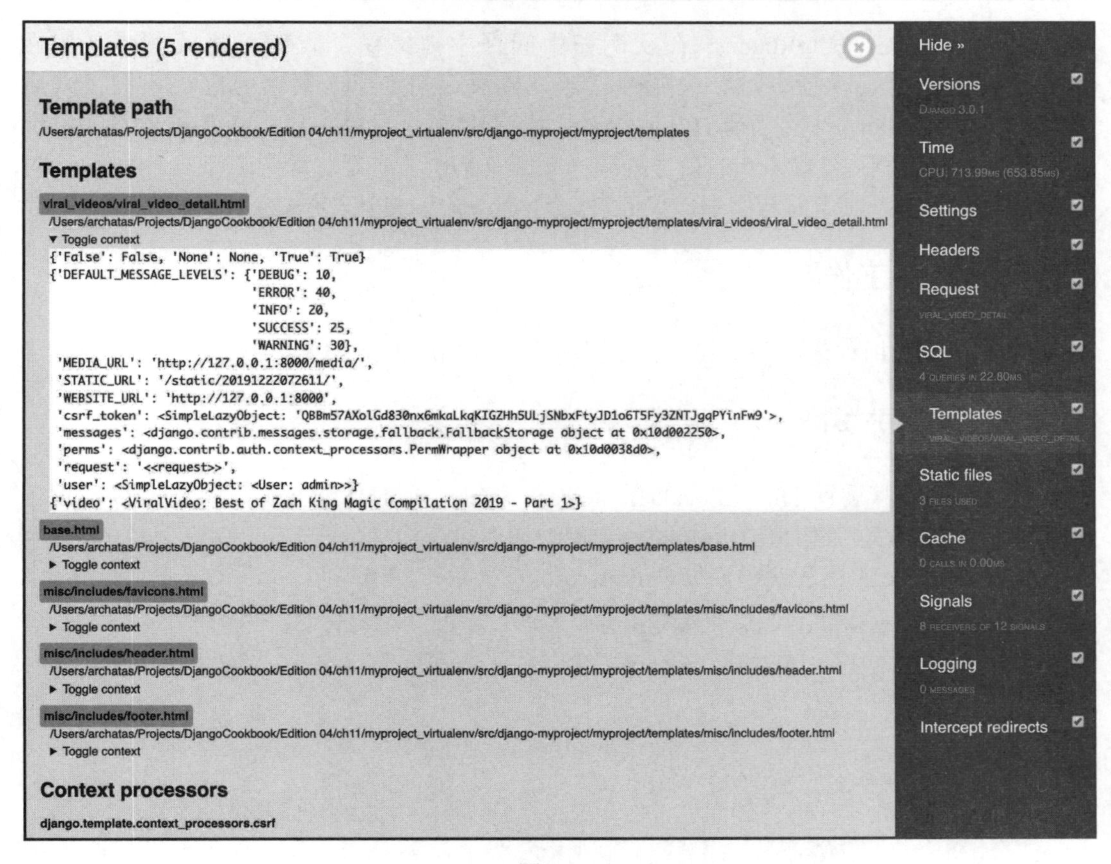

图 10.14

单击第二个书签小工具 Debug Toolbar Off，将 DebugToolbar Cookie 设置为 0。刷新页面后将再次隐藏工具栏。

10.6.4　延伸阅读

读者还可参考第 13 章中的“通过电子邮件获取详细的错误报告”示例。

10.7　使用 ThreadLocalMiddleware

HttpRequest 对象包含了与当前用户、语言、服务器变量、cookie、会话等有用的信息。实际上，HttpRequest 在视图和中间件中被提供，我们可将其（或其属性值）传递至表单、模型方法、模型管理器、模板等中。出于简单考虑，我们可采用 ThreadLocalMiddleware 将当前 HttpRequest 对象存储于全局可访问的 Python 线程中。因此，我们可从模型方法、表单、信号处理程序以及其他之前无法直接访问 HttpRequest 对象的地方访问它。在当前示例中，我们将定义 ThreadLocalMiddleware 中间件。

10.7.1　准备工作

创建 core 应用程序，并将其置于设置项的 INSTALLED_APPS 中。

10.7.2　实现方式

执行下列步骤设置 ThreadLocalMiddleware。ThreadLocalMiddleware 可用于获取项目代码的函数或方法中的当前 HttpRequest 或用户。

（1）向 core 应用程序中添加 middleware.py 文件，如下所示。

```
# myproject/apps/core/middleware.py
from threading import local

_thread_locals = local()

def get_current_request():
    """
    :returns the HttpRequest object for this thread
    """
    return getattr(_thread_locals, "request", None)

def get_current_user():
    """
    :returns the current user if it exists or None otherwise """
    request = get_current_request()
    if request:
        return getattr(request, "user", None)

class ThreadLocalMiddleware(object):
    """
    Middleware to add the HttpRequest to thread local storage
    """

    def __init__(self, get_response):
```

```
        self.get_response = get_response

    def __call__(self, request):
        _thread_locals.request = request
        return self.get_response(request)
```

（2）向设置项中的 MIDDLEWARE 添加 ThreadLocalMiddleware 中间件。

```
# myproject/settings/_base.py
MIDDLEWARE = [
    "django.middleware.security.SecurityMiddleware",
    "django.contrib.sessions.middleware.SessionMiddleware",
    "django.middleware.common.CommonMiddleware",
    "django.middleware.csrf.CsrfViewMiddleware",
    "django.contrib.auth.middleware.AuthenticationMiddleware",
    "django.contrib.messages.middleware.MessageMiddleware",
    "django.middleware.clickjacking.XFrameOptionsMiddleware",
    "django.middleware.locale.LocaleMiddleware",
    "debug_toolbar.middleware.DebugToolbarMiddleware",
    "myproject.apps.core.middleware.ThreadLocalMiddleware",
]
```

10.7.3　工作方式

ThreadLocalMiddleware 处理每个请求，并将当前 HttpRequest 存储于当前线程中。Django 中的每个请求-响应周期均为单一线程。这里，我们创建了两个函数，即 get_current_request()和 get_current_user()，这两个函数可分别获取当前 HttpRequest 对象和读取用户。

如我们可使用 ThreadLocalMiddleware 中间件开发和使用 CreatorMixin，它将当前用户保存为新模型对象的创建者，如下所示。

```
# myproject/apps/core/models.py
from django.conf import settings
from django.db import models
from django.utils.translation import gettext_lazy as _

class CreatorBase(models.Model):
    """
    Abstract base class with a creator
    """

    creator = models.ForeignKey(
        settings.AUTH_USER_MODEL,
```

```
        verbose_name=_("creator"),
        editable=False,
        blank=True,
        null=True,
        on_delete=models.SET_NULL,
    )

    class Meta:
        abstract = True

    def save(self, *args, **kwargs):
        from .middleware import get_current_user

        if not self.creator:
            self.creator = get_current_user()
        super().save(*args, **kwargs)

    save.alters_data = True
```

10.7.4 延伸阅读

- 第 2 章中的“利用与 URL 相关的方法创建一个模型混入”示例。
- 第 2 章中的“创建一个模型混入以处理日期的创建和修改”示例。
- 第 2 章中的“创建一个模型混入以关注元标签”示例。
- 第 2 章中的“创建一个模型混入以处理 Generic Relation”示例。

10.8 使用信号通知管理员有关新的条目

Django 框架中包含了信号这一概念，这类似于 JavaScript 中的事件。相应地，存在一些内建的信号，并以此在模型初始化、保存或删除一个实例、迁移数据库模式、处理请求等前后触发相关动作。不仅如此，我们还可在可复用的应用程序中创建自己的信号，并在其他应用程序中处理这些信号。在当前示例中，我们将学习如何在保存特定模型时使用信号将电子邮件发送至管理员。

10.8.1 准备工作

当前示例将在之前 viral_videos 应用程序示例的基础上完成。

10.8.2　实现方式

下列步骤将创建管理员的通知。

（1）创建 signals.py 文件，如下所示。

```
# myproject/apps/viral_videos/signals.py
from django.db.models.signals import post_save
from django.dispatch import receiver
from django.template.loader import render_to_string

from .models import ViralVideo

@receiver(post_save, sender=ViralVideo)
def inform_administrators(sender, **kwargs):
    from django.core.mail import mail_admins

    instance = kwargs["instance"]
    created = kwargs["created"]

    if created:
        context = {"title": instance.title, "link": instance.get_url()}
        subject = render_to_string(
            "viral_videos/email/administrator/subject.txt", context
        )

        plain_text_message = render_to_string(
            "viral_videos/email/administrator/message.txt", context
        )
        html_message = render_to_string(
            "viral_videos/email/administrator/message.html", context
        )

        mail_admins(
            subject=subject.strip(),
            message=plain_text_message,
            html_message=html_message,
            fail_silently=True,
        )
```

（2）创建模板。首先是电子邮件主题的模板。

```
{# viral_videos/email/administrator/subject.txt #}
```

```
New Viral Video Added
```

（3）创建一个纯文本消息的模板，如下所示。

```
{# viral_videos/email/administrator/message.txt #}
A new viral video called "{{ title }}" has been created.
You can preview it at {{ link }}.
```

（4）创建 HTML 消息模板，如下所示。

```
{# viral_videos/email/administrator/message.html #}
<p>A new viral video called "{{ title }}" has been created.</p>
<p>You can <a href="{{ link }}">preview it here</a>.</p>
```

（5）创建 apps.py 文件，如下所示。

```
# myproject/apps/viral_videos/apps.py
from django.apps import AppConfig
from django.utils.translation import ugettext_lazy as _

class ViralVideosAppConfig(AppConfig):
    name = "myproject.apps.viral_videos"
    verbose_name = _("Viral Videos")

    def ready(self):
        from .signals import inform_administrators
```

（6）更新__init__.py 文件，如下所示。

```
# myproject/apps/viral_videos/__init__.py
default_app_config =
"myproject.apps.viral_videos.apps.ViralVideosAppConfig"
```

确保在项目的设置项中设置了 ADMINS，如下所示。

```
# myproject/settings/_base.py
ADMINS = [("Administrator", "admin@example.com")]
```

10.8.3　工作方式

ViralVideosAppConfig 应用程序配置类包含一个 ready()方法，该方法在全部项目模型载入至内存中后被调用。根据 Django 文档，信号允许特定的发送者通知一组执行某种动作的接收者。因此，在 ready()方法中，我们导入了 inform_administrators()函数。

通过@receiver 装饰器，inform_administrators()针对 post_save 信号而注册，我们将其

限制于仅处理 ViralVideo 模型是 sender 的信号。因此，当保存一个 ViralVideo 对象时，将调用 receiver 函数。inform_administrators()函数检查视频是否为新创建的。若是，则向系统管理员发送一封电子邮件，该系统管理员列于设置项的 ADMINS 中。

我们使用模板生成 subject、plain_text_message 和 html_message 的内容，以便能够在应用程序中定义各自的默认模板。如果使 viral_videos 应用程序公开、有效，那些将其放置到自己项目的用户可根据自己的需要定制模板，也可以把它们封装在公司的电子邮件模板封装器中。

注意：

关于 Django 信号的更多内容，读者可查看官方文档，对应网址为 https://docs.djangoproject.com/en/3.0/topics/signals/。

10.8.4　延伸阅读

- 第 1 章中的“创建应用程序配置”示例。
- “使用数据库查询表达式”示例。
- “检查缺失设置项”示例。

10.9　检查缺失设置项

自 Django 1.7 起，我们可使用一个可扩展的系统检查框架，该框架替代了早期的 validate 管理命令。在当前示例中，我们将学习如何在设置了 ADMINS 设置项后创建一项检查。类似地，我们将能够检查不同的密钥或访问令牌是否已针对所用的 API 进行设置。

10.9.1　准备工作

当前示例将在之前 viral_videos 应用程序的基础上完成。

10.9.2　实现方式

下列步骤将使用系统检查框架。

（1）创建 checks.py 文件，如下所示。

```
# myproject/apps/viral_videos/checks.py
from textwrap import dedent
```

```
from django.core.checks import Warning, register, Tags

@register(Tags.compatibility)
def settings_check(app_configs, **kwargs):
    from django.conf import settings

    errors = []

    if not settings.ADMINS:
        errors.append(
            Warning(
                dedent("""
                    The system admins are not set in the project settings
                """),
                obj=settings,
                hint=dedent("""
                    In order to receive notifications when new
                    videos are created, define system admins
                    in your settings, like:
                    ADMINS = (
                        ("Admin", "administrator@example.com"),
                    )
                """),
                id="viral_videos.W001",
            )
        )

    return errors
```

（2）导入应用程序配置的 ready()方法中的检查，如下所示。

```
# myproject/apps/viral_videos/apps.py
from django.apps import AppConfig
from django.utils.translation import ugettext_lazy as _

class ViralVideosAppConfig(AppConfig):
    name = "myproject.apps.viral_videos"
    verbose_name = _("Viral Videos")

    def ready(self):
        from .signals import inform_administrators
        from .checks import settings_check
```

（3）尝试使用刚刚创建的检查，移除或注释出 ADMINS 设置项，随后在虚拟环境中运行 check 管理命令。

```
(env)$ python manage.py check
System check identified some issues:

WARNINGS:
<Settings "myproject.settings.dev">: (viral_videos.W001)
The system admins are not set in the project settings

HINT:
In order to receive notifications when new videos are
created, define system admins in your settings, like:

ADMINS = (
    ("Admin", "administrator@example.com"),
)

System check identified 1 issue (0 silenced).
```

10.9.3 工作方式

系统检查框架在模型、字段、数据库、管理验证配置、内容类型和安全设置中包含了一些检查，如果项目中的某些内容未正确设置，检查将会生成错误或警告信息。除此之外，我们还可创建自己的检查。

这里，我们注册了 settings_check()函数，如果不存在针对项目定义的 ADMINS 设置项，该函数将返回一个 Warning 列表。

除了 django.core.checks 模块中的 Warning 实例，返回的列表还可能包含 Debug、Info、Error 和 Critical 内建类实例，或者继承自 django.core.checks.CheckMessage 的其他类实例。调试、信息和警告级别的日志将呈现静默方式的失败，而错误和严重级别的日志将阻止项目的运行。

在当前示例中，检查通过传递至@register 装饰器的 Tags.compatibility 参数标记为兼容型检查。Tags 中提供的其他选项还包括：

- admin 用于与管理站点相关的检查。
- caches 用于与服务器缓存相关的检查。
- database 用于与数据库配置相关的检查。
- models 用于与模型、模型字段和管理器相关的检查。

- security 用于与安全相关的检查。
- signals 用于与信号声明和处理程序相关的检查。
- staticfiles 用于静态文件检查。
- templates 用于与模板相关的检查。
- translation 用于与字符串翻译相关的检查。
- url 用于与 URL 配置相关的检查。

注意：

关于系统检查的更多信息，读者可查看官方文档，对应网址为 https://docs.djangoproject.com/en/3.0/topics/checks/。

10.9.4 延伸阅读

- 第 1 章中的“创建应用程序配置”示例。
- “使用数据库查询表达式”示例。
- “使用信号通知管理员有关新的条目”示例。

第 11 章　测　试

本章主要涉及下列主题。

- 利用 Mock 测试视图。
- 利用 Selenium 测试用户界面。
- 利用 Django REST 框架测试 API。
- 确保测试覆盖范围。

11.1　简　介

为了确保代码的质量和准确性，应实现自动化软件测试。Django 提供了相应的工具可编写测试网站套件。测试套件自动检查网站及其组件，以确保一切工作正常。当调整代码时，我们可运行测试以检查变化是否对应用程序的行为带来负面影响。

自动软件测试涵盖了较为广泛的内容和术语。就本书而言，我们将测试划分为以下几类。

（1）单元测试是指严格针对代码的个体部分或单元的测试。大多数时候，单元对应于一个文件或模块，而单元测试则验证逻辑和行为是否按照期望方式进行。

（2）集成测试则更进一步处理两个或多个单元之间的彼此协同工作方式。其粒度与单元测试有所不同，通常假设在验证集成测试时，所有的单元测试均已通过。因此，集成测试仅覆盖了一组行为，这些行为对于单元之间的协同工作必须是真实可行的。

（3）组件接口测试则是集成测试的高阶形式，其中单个组件从一端到另一端进行验证。此类测试将忽略底层逻辑（用于提供组件行为），因而逻辑可在不调整行为的情况下修改，测试仍然会通过。

（4）系统测试验证构成系统的全部组件的端到端集成，通常与完整的用户流相对应。

（5）运行接受度测试检查系统所有非功能方面的内容是否正确运行。接受度测试检查业务逻辑，并从最终用户的角度来看，项目是否按照预期的方式工作。

11.2　需求条件

当与本章代码协同工作时，需要安装最新稳定版本的 Python、MySQL 或 PostgreSQL

数据库，以及虚拟环境下的 Django 项目。

读者可访问 GitHub 储存库的 ch11 目录查看本章的全部代码，对应网址为 https://github.com/PacktPublishing/Django-3-Web-Development-Cookbook-Fourth-Edition。

11.3　利用 Mock 测试视图

在当前示例中，我们将探讨如何编写单元测试。单元测试检查单个函数或方法是否返回正确的结果。这里，我们将探讨 likes 应用程序，并编写测试检查发送至 json_set_like() 视图是否会对未经身份验证的用户返回失败的响应，并对经过身份验证的用户返回成功的结果。其间，我们将使用 Mock 对象模拟 HttpRequest 和 AnonymousUser 对象。

11.3.1　准备工作

当前示例将在 Locations 和 likes 应用程序的基础上完成。

随后将使用 mock 库，该库自 Python 3.3 以来一直作为内建的 unittest.mock 使用。

11.3.2　实现方式

执行以下步骤，使用 mock 来测试 liking 动作。

（1）在 likes 应用程序中创建 tests 模块。

（2）在该模块中，创建 test_views.py 文件，如下所示。

```
# myproject/apps/likes/tests/test_views.py
import json
from unittest import mock
from django.contrib.auth.models import User
from django.contrib.contenttypes.models import ContentType
from django.test import TestCase
from myproject.apps.locations.models import Location

class JSSetLikeViewTest(TestCase):
    @classmethod
    def setUpClass(cls):
        super(JSSetLikeViewTest, cls).setUpClass()

        cls.location = Location.objects.create(
            name="Park Güell",
```

```
            description="If you want to see something spectacular,
            come to Barcelona, Catalonia, Spain and visit Park
            Güell. Located on a hill, Park Güell is a public
            park with beautiful gardens and organic
            architectural elements.",
            picture="locations/2020/01/20200101012345.jpg",
            # dummy path
        )
        cls.content_type = ContentType.objects.get_for_model(Location)
        cls.superuser = User.objects.create_superuser(
            username="admin", password="admin",
             email="admin@example.com"
        )

    @classmethod
    def tearDownClass(cls):
        super(JSSetLikeViewTest, cls).tearDownClass()
        cls.location.delete()
        cls.superuser.delete()

    def test_authenticated_json_set_like(self):
        from ..views import json_set_like

        mock_request = mock.Mock()
        mock_request.user = self.superuser
        mock_request.method = "POST"

        response = json_set_like(mock_request,
         self.content_type.pk, self.location.pk)
        expected_result = json.dumps(
            {"success": True, "action": "add", "count":
             Location.objects.count()}
        )
        self.assertJSONEqual(response.content, expected_result)

    @mock.patch("django.contrib.auth.models.User")
    def test_anonymous_json_set_like(self, MockUser):
        from ..views import json_set_like

        anonymous_user = MockUser()
        anonymous_user.is_authenticated = False
```

```
        mock_request = mock.Mock()
        mock_request.user = anonymous_user
        mock_request.method = "POST"

        response = json_set_like(mock_request,
        self.content_type.pk, self.location.pk)
        expected_result = json.dumps({"success": False})
        self.assertJSONEqual(response.content, expected_result)
```

（3）针对 likes 应用程序进行测试，如下所示。

```
(env)$ python manage.py test myproject.apps.likes --
settings=myproject.settings.test
Creating test database for alias 'default'...
System check identified no issues (0 silenced).
..
-----------------------------------------------------------------
---
Ran 2 tests in 0.268s
OK
Destroying test database for alias 'default'...
```

11.3.3　工作方式

当针对 likes 应用程序运行测试时，首先需要创建一个临时的测试数据库，随后调用 setUpClass()方法。接下来执行名称始于 test 的方法，最后调用 tearDownClass()方法。对于每个通过的测试，我们可以在命令行工具中看到一个“.”；对于每个失败的测试，将会看到字母 F；而对于测试中的每个错误，则会看到字母 E。最终，我们将会看到与失败和错误测试相关的提示信息。因为当前在 likes 应用程序的套件中仅包含两个测试，因而可在结果中看到两个“.”。

在 setUpClass()中，我们创建了一个位置和一个超级用户。此外，我们还针对 Location 模型查找 ContentType 对象，以供 json_set_like()视图使用，该视图用于设置和移除不同对象的点赞行为。注意，该视图如下所示，并返回一个 JSON 字符串作为结果。

```
def json_set_like(request, content_type_id, object_id):
    # all the view logic goes here…
    return JsonResponse(result)
```

在 test_authenticated_json_set_like()和 test_anonymous_json_set_like()方法中，我们使用了 Mock 对象，此类对象可包含任意属性或方法。Mock 对象的每个未定义的属性或方

法表示为另一个 Mock 对象。因此，在 shell 中，我们可尝试链接属性，如下所示。

```
>>> from unittest import mock
>>> m = mock.Mock()
>>> m.whatever.anything().whatsoever
<Mock name='mock.whatever.anything().whatsoever' id='4320988368'>
```

在当前测试中，我们使用 Mock 对象模拟 HttpRequest 对象。对于匿名用户，MockUser 是通过@mock.patch()装饰器作为标准 Django User 对象的补丁生成的。对于经过身份验证的用户，我们仍然需要真正的 User 对象，因为视图将用户的 ID 用于 Like 对象。

因此，我们调用 json_set_like()函数，并检查返回的 JSON 响应结果是否正确。

- 如果访问者未经过身份验证，则在响应结果中返回{"success": false}。
- 对于经过身份验证的用户，则返回{"action": "add", "count": 1, "success":true}。

最后调用 tearDownClass()类方法，并从测试数据库中删除位置和超级用户。

11.3.4　更多内容

当采用 HttpRequest 对象进行测试时，还可使用 Django Request Factory，读者可访问 https://docs.djangoproject.com/en/3.0/topics/testing/advanced/#the-request-factory 查看其应用方式。

11.3.5　延伸阅读

- 第 4 章中的“实现 Like 微件”示例。
- “利用 Selenium 测试用户界面”示例。
- “利用 Django REST 框架测试 API”示例。
- “确保测试覆盖率”示例。

11.4　利用 Selenium 测试用户界面

运行可接受测试检查业务逻辑，进而查看项目是否以期望方式工作。在当前示例中，我们将学习如何通过 Selenium 编写可接受度测试，进而模拟前端行为，如填写表单或单击浏览器中特定的 DOM 元素。

11.4.1 准备工作

当前示例将在之前的 locations 和 likes 应用程序的基础上完成。

针对当前示例，我们将使用 Selenium 库、Chrome 浏览器和 ChromeDriver 进行控制。

（1）访问 https://www.google.com/chrome/下载并安装 Chrome 浏览器。

（2）在 Django 项目中创建 drivers 目录。访问 https://sites.google.com/a/chromium.org/chromedriver/下载最终稳定版本的 ChromeDriver，解压后将其置于新创建的 drivers 目录中。

（3）在虚拟环境中安装 Selenium，如下所示。

```
(env)$ pip install selenium
```

11.4.2 实现方式

下列步骤将利用 Selenium 测试基于 Ajax 的点赞功能。

（1）在项目设置项中，添加 TESTS_SHOW_BROWSER 设置项。

```
# myproject/settings/_base.py
TESTS_SHOW_BROWSER = True
```

（2）在 locations 应用程序中创建 tests 模块，并向其中添加一个 test_frontend.py 文件，如下所示。

```
# myproject/apps/locations/tests/test_frontend.py
import os
from io import BytesIO
from time import sleep

from django.core.files.storage import default_storage
from django.test import LiveServerTestCase
from django.contrib.contenttypes.models import ContentType
from django.contrib.auth.models import User
from django.conf import settings
from django.test import override_settings
from django.urls import reverse
from selenium import webdriver
from selenium.webdriver.chrome.options import Options
from selenium.webdriver.support.ui import WebDriverWait
from myproject.apps.likes.models import Like
```

```
from ..models import Location

SHOW_BROWSER = getattr(settings, "TESTS_SHOW_BROWSER", False)

@override_settings(DEBUG=True)
class LiveLocationTest(LiveServerTestCase):
    @classmethod
    def setUpClass(cls):
        super(LiveLocationTest, cls).setUpClass()
        driver_path = os.path.join(settings.BASE_DIR, "drivers",
        "chromedriver")
        chrome_options = Options()
        if not SHOW_BROWSER:
            chrome_options.add_argument("--headless")
        chrome_options.add_argument("--window-size=1200,800")

        cls.browser = webdriver.Chrome(
            executable_path=driver_path, options=chrome_options
        )
        cls.browser.delete_all_cookies()

        image_path = cls.save_test_image("test.jpg")
        cls.location = Location.objects.create(
            name="Park Güell",
            description="If you want to see something spectacular,
             come to Barcelona, Catalonia, Spain and visit Park
             Güell. Located on a hill, Park Güell is a public
             park with beautiful gardens and organic
             architectural elements.",
            picture=image_path, # dummy path
        )
        cls.username = "admin"
        cls.password = "admin"
        cls.superuser = User.objects.create_superuser(
            username=cls.username, password=cls.password,
             email="admin@example.com"
        )

    @classmethod
    def tearDownClass(cls):
        super(LiveLocationTest, cls).tearDownClass()
        cls.browser.quit()
```

```
        cls.location.delete()
        cls.superuser.delete()

    @classmethod
    def save_test_image(cls, filename):
        from PIL import Image

        image = Image.new("RGB", (1, 1), 0)
        image_buffer = BytesIO()
        image.save(image_buffer, format="JPEG")

        path = f"tests/{filename}"
        default_storage.save(path, image_buffer)
        return path

    def wait_a_little(self):
        if SHOW_BROWSER:
            sleep(2)

    def test_login_and_like(self):
        # login
        login_path = reverse("admin:login")
        self.browser.get(
            f"{self.live_server_url}{login_path}?next=
            {self.location.get_url_path()}"
        )
        username_field =
        self.browser.find_element_by_id("id_username")
        username_field.send_keys(self.username)
        password_field =
        self.browser.find_element_by_id("id_password")
        password_field.send_keys(self.password)
        self.browser.find_element_by_css_selector
        ('input[type="submit"]').click()
        WebDriverWait(self.browser, timeout=10).until(
            lambda x:
        self.browser.find_element_by_css_selector(".like-button")
        )
        # click on the "like" button
        like_button =
        self.browser.find_element_by_css_selector(".like-button")
        is_initially_active = "active" in
```

```
like_button.get_attribute("class")
initial_likes = int(
    self.browser.find_element_by_css_selector
     (".like-badge").text
)

self.assertFalse(is_initially_active)
self.assertEqual(initial_likes, 0)

self.wait_a_little()

like_button.click()
WebDriverWait(self.browser, timeout=10).until(
    lambda x:
    int(self.browser.find_element_by_css_selector
     (".like-badge").text) != initial_likes
)

likes_in_html = int(
    self.browser.find_element_by_css_selector
     (".like-badge").text
)
likes_in_db = Like.objects.filter(
content_type=ContentType.objects.get_for_model(Location),
    object_id=self.location.pk,
).count()
self.assertEqual(likes_in_html, 1)
self.assertEqual(likes_in_html, likes_in_db)

self.wait_a_little()

self.assertGreater(likes_in_html, initial_likes)

# click on the "like" button again to switch back to the
# previous state
like_button.click()
WebDriverWait(self.browser, timeout=10).until(
    lambda x: int(self.browser.find_element_by_css_selector
     (".like-badge").text) == initial_likes
)

self.wait_a_little()
```

（3）运行 locations 应用程序的测试，如下所示。

```
(env)$ python manage.py test myproject.apps.locations --
settings=myproject.settings.test
Creating test database for alias 'default'...
System check identified no issues (0 silenced).
.
----------------------------------------------------------------------
---
Ran 1 test in 4.284s

OK
Destroying test database for alias 'default'...
```

11.4.3　工作方式

当运行上述测试时，我们将看到一个打开的 Chrome 窗口，在 URL（对应 http://localhost:63807/en/admin/login/?next=/en/locations/176255a9-9c07-4542-8324-83ac0d21b7c3/）下管理登录界面。

用户名和密码字段将通过 admin 填写，用户将重定向至 Park Güell 位置的详细页面，如 http://localhost:63807/en/locations/176255a9-9c07-4542-8324-83ac0d21b7c3/。其中，我们可以看到 Like 按钮被单击两次，分别表示点赞和取消点赞行为。

如果将 TESTS_SHOW_BROWSER 设置为 False（或全部移除它）并再次运行测试，测试将以最小等待时间，并且于后台（不会打开浏览器的窗口）进行。

下面探讨在测试套件中的工作方式。我们定义了一个扩展 LiveServerTestCase 的类，该类创建一个测试套件并在一个随机未使用端口（如 63807）下运行本地服务器。默认情况下，LiveServerTestCase 以非调试模式运行一个服务器。但是我们利用 override_settings() 装饰器将其切换至调试模式，以使静态文件可访问且无须收集这些文件，并在任何页面上发生错误时显示错误回溯。setUpClass()类方法将在所有测试开始时执行，tearDownClass()类方法则在测试运行之后执行。在中间阶段，测试执行名称始于 test 的所有套件方法。

当开始测试时，将创建一个新的测试数据库。在 setUpClass()中，我们创建了一个浏览器对象、一个位置和一个超级用户。随后，test_login_and_like()方法将被执行，该方法打开管理登录页面、查找用户名字段、输入管理员用户名、查找密码字段、输入管理员密码、查找提交按钮并单击该按钮。接下来，该方法最多等待 10 秒，直至页面上查找到带有.like-button CSS 类的 DOM 元素。

在第 4 章曾有所介绍，微件由两个元素构成。

（1）一个 Like 按钮。

（2）一个显示全部点赞数量的标记。

如果单击按钮，Like 实例将通过 Ajax 调用添加至数据库中，或从数据库中被删除。而且，标记的计数将被更新，以反映数据库中点赞的数量。

进一步的测试中，我们检查了按钮的初始状态（是否包含.active CSS 类）、检查了初始的点赞数量，并模拟了按钮上的单击操作。我们最多等待 10 秒，直至标记中的计数发生变化。随后，我们检查标记中的计数是否匹配于数据库中位置的总点赞数。除此之外，我们还将检查标记中计数的变化方式（递增）。最后，我们再次模拟按钮上的单击操作，以切换到上一个状态。

随后调用 tearDownClass()方法，这将关闭浏览器并从测试数据库中移除位置和超级用户。

11.4.4　延伸阅读

- 第 4 章中的“实现 Like 微件”示例。
- “利用 Mock 测试视图”示例。
- “利用 Django REST 框架测试 API”示例。
- “确保测试覆盖率”示例。

11.5　利用 Django REST 框架测试 API

截至目前，相信读者已经理解了如何编写单元测试和运行可接受度测试。在当前示例中，我们将讨论 RESTful API 的组件接口测试。

如果读者尚不了解 RESTful API 以及 API 的应用方式，可先期访问 http://www.restapitutorial.com/以查看相关内容。

11.5.1　准备工作

当前示例将在第 9 章 music 应用程序的基础上完成。

11.5.2　实现方式

下列步骤将测试 RESTful API。

（1）在 music 应用程序中创建 tests 模块。在 tests 模块中，利用 SongTests 类创建一个 test_api.py 文件。该类包含 setUpClass()和 tearDownClass()方法，如下所示。

```
# myproject/apps/music/tests/test_api.py
from django.contrib.auth.models import User
from django.urls import reverse

from rest_framework import status
from rest_framework.test import APITestCase
from ..models import Song

class SongTests(APITestCase):
    @classmethod
    def setUpClass(cls):
        super().setUpClass()

        cls.superuser = User.objects.create_superuser(
            username="admin", password="admin",
             email="admin@example.com"
        )

        cls.song = Song.objects.create(
            artist="Lana Del Rey",
            title="Video Games - Remastered",
            url="https://open.spotify.com/track/5UOo694cVvj
             cPFqLFiNWGU?si=maZ7JCJ7Rb6WzESLXg1Gdw",
        )

        cls.song_to_delete = Song.objects.create(
            artist="Milky Chance",
            title="Stolen Dance",
            url="https://open.spotify.com/track/3miMZ2IlJ
             iaeSWo1DohXlN?si=g-xMM4m9S_yScOm02C2MLQ",
        )

    @classmethod
    def tearDownClass(cls):
        super().tearDownClass()

        cls.song.delete()
        cls.superuser.delete()
```

（2）添加一个检查列表歌曲的 API 测试。

```
def test_list_songs(self):
    url = reverse("rest_song_list")
    data = {}
    response = self.client.get(url, data, format="json")

    self.assertEqual(response.status_code, status.HTTP_200_OK)
    self.assertEqual(response.data["count"], Song.objects.count())
```

（3）添加一个检查单首歌曲详细信息的 API 测试。

```
def test_get_song(self):
    url = reverse("rest_song_detail", kwargs={"pk": self.song.pk})
    data = {}
    response = self.client.get(url, data, format="json")

    self.assertEqual(response.status_code, status.HTTP_200_OK)
    self.assertEqual(response.data["uuid"], str(self.song.pk))
    self.assertEqual(response.data["artist"], self.song.artist)
    self.assertEqual(response.data["title"], self.song.title)
    self.assertEqual(response.data["url"], self.song.url)
```

（4）添加一个检查成功创建一首新歌曲的 API 测试。

```
def test_create_song_allowed(self):
        # login
        self.client.force_authenticate(user=self.superuser)

        url = reverse("rest_song_list")
        data = {
            "artist": "Capital Cities",
            "title": "Safe And Sound",
            "url": "https://open.spotify.com/track/40Fs0YrUGu
                wLNQSaHGVfqT?si=2OUawusIT-evyZKonT5GgQ",
        }
        response = self.client.post(url, data, format="json")

        self.assertEqual(response.status_code, status.HTTP_201_CREATED)

        song = Song.objects.filter(pk=response.data["uuid"])
        self.assertEqual(song.count(), 1)

        # logout
        self.client.force_authenticate(user=None)
```

（5）添加一个测试，该测试尝试在未经验证和失败的情况下创建一首歌曲，如下所示。

```
def test_create_song_restricted(self):
        # make sure the user is logged out
        self.client.force_authenticate(user=None)

        url = reverse("rest_song_list")
        data = {
            "artist": "Men I Trust",
            "title": "Tailwhip",
            "url": "https://open.spotify.com/track/2DoO0sn4S
              bUrz7Uay9ACTM?si=SC_MixNKSnuxNvQMf3yBBg",
        }
        response = self.client.post(url, data, format="json")

        self.assertEqual(response.status_code,
          status.HTTP_403_FORBIDDEN)
```

（6）添加一个测试，检查成功地修改了一首歌曲。

```
def test_change_song_allowed(self):
        # login
        self.client.force_authenticate(user=self.superuser)

        url = reverse("rest_song_detail", kwargs={"pk": self.song.pk})

        # change only title
        data = {
            "artist": "Men I Trust",
            "title": "Tailwhip",
            "url": "https://open.spotify.com/track/2DoO0sn4S
                bUrz7Uay9ACTM?si=SC_MixNKSnuxNvQMf3yBBg",
        }
        response = self.client.put(url, data, format="json")

        self.assertEqual(response.status_code, status.HTTP_200_OK)
        self.assertEqual(response.data["uuid"], str(self.song.pk))
        self.assertEqual(response.data["artist"], data["artist"])
        self.assertEqual(response.data["title"], data["title"])
        self.assertEqual(response.data["url"], data["url"])

        # logout
        self.client.force_authenticate(user=None)
```

（7）添加一个测试，检查缺失验证而导致的无效修改。

```
def test_change_song_restricted(self):
    # make sure the user is logged out
    self.client.force_authenticate(user=None)

    url = reverse("rest_song_detail", kwargs=
     {"pk": self.song.pk})

    # change only title
    data = {
        "artist": "Capital Cities",
        "title": "Safe And Sound",
        "url": "https://open.spotify.com/track/40Fs0YrU
          GuwLNQSaHGVfqT?si=2OUawusIT-evyZKonT5GgQ",
    }
    response = self.client.put(url, data, format="json")

    self.assertEqual(response.status_code,
     status.HTTP_403_FORBIDDEN)
```

（8）添加一个测试，检查一首歌曲的无效删除。

```
def test_delete_song_restricted(self):
    # make sure the user is logged out
    self.client.force_authenticate(user=None)

    url = reverse("rest_song_detail", kwargs=
     {"pk": self.song_to_delete.pk})

    data = {}
    response = self.client.delete(url, data, format="json")

    self.assertEqual(response.status_code,
     status.HTTP_403_FORBIDDEN)
```

（9）添加一个测试，检查成功地删除一首歌曲。

```
def test_delete_song_allowed(self):
    # login
    self.client.force_authenticate(user=self.superuser)

    url = reverse("rest_song_detail", kwargs=
     {"pk": self.song_to_delete.pk})
```

```
        data = {}
        response = self.client.delete(url, data, format="json")

        self.assertEqual(response.status_code,
         status.HTTP_204_NO_CONTENT)

        # logout
        self.client.force_authenticate(user=None)
```

（10）运行 music 应用程序的测试，如下所示。

```
(env)$ python manage.py test myproject.apps.music --
settings=myproject.settings.test
Creating test database for alias 'default'...
System check identified no issues (0 silenced).
........
-------------------------------------------------------------------
---
Ran 8 tests in 0.370s

OK
Destroying test database for alias 'default'...
```

11.5.3　工作方式

这里，RESTful API 测试套件扩展了 APITestCase 类。再次说明，setUpClass()和 tearDownClass()类方法在不同的测试之前或之后执行。另外，测试套件包含了一个 APIClient 类型的客户端属性，可用于模拟 API 调用。客户端针对所有的标准 HTTP 调用提供了相应的方法，包括 get()、post()、put()、patch()、delete()、head()和 options()。

在当前测试中，我们使用了 GET、POST 和 DELETE 请求。另外，客户端包含了相关方法，并根据登录证书、令牌或 User 对象强制用户进行身份验证，在测试过程中，我们验证了第三种方法：将用户直接传递至 force_authenticate()方法中。

代码的其余部分具有自解释性，此处不再赘述。

11.5.4　延伸阅读

- 第 9 章中的“使用 Django REST 框架创建一个 API”示例。
- “利用 Mock 测试视图”示例。
- “利用 Selenium 测试用户界面”示例。

- ❑ “确保测试覆盖率”示例。

11.6　确保测试覆盖率

Django 支持项目的快速原型和构建。但是，为了确保项目的稳定性并满足产品升级要求，我们应尽可能地测试各项功能。根据测试覆盖率，我们可以检查项目代码的测试量。

11.6.1　准备工作

为项目准备好测试。

在虚拟环境中安装 coverage 实用程序。

```
(env)$ pip install coverage~=5.0.1
```

11.6.2　实现方式

下列步骤将检查项目的覆盖率。

（1）针对 coverage 实用程序，创建 setup.cfg 配置文件，如下所示。

```
# setup.cfg
[coverage:run]
source = .
omit =
    media/*
    static/*
    tmp/*
    drivers/*
    locale/*
    myproject/site_static/*
    myprojext/templates/*
```

（2）当采用 Git 版本控制时，确保.gitignore 文件中包含下列内容。

```
# .gitignore
htmlcov/
.coverage
.coverage.*
coverage.xml
*.cover
```

（3）利用下列命令创建 shell 脚本 run_tests_with_coverage.sh，运行基于覆盖率的测试并报告结果。

```
# run_tests_with_coverage.sh
#!/usr/bin/env bash
coverage erase
coverage run manage.py test --settings=myproject.settings.test
coverage report
```

（4）针对脚本添加执行授权许可。

```
(env)$ chmod +x run_tests_with_coverage.sh
```

（5）运行脚本。

```
(env)$ ./run_tests_with_coverage.sh
Creating test database for alias 'default'...
System check identified no issues (0 silenced).
...........
-------------------------------------------------------------------
---
Ran 11 tests in 12.940s

OK
Destroying test database for alias 'default'...
Name Stmts Miss Cover
-------------------------------------------------------------------
----------------------------
manage.py 12 2 83%
myproject/__init__.py 0 0 100%
myproject/apps/__init__.py 0 0 100%
myproject/apps/core/__init__.py 0 0 100%
myproject/apps/core/admin.py 16 10 38%
myproject/apps/core/context_processors.py 3 0 100%
myproject/apps/core/model_fields.py 48 48 0%
myproject/apps/core/models.py 87 29 67%
myproject/apps/core/templatetags/__init__.py 0 0 100%
myproject/apps/core/templatetags/utility_tags.py 171 135 21%

the statistics go on…

myproject/settings/test.py 5 0 100%
myproject/urls.py 10 0 100%
myproject/wsgi.py 4 4 0%
```

```
----------------------------------------------------------------
----------------------------
TOTAL 1363 712 48%
```

11.6.3　工作方式

coverage 实用程序运行测试，并检查测试覆盖的代码行数。在当前示例中，所编写的测试覆盖了 48%的代码。如果项目的稳定性十分重要，建议覆盖率接近 100%。

在覆盖率配置中，我们忽略了静态数据资源、模板和其他非 Python 文件。

11.6.4　延伸阅读

- “利用 Mock 测试视图”示例。
- “利用 Selenium 测试用户界面”示例。
- “利用 Django REST 框架测试 API”示例。

第 12 章　部　　署

本章主要涉及下列主题。

- ❑ 发布可复用的 Django 应用程序。
- ❑ 针对预发布环境利用 mod_wsgi 在 Apache 上部署。
- ❑ 针对产品环境利用 mod_wsgi 在 Apache 上部署。
- ❑ 针对预发布环境在 Nginx 和 Gunicorn 上部署。
- ❑ 针对产品环境在 Nginx 和 Gunicorn 上部署。

12.1　简　　介

一旦拥有了可工作的网站和可复用的应用程序，将其向用户公开。部署网站是 Django 开发过程中较为困难的一项任务，因为存在很多变化的内容需要我们处理。

- ❑ 管理 Web 服务器。
- ❑ 配置数据库。
- ❑ 处理静态和媒体文件。
- ❑ 处理 Django 项目。
- ❑ 配置缓存。
- ❑ 设置电子邮件发送机制。
- ❑ 管理域。
- ❑ 安排后台任务和定时任务。
- ❑ 设置连续集成。
- ❑ 其他任务（取决于项目的规模和复杂度）。

在较大的团队中，所有任务均由 DevOps 工程师完成，这些工程师具备深入理解网络和计算机体系结构、管理 Linux 服务器、bash 脚本、使用 vim 等技能。

专业的网站通常具备开发、预发布和产品环境，且每种环境包含特定的目标。其中，开发环境用于创建项目；产品环境则意味着公共站点所托管的服务器（或多台服务器）；预发布环境为技术上类似于产品的系统，但用于检查发布前的新特性和优化措施。

12.2　技术需求

当与本章代码协同工作时，需要安装最新稳定版本的 Python、MySQL 或 PostgreSQL 数据库，以及虚拟环境下的 Django 项目。

读者可访问 GitHub 储存库的 ch12 目录查看本章的所有代码，对应网址为 https://github.com/PacktPublishing/Django-3-Web-Development-Cookbook-Fourth-Edition。

12.3　发布可复用的 Django 应用程序

Django 文档阐述了如何打包可复用的应用程序，以便后续在虚拟环境中利用 pip 进行安装，对应网址为 https://docs.djangoproject.com/en/3.0/intro/reusable-apps/。

然而，还存在另一种方式并通过工具打包和发布一个可复用的 Django 应用程序，并针对不同的代码项目创建模板，如新的 Django CMS 网站、Flask 网站或 jQuery 插件。其中一个有效的项目模板是 cookiecutter-djangopackage。在当前示例中，我们将学习如何使用该模板发布可复用的 likes 应用程序。

12.3.1　准备工作

利用虚拟环境创建一个新项目，并安装 cookiecutter，如下所示。

```
(env)$ pip install cookiecutter~=1.7.0
```

12.3.2　实现方式

下列步骤将发布 likes 应用程序。

（1）启用新的 Django 应用程序项目，如下所示。

```
(env)$ cookiecutter
https://github.com/pydanny/cookiecutter-djangopackage.git
```

由于这是 GitHub 托管的 cookiecutter 模板，因此可使用简洁方式的语法，如下所示。

```
(env)$ cookiecutter gh:pydanny/cookiecutter-djangopackage
```

（2）回答问题并创建应用程序模板，如下所示。

```
full_name [Your full name here]: Aidas Bendoraitis
email [you@example.com]: aidas@bendoraitis.lt
github_username [yourname]: archatas
project_name [Django Package]: django-likes
repo_name [dj-package]: django-likes
app_name [django_likes]: likes
app_config_name [LikesConfig]:
project_short_description [Your project description goes here]:
Django app for liking anything on your website.
models [Comma-separated list of models]: Like
django_versions [1.11,2.1]: master
version [0.1.0]:
create_example_project [N]:
Select open_source_license:
1 - MIT
2 - BSD
3 - ISCL
4 - Apache Software License 2.0
5 - Not open source
Choose from 1, 2, 3, 4, 5 [1]:
```

这将生成可发布的 Django 包的基本的文件结构，如下所示。

```
django-likes/
├── docs/
│   ├── Makefile
│   ├── authors.rst
│   ├── conf.py
│   ├── contributing.rst
│   ├── history.rst
│   ├── index.rst
│   ├── installation.rst
│   ├── make.bat
│   ├── readme.rst
│   └── usage.rst
├── likes/
│   ├── static/
│   │   ├── css/
│   │   │   └── likes.css
│   │   ├── img/
│   │   └── js/
│   │       └── likes.js
│   ├── templates/
```

```
│   │   └── likes/
│   │       └── base.html
│   └── test_utils/
│       ├── test_app/
│       │   ├── migrations/
│       │   │   └── __init__.py
│       │   ├── __init__.py
│       │   ├── admin.py
│       │   ├── apps.py
│       │   └── models.html
│       ├── __init__.py
│       ├── admin.py
│       ├── apps.py
│       ├── models.py
│       ├── urls.py
│       └── views.py
├── tests/
│   ├── __init__.py
│   ├── README.md
│   ├── requirements.txt
│   ├── settings.py
│   ├── test_models.py
│   └── urls.py
├── .coveragerc
├── .editorconfig
├── .gitignore
├── .travis.yml
├── AUTHORS.rst
├── CONTRIBUTING.rst
├── HISTORY.rst
├── LICENSE
├── MANIFEST.in
├── Makefile
├── README.rst
├── manage.py
├── requirements.txt
├── requirements_dev.txt
├── requirements_test.txt
├── runtests.py
├── setup.cfg
├── setup.py*
└── tox.ini
```

（3）将 Django 项目中使用的 likes 应用程序文件复制至 django-likes/likes 目录中。在 cookiecutter 创建相同文件的情况下，内容需要被合并而非覆写。如 likes/__init__.py 文件需要包含一个版本字符串，以正确地与后续步骤中的 setup.py 协同工作，如下所示。

```
# django-likes/likes/__init__.py
__version__ = '0.1.0'
```

（4）重置依赖项，以便不会从 Django 项目中导入任何内容，所使用的函数和类均位于当前应用程序中。如在 likes 应用程序中，存在 core 应用程序中某些混入的依赖项。我们需要将相关代码直接复制至 django-likes 应用程序中的文件中。

提示：

如果存在大量的依赖代码，则可将 core 应用程序作为一个未耦合的包发布，但之后需要单独对其进行维护。

（5）使用之前输入的 repo_name，将可复用的应用程序项目添加至 GitHub 中的 Git 储存库中。

（6）查看不同的文件并完成证书、README、文档、配置和其他文件。

（7）确保应用程序通过 cookiecutter 模板测试。

```
(env)$ pip install -r requirements_test.txt
(env)$ python runtests.py
Creating test database for alias 'default'...
System check identified no issues (0 silenced).
.
-----------------------------------------------------------------
---
Ran 1 test in 0.001s

OK
Destroying test database for alias 'default'...
```

（8）如果包是闭源的，那么创建一个可共享的 ZIP 归档版本，如下所示。

```
(env)$ python setup.py sdist
```

这将创建一个 django-likes/dist/django-likes-0.1.0.tar.gz 文件，并于随后利用 pip 在项目的虚拟环境中进行安装或卸载，如下所示。

```
(env)$ pip install django-likes-0.1.0.tar.gz
(env)$ pip uninstall django-likes
```

（9）如果包是开源的，则可将应用程序注册并发布至 Python Package Index（PyPI）。

```
(env)$ python setup.py register
(env)$ python setup.py publish
```

（10）此外，为了传播单词，通过在 https://www.djangopackages.com/packages/add/提交一个表单，将应用程序添加至 Django 包中。

12.3.3　工作方式

如果只是按下 Enter 键而不输入任何内容，通过方括号给出的默认值，Cookiecutter 在 Django 应用程序项目模板的不同部分填充请求的数据。最终，我们得到了 setup.py 文件，并分发至 Python Package Index、Sphinx 文档、作为默认证书的 MIT、项目的通用文本编辑器配置、应用程序中包含的静态文件和模板等。

12.3.4　延伸阅读

- 第 1 章中的“创建一个项目文件结构”示例。
- 第 1 章中的“针对 Django、Dunicorn、Nginx 和 PostgreSQL，与 Docker 容器协同工作”示例。
- 第 1 章中的“利用 pip 处理项目依赖项”示例。
- 第 4 章中的“实现 Like 微件”示例。
- 第 11 章中的“利用 Mock 测试视图”示例。

12.4　针对预发布环境利用 mod_wsgi 在 Apache 上部署

在当前示例中，我们将展示如何创建一个脚本，并将项目部署至计算机虚拟机上的预发布环境中。该项目使用基于 mod_wsgi 模块的 Apache Web 服务器。针对安装过程，我们将使用 Ansible、Vagrant 和 VirtualBox。如前所述，由于存在大量细节内容需要被关注，因而开发一个较优的部署脚本可能会花费几天的时间。

12.4.1　准备工作

访问部署检查列表，并确保配置通过全部安全推荐规范，对应网址为 https://docs.djangoproject.com/en/3.0/howto/deployment/checklist/。至少应确保在运行下列内容时项目配置不会产生警告消息。

```
(env)$ python manage.py check --deploy --
settings=myproject.settings.staging
```

安装最新稳定版本的 Ansible、Vagrant 和 VirtualBox，具体的官方网站如下所示。

- Ansible：https://docs.ansible.com/ansible/latest/installation_guide/intro_installation.html。
- VirtualBox：https://www.virtualbox.org/wiki/Downloads。
- Vagrant：https://www.vagrantup.com/downloads.html。

在 macOS X 环境下，可利用 HomeBrew 安装 Ansible、Vagrant 和 VirtualBox。

```
$ brew install ansible
$ brew cask install virtualbox
$ brew cask install vagrant
```

12.4.2 实现方式

首先需要针对服务器上使用的不同服务创建一些配置模板，以供预发布和生产部署使用。

（1）在 Django 项目中，创建一个 deployment 目录，并于其中创建一个 ansible_templates 目录。

（2）创建一个 Jinja 模板文件以供配置时区使用。

```
{# deployment/ansible_templates/timezone.j2 #}
{{ timezone }}
```

（3）在设置 SSL 证书之前，创建一个 Jinja 模板文件以供 Apache 域配置使用。

```
{# deployment/ansible_templates/apache_site-pre.conf.j2 #}
<VirtualHost *:80>
   ServerName {{ domain_name }}
   ServerAlias {{ domain_name }} www.{{ domain_name }}

   DocumentRoot {{ project_root }}/public_html
   DirectoryIndex index.html

   ErrorLog ${APACHE_LOG_DIR}/error.log
   CustomLog ${APACHE_LOG_DIR}/access.log combined

   AliasMatch ^/.well-known/(.*) "/var/www/letsencrypt/$1"
```

```
    <Directory "/var/www/letsencrypt">
        Require all granted
    </Directory>

    <Directory "/">
        Require all granted
    </Directory>

</VirtualHost>
```

（4）创建一个 Jinja 模板文件 deployment/ansible_templates/apache_site.conf.j2 以供域配置使用，同时还包括 SSL 证书。针对该文件，复制 https://raw.githubusercontent.com/PacktPublishing/Django-3-Web-Development-Cookbook-Fourth-Edition/master/ch12/myproject_virtualenv/src/django-myproject/deployment-apache/ansible_templates/apache_site.conf.j2 中的内容。

（5）针对 PostgreSQL 配置文件创建一个模板 deployment/ansible_templates/postgresql.j2，并包含 https://github.com/postgres/postgres/blob/REL_10_STABLE/src/backend/utils/misc/ postgresql.conf.sample 中的内容。稍后可调整配置内容以满足服务器的需求。

（6）针对 PostgreSQL 授权配置文件（稍后可根据需求进行调整）创建一个模板。

```
{# deployment/ansible_templates/pg_hba.j2 #}
# TYPE      DATABASE        USER            CIDR-ADDRESS        METHOD
local       all             all                                 ident
host        all             all             ::0/0               md5
host        all             all             0.0.0.0/32          md5
host        {{ db_name }}   {{ db_user }}   127.0.0.1/32        md5
```

（7）创建一个 Postfix 电子邮件服务器配置模板。

```
{# deployment/ansible_templates/postfix.j2 #}
# See /usr/share/postfix/main.cf.dist for a commented, more
# complete version

# Debian specific: Specifying a file name will cause the first
# line of that file to be used as the name. The Debian default
# is /etc/mailname.
# myorigin = /etc/mailname

smtpd_banner = $myhostname ESMTP $mail_name (Ubuntu)
biff = no
```

```
# appending .domain is the MUA's job.
append_dot_mydomain = no

# Uncomment the next line to generate "delayed mail" warnings
# delay_warning_time = 4h

readme_directory = no

# TLS parameters
smtpd_tls_cert_file=/etc/ssl/certs/ssl-cert-snakeoil.pem
smtpd_tls_key_file=/etc/ssl/private/ssl-cert-snakeoil.key
smtpd_use_tls=yes
smtpd_tls_session_cache_database = btree:${data_directory}/smtpd_scache
smtp_tls_session_cache_database = btree:${data_directory}/smtp_scache

# See /usr/share/doc/postfix/TLS_README.gz in the postfix-doc
# package for information on enabling SSL in
# the smtp client.

smtpd_relay_restrictions = permit_mynetworks
permit_sasl_authenticated defer_unauth_destination
myhostname = {{ domain_name }}
alias_maps = hash:/etc/aliases
alias_database = hash:/etc/aliases
mydestination = $myhostname, localhost, localhost.localdomain, ,
 localhost
relayhost =
mynetworks = 127.0.0.0/8 [::ffff:127.0.0.0]/104 [::1]/128
mailbox_size_limit = 0
recipient_delimiter = +
inet_interfaces = all
inet_protocols = all
virtual_alias_domains = {{ domain_name }}
virtual_alias_maps = hash:/etc/postfix/virtual
```

（8）创建一个电子邮件转发配置模板。

```
{# deployment/ansible_templates/virtual.j2 #}
# /etc/postfix/virtual

hello@{{ domain_name }} admin@example.com
@{{ domain_name }} admin@example.com
```

（9）创建一个 memcached 配置模板。

```
{# deployment/ansible_templates/memcached.j2 #}
# memcached default config file
# 2003 - Jay Bonci <jaybonci@debian.org>
# This configuration file is read by the start-memcached script
# provided as part of the Debian GNU/Linux
# distribution.

# Run memcached as a daemon. This command is implied, and is not
# needed for the daemon to run. See the README.Debian that

# comes with this package for more information.
-d

# Log memcached's output to /var/log/memcached
logfile /var/log/memcached.log

# Be verbose
# -v

# Be even more verbose (print client commands as well)
# -vv

# Use 1/16 of server RAM for memcached
-m {{ (ansible_memtotal_mb * 0.0625) | int }}

# Default connection port is 11211
-p 11211

# Run the daemon as root. The start-memcached will default to
# running as root if no -u command is present
# in this config file
-u memcache

# Specify which IP address to listen on. The default is to
# listen on all IP addresses
# This parameter is one of the only security measures that
# memcached has, so make sure it's listening on
# a firewalled interface.
-l 127.0.0.1

# Limit the number of simultaneous incoming connections.
```

```
# The daemon default is 1024
# -c 1024

# Lock down all paged memory. Consult with the README and
# homepage before you do this
# -k

# Return error when memory is exhausted (rather than
# removing items)
# -M

# Maximize core file limit
# -r
```

（10）创建 secrets.json 文件的 Jinja 模板。

```
{# deployment/ansible_templates/secrets.json.j2 #}
{
   "DJANGO_SECRET_KEY": "{{ django_secret_key }}",
   "DATABASE_ENGINE": "django.contrib.gis.db.backends.postgis",
   "DATABASE_NAME": "{{ db_name }}",
   "DATABASE_USER": "{{ db_user }}",
   "DATABASE_PASSWORD": "{{ db_password }}",
   "EMAIL_HOST": "{{ email_host }}",
   "EMAIL_PORT": "{{ email_port }}",
   "EMAIL_HOST_USER": "{{ email_host_user }}",
   "EMAIL_HOST_PASSWORD": "{{ email_host_password }}"
}
```

接下来处理与特定的预发布环境相关的 Vagrant 和 Ansible 脚本。

（1）在.gitignore 文件中，忽略某些与 Vagrant 和 Ansible 相关的文件。

```
# .gitignore
# Secrets
secrets.json
secrets.yml

# Vagrant / Ansible
.vagrant
*.retry
```

（2）创建两个目录 deployment/staging 和 deployment/staging/ansible。

（3）创建 Vagrantfile 文件，并利用下列脚本设置基于 Ubuntu 18 的虚拟机，并于其中运行 Ansible 脚本。

```
# deployment/staging/ansible/Vagrantfile
VAGRANTFILE_API_VERSION = "2"

Vagrant.configure(VAGRANTFILE_API_VERSION) do |config|
    config.vm.box = "bento/ubuntu-18.04"
    config.vm.box_version = "201912.14.0"
    config.vm.box_check_update = false
    config.ssh.insert_key=false
    config.vm.provider "virtualbox" do |v|
        v.memory = 512
        v.cpus = 1
        v.name = "myproject"
    end
    config.vm.network "private_network", ip: "192.168.50.5"
    config.vm.provision "ansible" do |ansible|
        ansible.limit = "all"
        ansible.playbook = "setup.yml"
        ansible.inventory_path = "./hosts/vagrant"
        ansible.host_key_checking = false
        ansible.verbose = "vv"
        ansible.extra_vars = { ansible_python_interpreter:
        "/usr/bin/python3" }
    end
end
```

（4）创建包含 vagrant 文件的 hosts 目录。

```
# deployment/staging/ansible/hosts/vagrant
[servers]
192.168.50.5
```

（5）创建一个 vars.yml 文件，其中包含在安装脚本和 Jinja 模板中使用的变量。

```
# deployment/staging/ansible/vars.yml
---
# a unix path-friendly name (IE, no spaces or special characters)
project_name: myproject

user_username: "{{ project_name }}"

# the base path to install to. You should not need to change this.
install_root: /home
```

```
project_root: "{{ install_root }}/{{ project_name }}"

# the python module path to your project's wsgi file
wsgi_module: myproject.wsgi

# any directories that need to be added to the PYTHONPATH.
python_path: "{{ project_root }}/src/{{ project_name }}"

# the git repository URL for the project
project_repo: git@github.com:archatas/django-myproject.git

# The value of your django project's STATIC_ROOT settings.
static_root: "{{ python_path }}/static"
media_root: "{{ python_path }}/media"

locale: en_US.UTF-8
timezone: Europe/Berlin

domain_name: myproject.192.168.50.5.xip.io
django_settings: myproject.settings.staging

letsencrypt_email: ""
wsgi_file_name: wsgi_staging.py
```

此外，我们还需要一个 secrets.yml 文件，其密码值包含密码和身份验证密钥。对此，首先创建一个 sample_secrets.yml 文件，该文件不包含任何敏感信息，且仅包含变量名，随后将其复制至 secrets.yml 文件中并填写密码。这里，前一个文件处于版本控制下，而后一个文件将被忽略。

```
# deployment/staging/ansible/sample_secrets.yml
# Django Secret Key
django_secret_key: "change-this-to-50-characterslong-random-string"

# PostgreSQL database settings
db_name: "myproject"
db_user: "myproject"
db_password: "change-this-to-a-secret-password"
db_host: "localhost"
db_port: "5432"

# Email SMTP settings
```

```
email_host: "localhost"
email_port: "25"
email_host_user: ""
email_host_password: ""

# a private key that has access to the repository URL
ssh_github_key: ~/.ssh/id_rsa_github
```

（6）在 deployment/staging/ansible/setup.yml 处创建一个 Ansible 脚本，并安装所有的依赖项和配置服务。从 https://raw.githubusercontent.com/PacktPublishing/Django-3-Web-Development-Cookbook-Fourth-Edition/master/ch12/myproject_virtualenv/src/djangomyproject/deployment-apache/staging/ansible/setup.yml 处复制该文件的内容。

（7）在 deployment/staging/ansible/deploy.yml 处创建另一个 Ansible 脚本，以处理 Django 项目。从 https://raw.githubusercontent.com/PacktPublishing/Django-3-Web-Development-book-CookFourth-Edition/master/ch12/myproject_virtualenv/src/django-myproject/deploymentapache/staging/ansible/deploy.yml 处复制该文件的内容。

（8）创建启动部署的 bash 脚本。

```
# deployment/staging/ansible/setup_on_virtualbox.sh
#!/usr/bin/env bash
echo "=== Setting up the local staging server ==="
date

cd "$(dirname "$0")"
vagrant up --provision
```

（9）添加 bash 脚本的执行权限并运行该脚本。

```
$ chmod +x setup_on_virtualbox.sh
$ ./setup_on_virtualbox.sh
```

（10）如果脚本出现错误，很可能是虚拟机需要针对变化内容而重启。对此，可通过 ssh 连接虚拟机，并更改为根用户随后重启，如下所示。

```
$ vagrant ssh
Welcome to Ubuntu 18.04.3 LTS (GNU/Linux 4.15.0-72-generic x86_64)

* Documentation:    https://help.ubuntu.com
* Management:       https://landscape.canonical.com
* Support:          https://ubuntu.com/advantage

System information as of Wed Jan 15 04:44:42 CET 2020
```

```
System load:    0.21              Processes:              126
Usage of /:     4.0% of 61.80GB   Users logged in:        1
Memory usage:   35%               IP address for eth0:    10.0.2.15
Swap usage:     4%                IP address for eth1:    192.168.50.5

0 packages can be updated.
0 updates are security updates.

*** System restart required ***

This system is built by the Bento project by Chef Software
More information can be found at https://github.com/chef/bento
Last login: Wed Jan 15 04:43:32 2020 from 192.168.50.1
vagrant@myproject:~$ sudo su
root@myproject:/home/vagrant#
reboot
Connection to 127.0.0.1 closed by remote host.
Connection to 127.0.0.1 closed.
```

（11）当浏览 Django 项目目录时，可 ssh 至虚拟机并将用户更改为 myproject，如下所示。

```
$ vagrant ssh
Welcome to Ubuntu 18.04.3 LTS (GNU/Linux 4.15.0-74-generic x86_64)
# …
vagrant@myproject:~$ sudo su - myproject
(env) myproject@myproject:~$ pwd
/home/myproject
(env) myproject@myproject:~$ ls
commands db_backups logs public_html src env
```

12.4.3　工作方式

VirtualBox 可在包含不同操作系统的计算机上设置多个虚拟机。Vagrant 工具可创建这些虚拟机，并通过脚本在这些虚拟机上下载和安装操作系统。Ansible 是一个基于 Python 的实用程序，可读取.yaml 配置文件中的指令，并在远程服务器上执行这些指令。

我们刚刚完成的部署脚本执行下列操作。

- 在 VirtualBox 中创建虚拟机并安装 Ubuntu 18。
- 将 IP 192.168.50.5 分配于虚拟机。

- 设置虚拟机的主机名。
- 更新 Linux 包。
- 设置服务器的本地化设置项。
- 安装 Linux 依赖项，包括 Python、Apache、PostgreSQL、Postfix、Memcached 等。
- 创建 Linux 用户和 Django 项目的 home 目录。
- 创建 Django 项目的虚拟环境。
- 创建 PostgreSQL 数据库用户和数据库。
- 配置 Apache Web 服务器。
- 安装自签名的 SSL 证书。
- 配置 Memcached 缓存服务。
- 配置 Postfix 电子邮件服务器。
- 复制 Django 项目储存库。
- 安装 Python 依赖项。
- 创建 secrets.json 文件。
- 迁移数据库。
- 收集静态文件。
- 重启 Apache。

当前，可登录 https://www.myproject.192.168.50.5.xip.io 并访问 Django 网站，此时将显示一个 Hello,World!页面。注意，某些浏览器（如 Chrome）可能不会打开包含自签名 SSL 证书的网站，并作为一项安全措施阻塞该网站。

提示：

xip.io 是一个通配符 DNS 服务，并将特定于 IP 的子域指向 IP，并可用于 SSL 证书或需要域名的其他网站特性中。

如果打算尝试不同的配置或额外的命令，可通过较小的幅度实现渐进式修改。对于某些部分，还需要直接在虚拟机上进行测试，随后将相关任务转换为 Ansible 指令。

提示：

关于如何使用 Ansible，读者可访问官方文档，其中展示了大多数用例可用的指令示例，对应网址为 https://docs.ansible.com/ansible/latest/index.html。

如果任何服务出现了错误，可 ssh 至虚拟机，切换至根用户，并查看对应服务的日志。同时，我们还可以在 Google 上查询错误，从而进一步了解当前工作的系统。

当重建虚拟机时，可使用下列命令。

```
$ vagrant destroy
$ vagrant up --provision
```

12.4.4　延伸阅读

- ❑ 第 1 章中的“创建一个项目文件结构”示例。
- ❑ 第 1 章中的“利用 pip 处理项目依赖项”示例。
- ❑ 第 1 章中的“以动态方式设置 STATIC_URL”示例。
- ❑ “针对产品环境利用 mod_wsgi 在 Apache 上部署”示例。
- ❑ “针对预发布环境在 Nginx 和 Gunicorn 上部署”示例。
- ❑ “针对产品环境在 Nginx 和 Gunicorn 上部署”示例。
- ❑ 第 13 章中的“创建和恢复 PostgreSQL 数据库备份”示例。
- ❑ 第 13 章中的“设置常规作业的定时任务”示例。

12.5　针对产品环境利用 mod_wsgi 在 Apache 上部署

Apache 是较为流行的 Web 服务器之一，如果还必须运行服务管理、监测、分析、博客、电子商务等需要在同一服务器上运行 Apache 的服务，那么在 Apache 下部署 Django 项目是十分有意义的。

当前示例将在之前示例的基础上完成，并实现一个 Ansible 脚本，以利用 mod_wsgi 模块在 Apache 上设置产品环境。

12.5.1　准备工作

确保在运行下列内容时项目配置不会产生警告消息。

```
(env)$ python manage.py check --deploy --
settings=myproject.settings.production
```

此外，还应安装最新稳定版本的 Ansible。

选择一家服务器提供商并创建一个专用服务器，通过 SSH 以及私钥和公钥身份验证实现根访问。笔者选择的提供商是 DigitalOcean（https://www.digitalocean.com/），并通过 Ubuntu 18 建立了一个专用服务器（Droplet）。利用新的 SSH 私钥和公钥对（~/.ssh/id_rsa_django_cookbook 和~/.ssh/id_rsa_django_cookbook.pub）以及 IP 地址 142.93.167.30，笔者可连接至服务器上。

在本地处，我们需要配置 SSH 连接，即创建或调整~/.ssh/config 文件，如下所示。

```
# ~/.ssh/config
Host *
    ServerAliveInterval 240
    AddKeysToAgent yes
    UseKeychain yes

Host github
    Hostname github.com
    IdentityFile ~/.ssh/id_rsa_github

Host myproject-apache
    Hostname 142.93.167.30
    User root
    IdentityFile ~/.ssh/id_rsa_django_cookbook
```

通过下列命令，应能够利用 SSH 并以根用户身份连接至专用服务器。

```
$ ssh myproject-apache
```

在域配置中，将域的 DNS A record 指向专用服务器的 IP 地址。在当前示例中，我们仅使用 myproject.142.93.167.30.xip.io 以展示如何利用 SSL 证书设置 Django 网站的服务器。

注意：

如前所述，xip.iox 是一个通配符 DNS 服务，并将特定于 IP 的子域指向对应 IP，并可用于 SSL 证书或需要域名的其他网站特性中。

12.5.2　实现方式

下列步骤将创建用于产品环境的部署脚本。

（1）确保 deployment/ansible_templates 目录中包含用于服务配置的 Jinja 模板。

（2）创建 Ansible 脚本的 deployment/production 和 deployment/production/ansible 目录。

（3）创建包含 remote 文件的 hosts 目录，如下所示。

```
# deployment/production/ansible/hosts/remote
[servers]
myproject-apache

[servers:vars]
ansible_python_interpreter=/usr/bin/python3
```

（4）创建 vars.yml 文件，包含安装脚本中使用的变量，以及配置的 Jinja 模板。

```
# deployment/production/ansible/vars.yml
---
# a unix path-friendly name (IE, no spaces or special characters)
project_name: myproject

user_username: "{{ project_name }}"

# the base path to install to. You should not need to change this.
install_root: /home

project_root: "{{ install_root }}/{{ project_name }}"

# the python module path to your project's wsgi file
wsgi_module: myproject.wsgi

# any directories that need to be added to the PYTHONPATH.
python_path: "{{ project_root }}/src/{{ project_name }}"

# the git repository URL for the project
project_repo: git@github.com:archatas/django-myproject.git

# The value of your django project's STATIC_ROOT settings.
static_root: "{{ python_path }}/static"
media_root: "{{ python_path }}/media"

locale: en_US.UTF-8
timezone: Europe/Berlin

domain_name: myproject.142.93.167.30.xip.io
django_settings: myproject.settings.production

# letsencrypt settings
letsencrypt_email: hello@myproject.com
wsgi_file_name: wsgi_production.py
```

（5）此外还需要一个带有秘密值的 secrets.yml 文件，其中包含密码和身份验证密钥。对此，首先创建一个不包含敏感信息的 sample_secrets.yml 文件，其中仅包含变量名，随后将其复制至 secrets.yml 文件中。这里，前一个文件处于版本控制下，后一个文件则被忽略。

```
# deployment/production/ansible/sample_secrets.yml
# Django Secret Key
django_secret_key: "change-this-to-50-characterslong-random-string"

# PostgreSQL database settings
db_name: "myproject"
db_user: "myproject"
db_password: "change-this-to-a-secret-password"
db_host: "localhost"
db_port: "5432"

# Email SMTP settings
email_host: "localhost"
email_port: "25"
email_host_user: ""
email_host_password: ""

# a private key that has access to the repository URL
ssh_github_key: ~/.ssh/id_rsa_github
```

（6）在 deployment/production/ansible/setup.yml 处创建一个 Ansible 脚本，用于安装所有的依赖项并配置服务。将 https://raw.githubusercontent.com/PacktPublishing/Django-3-Web-Development-Cookbook-Fourth-Edition/master/ch12/myproject_virtualenv/src/django-myproject/deployment-apache/production/ansible/setup.yml.中的内容复制至该文件中。

（7）创建另一个 Ansible 脚本 deployment/production/ansible/deploy.yml，用于处理 Django 项目。将 https://raw.githubusercontent.com/PacktPublishing/Django-3-Web-Development-Cookbook-Fourth-Edition/master/ch12/myproject_virtualenv/src/djangomyproject/deployment-apache/production/ansible/deploy.yml 中的内容复制至该文件中。

（8）创建一个 bash 脚本以启用部署。

```
# deployment/production/ansible/setup_remotely.sh
#!/usr/bin/env bash
echo "=== Setting up the production server ==="
date

cd "$(dirname "$0")"
ansible-playbook setup.yml -i hosts/remote
```

（9）添加 bash 脚本的执行许可并运行该脚本。

```
$ chmod +x setup_remotely.sh
$ ./setup_remotely.sh
```

（10）如果脚本出现错误，很可能是内容发生变化需要重启专用服务器。对此，可通过 ssh 连接至服务器并重启，如下所示。

```
$ ssh myproject-apache
Welcome to Ubuntu 18.04.3 LTS (GNU/Linux 4.15.0-74-generic x86_64)

* Documentation: https://help.ubuntu.com
* Management: https://landscape.canonical.com
* Support: https://ubuntu.com/advantage

System information as of Wed Jan 15 11:39:51 CET 2020

System load: 0.08 Processes: 104
Usage of /: 8.7% of 24.06GB Users logged in: 0
Memory usage: 35% IP address for eth0: 142.93.167.30
Swap usage: 0%

* Canonical Livepatch is available for installation.
- Reduce system reboots and improve kernel security. Activate at:
https://ubuntu.com/livepatch

0 packages can be updated.
0 updates are security updates.

*** System restart required ***

Last login: Sun Jan 12 12:23:35 2020 from 178.12.115.146
root@myproject:~# reboot
Connection to 142.93.167.30 closed by remote host.
Connection to 142.93.167.30 closed.
```

（11）创建另一个 bash 脚本，用于更新 Django 项目。

```
# deployment/production/ansible/deploy_remotely.sh
#!/usr/bin/env bash
echo "=== Deploying project to production server ==="
date

cd "$(dirname "$0")"
ansible-playbook deploy.yml -i hosts/remote
```

（12）添加 bash 脚本的运行许可。

```
$ chmod +x deploy_remotely.sh
```

12.5.3　工作方式

Ansible 脚本是幂等的，这意味着，我们可多次执行该脚本，并且总是会得到相同的结果，即一个最新的专用服务器，同时安装并运行 Django 网站。如果服务器存在任何技术硬件问题，并且持有数据库和媒体文件备份，那么我们可较快地在另一台专用服务器上安装相同的配置。

产品部署脚本执行下列操作。

- 设置虚拟机的主机名。
- 更新 Linux 包。
- 设置服务器的本地化设置项。
- 安装 Linux 依赖项，包括 Python、Apache、PostgreSQL、Postfix、Memcached 等。
- 创建 Linux 用户和 Django 项目的 home 目录。
- 创建 Django 项目的虚拟环境。
- 创建 PostgreSQL 数据库用户和数据库。
- 配置 Apache Web 服务器。
- 安装 Let's Encrypt SSL 证书。
- 配置 Memcached 缓存服务。
- 配置 Postfix 电子邮件服务。
- 复制 Django 项目储存库。
- 安装 Python 依赖项。
- 创建 secrets.json 文件。
- 迁移数据库。
- 收集静态文件。
- 重启 Apache。

当首次安装服务和依赖项时，可运行 setup_remotely.sh 脚本。稍后，如果仅需要更新 Django 项目，则可使用 deploy_remotely.sh。可以看到，安装过程与预发布服务器上的安装十分类似，但为了更具灵活性和可调性，我们将该文件单独保存在一个 deployment/production 目录中。

从理论上讲，我们可以省略预发布环境，但尝试在虚拟机上首次实施部署过程更具实际意义，而不是在远程服务器上直接安装。

12.5.4　延伸阅读

- 第 1 章中的“创建一个项目文件结构”示例。
- 第 1 章中的“利用 pip 处理项目依赖项”示例。
- 第 1 章中的“以动态方式设置 STATIC_URL”示例。
- “针对预发布环境利用 mod_wsgi 在 Apache 上部署”示例。
- “针对预发布环境在 Nginx 和 Gunicorn 上部署”示例。
- “针对产品环境在 Nginx 和 Gunicorn 上部署”示例。
- 第 13 章中的“创建和恢复 PostgreSQL 数据库备份”示例。
- 第 13 章中的“设置常规作业的定时任务”示例。

12.6　针对预发布环境在 Nginx 和 Gunicorn 上部署

基于 mod_wsgi 的 Apache 是一个稳定的部署方案。然而，当对性能提出更高要求时，建议使用 Nginx 和 Gunicorn 处理 Django 网站。Gunicorn 是一个运行 WSGI 脚本的 Python 服务器，而 Nginx 则是一个 Web 服务器，解析域配置并将请求传递至 Gunicorn。

在当前示例中，我们将展示如何创建一个脚本，并将项目部署至计算机虚拟机上的预发布环境中。对此，我们将使用 Ansible、Vagrant 和 VirtualBox。如前所述，开发较优的部署脚本涉及大量的细节问题，有时可能会花费几天的时间才可能完成任务。

12.6.1　准备工作

访问部署检查列表以确保配置通过 https://docs.djangoproject.com/en/3.0/howto/deployment/checklist/中的全部安全建议，至少应确保运行下列内容时，项目配置不会出现警告信息。

```
(env)$ python manage.py check --deploy --
settings=myproject.settings.staging
```

安装最新稳定版本的 Ansible、Vagrant 和 VirtualBox。

- Ansible：https://docs.ansible.com/ansible/latest/installation_guide/intro_installation.html。
- VirtualBox：https://www.virtualbox.org/wiki/Downloads。
- Vagrant：https://www.vagrantup.com/downloads.html。

在 macOS X 环境下，可利用 Homebrew 安装 Ansible、Vagrant 和 VirtualBox。

```
$ brew install ansible
$ brew cask install virtualbox
$ brew cask install vagrant
```

12.6.2　实现方式

首先需要针对服务器上使用的不同的服务创建一些配置模板。这些模板用于部署过程，即预发布环境和产品环境。

（1）在 Django 项目中，创建 deployment 目录并在其中创建一个 ansible_templates 目录。

（2）创建时区配置的 Jinja 模板文件。

```
{# deployment/ansible_templates/timezone.j2 #}
{{ timezone }}
```

（3）在设置 SSL 证书之前，创建 Nginx 域配置的 Jinja 模板文件。

```
{# deployment/ansible_templates/nginx-pre.j2 #}
server{
    listen 80;
    server_name {{ domain_name }};

    location /.well-known/acme-challenge {
        root /var/www/letsencrypt;
        try_files $uri $uri/ =404;
    }
    location / {
        root /var/www/letsencrypt;
    }
}
```

（4）在 deployment/ansible_templates/nginx.j2 中创建 Nginx 域配置 Jinja 模板文件，包括 SSL 证书。对于该文件，复制 https://raw.githubusercontent.com/PacktPublishing/Django-3-Web-Development-Cookbook-Fourth-Edition/master/ch12/myproject_virtualenv/src/django-myproject/deployment-nginx/ansible_templates/nginx.j2 中的内容。

（5）创建 Gunicorn 服务配置的模板。

```
# deployment/ansible_templates/gunicorn.j2
[Unit]
Description=Gunicorn daemon for myproject website
After=network.target
```

```
[Service]
PIDFile=/run/gunicorn/pid
Type=simple
User={{ user_username }}
Group=www-data
RuntimeDirectory=gunicorn
WorkingDirectory={{ python_path }}
ExecStart={{ project_root }}/env/bin/gunicorn --pid
/run/gunicorn/pid --log-file={{ project_root }}/logs/gunicorn.log -
-workers {{ ansible_processor_count | int }} --bind 127.0.0.1:8000
{{ project_name }}.wsgi:application --env DJANGO_SETTINGS_MODULE={{
django_settings }} --max-requests 1000
ExecReload=/bin/kill -s HUP $MAINPID
ExecStop=/bin/kill -s TERM $MAINPID
PrivateTmp=true

[Install]
WantedBy=multi-user.target
```

（6）在 deployment/ansible_templates/postgresql.j2 处创建 PostgreSQL 配置文件的模板，其内容来自 https://github.com/postgres/postgres/blob/REL_10_STABLE/src/backend/utils/misc/postgresql.conf.sample。稍后可调整该文件的配置内容。

（7）创建 PostgreSQL 许可配置的模板（稍后可根据需要对其进行调整）。

```
{# deployment/ansible_templates/pg_hba.j2 #}
# TYPE      DATABASE        USER            CIDR-ADDRESS        METHOD
local       all             all                                 ident
host        all             all             ::0/0               md5
host        all             all             0.0.0.0/32          md5
host        {{ db_name }}   {{ db_user }}   127.0.0.1/32        md5
```

（8）创建 Postfix 电子邮件服务器配置的模板。

```
{# deployment/ansible_templates/postfix.j2 #}
# See /usr/share/postfix/main.cf.dist for a commented, more
# complete version

# Debian specific: Specifying a file name will cause the first
# line of that file to be used as the name. The Debian default
# is /etc/mailname.
# myorigin = /etc/mailname
```

```
smtpd_banner = $myhostname ESMTP $mail_name (Ubuntu)
biff = no

# appending .domain is the MUA's job.
append_dot_mydomain = no

# Uncomment the next line to generate "delayed mail" warnings
#delay_warning_time = 4h

readme_directory = no

# TLS parameters
smtpd_tls_cert_file=/etc/ssl/certs/ssl-cert-snakeoil.pem
smtpd_tls_key_file=/etc/ssl/private/ssl-cert-snakeoil.key
smtpd_use_tls=yes
smtpd_tls_session_cache_database = btree:${data_directory}/smtpd_scache
smtp_tls_session_cache_database =btr ee:${data_directory}/smtp_scache

# See /usr/share/doc/postfix/TLS_README.gz in the postfix-doc
# package for information on enabling SSL
# in the smtp client.

smtpd_relay_restrictions = permit_mynetworks
permit_sasl_authenticated defer_unauth_destination
myhostname = {{ domain_name }}
alias_maps = hash:/etc/aliases
alias_database = hash:/etc/aliases
mydestination = $myhostname, localhost, localhost.localdomain, ,
  localhost
relayhost =
mynetworks = 127.0.0.0/8 [::ffff:127.0.0.0]/104 [::1]/128
mailbox_size_limit = 0
recipient_delimiter = +
inet_interfaces = all
inet_protocols = all
virtual_alias_domains = {{ domain_name }}
virtual_alias_maps = hash:/etc/postfix/virtual
```

（9）创建电子邮件转发配置模板。

```
{# deployment/ansible_templates/virtual.j2 #}
# /etc/postfix/virtual
```

```
hello@{{ domain_name }} admin@example.com
@{{ domain_name }} admin@example.com
```

（10）创建 memcached 配置模板。

```
{# deployment/ansible_templates/memcached.j2 #}
# memcached default config file
# 2003 - Jay Bonci <jaybonci@debian.org>
# This configuration file is read by the start-memcached script
# provided as part of the Debian GNU/Linux distribution.

# Run memcached as a daemon. This command is implied, and is not
# needed for the daemon to run. See the README.Debian
# that comes with this package for more information.
-d

# Log memcached's output to /var/log/memcached
logfile /var/log/memcached.log

# Be verbose
# -v

# Be even more verbose (print client commands as well)
# -vv

# Use 1/16 of server RAM for memcached
-m {{ (ansible_memtotal_mb * 0.0625) | int }}

# Default connection port is 11211
-p 11211

# Run the daemon as root. The start-memcached will default to
# running as root if no -u command is present
# in this config file
-u memcache

# Specify which IP address to listen on. The default is to
# listen on all IP addresses
# This parameter is one of the only security measures that
# memcached has, so make sure it's listening
# on a firewalled interface.
-l 127.0.0.1
```

```
# Limit the number of simultaneous incoming connections. The
# daemon default is 1024
# -c 1024

# Lock down all paged memory. Consult with the README and homepage
# before you do this
# -k

# Return error when memory is exhausted (rather than
# removing items)
# -M

# Maximize core file limit
# -r
```

（11）创建 secrets.json 文件的 Jinja 模板。

```
{# deployment/ansible_templates/secrets.json.j2 #}
{
    "DJANGO_SECRET_KEY": "{{ django_secret_key }}",
    "DATABASE_ENGINE": "django.contrib.gis.db.backends.postgis",
    "DATABASE_NAME": "{{ db_name }}",
    "DATABASE_USER": "{{ db_user }}",
    "DATABASE_PASSWORD": "{{ db_password }}",
    "EMAIL_HOST": "{{ email_host }}",
    "EMAIL_PORT": "{{ email_port }}",
    "EMAIL_HOST_USER": "{{ email_host_user }}",
    "EMAIL_HOST_PASSWORD": "{{ email_host_password }}"
}
```

接下来研究特定于预发布环境的 Vagrant 和 Ansible 脚本。

（1）在.gitignore 文件中，添加下列代码行并忽略某些特定于 Vagrant 和 Ansible 的文件。

```
# .gitignore
# Secrets
secrets.json
secrets.yml

# Vagrant / Ansible
.vagrant
*.retry
```

（2）创建 deployment/staging 和 deployment/staging/ansible 目录。

（3）在 deployment/staging/ansible 目录中，创建 Vagrantfile 文件并包含下列脚本以利用 Ubuntu 18 设置虚拟机，并在其中运行 Ansible 脚本。

```
# deployment/staging/ansible/Vagrantfile
VAGRANTFILE_API_VERSION = "2"

Vagrant.configure(VAGRANTFILE_API_VERSION) do |config|
    config.vm.box = "bento/ubuntu-18.04"
    config.vm.box_version = "201912.14.0"
    config.vm.box_check_update = false
    config.ssh.insert_key=false
    config.vm.provider "virtualbox" do |v|
        v.memory = 512
        v.cpus = 1
        v.name = "myproject"
    end
    config.vm.network "private_network", ip: "192.168.50.5"
    config.vm.provision "ansible" do |ansible|
        ansible.limit = "all"
        ansible.playbook = "setup.yml"
        ansible.inventory_path = "./hosts/vagrant"
        ansible.host_key_checking = false
        ansible.verbose = "vv"
        ansible.extra_vars = { ansible_python_interpreter:
        "/usr/bin/python3" }
    end
end
```

（4）创建包含 vagrant 文件的 hosts 目录，其中的 vagrant 文件包含下列内容。

```
# deployment/staging/ansible/hosts/vagrant
[servers]
192.168.50.5
```

（5）创建一个 vars.yml 文件，其中包含安装脚本和 Jinja 配置模板所用的变量。

```
# deployment/staging/ansible/vars.yml
---
# a unix path-friendly name (IE, no spaces or special characters)
project_name: myproject

user_username: "{{ project_name }}"
```

```
# the base path to install to. You should not need to change this.
install_root: /home

project_root: "{{ install_root }}/{{ project_name }}"

# the python module path to your project's wsgi file
wsgi_module: myproject.wsgi

# any directories that need to be added to the PYTHONPATH.
python_path: "{{ project_root }}/src/{{ project_name }}"

# the git repository URL for the project
project_repo: git@github.com:archatas/django-myproject.git

# The value of your django project's STATIC_ROOT settings.
static_root: "{{ python_path }}/static"
media_root: "{{ python_path }}/media"

locale: en_US.UTF-8
timezone: Europe/Berlin

domain_name: myproject.192.168.50.5.xip.io
django_settings: myproject.settings.staging

letsencrypt_email: ""
```

（6）此外还需要一个包含密码值（如密码和身份验证密钥）的 secrets.yml 文件，对此，创建一个不包含敏感信息的 sample_secrets.yml 文件，但只包含变量名，随后将其复制至 secrets.yml 文件中并填写密码。这里，前一个文件处于版本控制下，而后一个文件则被忽略。

```
# deployment/staging/ansible/sample_secrets.yml
# Django Secret Key
django_secret_key: "change-this-to-50-characters-long-randomstring"

# PostgreSQL database settings
db_name: "myproject"
db_user: "myproject"
db_password: "change-this-to-a-secret-password"
db_host: "localhost"
db_port: "5432"
```

```
# Email SMTP settings
email_host: "localhost"
email_port: "25"
email_host_user: ""
email_host_password: ""

# a private key that has access to the repository URL
ssh_github_key: ~/.ssh/id_rsa_github
```

（7）在 deployment/staging/ansible/setup.yml 处创建一个 Ansible 脚本，用于安装所有的依赖项和配置服务。从 https://raw.githubusercontent.com/PacktPublishing/Django-3-Web-Development-Cookbook-Fourth-Edition/master/ch12/myproject_virtualenv/src/django-myproject/deployment-nginx/staging/ansible/setup.yml 处复制该文件的内容。

（8）随后创建另一个 Ansible 脚本 deployment/staging/ansible/deploy.yml，用于处理 Django 项目。从 https://raw.githubusercontent.com/PacktPublishing/Django-3-Web- Development-Cookbook-Fourth-Edition/master/ch12/myproject_virtualenv/src/django-myproject/deploymentnginx/staging/ansible/deploy.yml 处复制该文件的内容。

（9）创建一个 bash 脚本用于以启动部署。

```
# deployment/staging/ansible/setup_on_virtualbox.sh
#!/usr/bin/env bash
echo "=== Setting up the local staging server ==="
date

cd "$(dirname "$0")"
vagrant up --provision
```

（10）添加 bash 脚本的运行许可并运行该脚本。

```
$ chmod +x setup_on_virtualbox.sh
$ ./setup_on_virtualbox.sh
```

（11）如果脚本出现任何错误，一般情况下是虚拟机需要针对变化内容重启。对此，可通过 ssh 连接至虚拟机，更改为根用户并于随后重启，如下所示。

```
$ vagrant ssh
Welcome to Ubuntu 18.04.3 LTS (GNU/Linux 4.15.0-72-generic x86_64)

* Documentation:    https://help.ubuntu.com
* Management:       https://landscape.canonical.com
* Support:          https://ubuntu.com/advantage
```

```
System information as of Wed Jan 15 04:44:42 CET 2020

System load:     0.21              Processes:             126
Usage of /:      4.0% of 61.80GB   Users logged in:       1
Memory usage:    35%               IP address for eth0:   10.0.2.15
Swap usage:      4%                IP address for eth1:   192.168.50.5

0 packages can be updated.
0 updates are security updates.

*** System restart required ***

This system is built by the Bento project by Chef Software
More information can be found at https://github.com/chef/bento
Last login: Wed Jan 15 04:43:32 2020 from 192.168.50.1
vagrant@myproject:~$ sudo su
root@myproject:/home/vagrant#
reboot
Connection to 127.0.0.1 closed by remote host.
Connection to 127.0.0.1 closed.
```

（12）浏览 Django 项目目录，ssh 至虚拟机并将用户更改为 myproject，如下所示。

```
$ vagrant ssh
Welcome to Ubuntu 18.04.3 LTS (GNU/Linux 4.15.0-74-generic x86_64)
# …
vagrant@myproject:~$ sudo su - myproject
(env) myproject@myproject:~$ pwd
/home/myproject
(env) myproject@myproject:~$ ls
commands db_backups logs public_html src env
```

12.6.3　工作方式

VirtualBox 可在包含不同操纵系统的计算机上持有多个虚拟机。Vagrant 作为一个工具，创建这些虚拟机，并在其上安装和下载操作系统。Ansible 是一个基于 Python 的实用程序，并从.yaml 配置文件中读取指令，进而在远程服务器上执行这些指令。

部署脚本执行下列操作。

- 在 VirtualBox 中创建一个虚拟机并安装 Ubuntu 18。
- 将 IP 值 192.168.50.5 分配于虚拟机。
- 设置虚拟机的主机名。

- 更新 Linux 包。
- 设置服务器的本地化设置项。
- 安装 Linux 依赖项，包括 Python、Nginx、PostgreSQL、Postfix、Memcached 等。
- 创建 Linux 用户和 Django 项目的 home 目录。
- 创建 Django 项目的虚拟环境。
- 创建 PostgreSQL 数据库用户和数据库。
- 配置 Nginx Web 服务器。
- 安装自签名的 SSL 证书。
- 配置 Memcached 缓存服务。
- 配置 Postfix 电子邮件服务器。
- 复制 Django 项目储存库。
- 安装 Python 依赖项。
- 设置 Gunicorn。
- 创建 secrets.json 文件。
- 迁移数据库。
- 收集静态文件。
- 重启 Nginx。

当前，可登录 https://www.myproject.192.168.50.5.xip.io 并访问 Django 网站。此时将显示一个 Hello,World!页面。注意，一些浏览器（包括 Chrome）并不会利用自签名的 SSL 证书打开网站，并作为安全措施阻塞该网站。

注意：

xip.io 是一个通配符 DNS 服务，并将特定于 IP 的子域指向 IP，从而可将其用于 SSL 证书或需要域的其他网站特性中。

如果打算尝试不同的配置或额外的命令，建议采用小幅度渐进方式进行修改。对于某些部分，还需要在将任务转换至 Ansible 指令之前直接在虚拟机上进行测试。

提示：

关于 Ansible 应用方式的更多信息，读者可查看官方文档，对应网址为 https://docs.ansible.com/ansible/latest/index.html，其中展示了大多数用例的指令示例。

如果服务出现错误，可 ssh 至虚拟机，切换至根用户，并查看对应服务的日志。随后在 Google 上查看错误的原因，以进一步了解当前系统。

当重新构建虚拟机时，可使用下列命令。

```
$ vagrant destroy
$ vagrant up --provision
```

12.6.4 延伸阅读

- 第 1 章中的“创建一个项目文件结构”示例。
- 第 1 章中的“利用 pip 处理项目依赖项”示例。
- 第 1 章中的“以动态方式设置 STATIC_URL”示例。
- “针对预发布环境利用 mod_wsgi 在 Apache 上部署”示例。
- “针对产品环境利用 mod_wsgi 在 Apache 上部署”示例。
- “针对产品环境在 Nginx 和 Gunicorn 上部署”示例。
- 第 13 章中的“创建和恢复 PostgreSQL 数据库备份”示例。
- 第 13 章中的“设置常规作业的定时任务”示例。

12.7 针对产品环境在 Nginx 和 Gunicorn 上部署

当前示例将在前述示例的基础上完成，实现一个 Ansible 脚本并利用 Nginx 和 Gunicorn 设置一个产品环境。

12.7.1 准备工作

当运行下列内容时，检查项目配置是否出现警告信息。

```
(env)$ python manage.py check --deploy --
settings=myproject.settings.production
```

确保安装了最新稳定版本的 Ansible。

选择一家服务器提供商，并通过私钥和公钥身份验证以及 ssh 创建一个基于根访问的专用服务器。本书选择的服务器供应商是 DigitalOcean（https://www.digitalocean.com/）。在 DigitalOcean 控制面板处，笔者通过 Ubuntu 18 创建了一个专用服务器（Droplet），并通过新的 SSH 私钥和公钥对（~/.ssh/id_rsa_django_cookbook 和~/.ssh/id_rsa_django_cookbook.pub）连接至服务器的 IP 地址 46.101.136.102。

在本地处，我们需要创建或调整~/.ssh/config 文件配置 SSH 连接，如下所示。

```
# ~/.ssh/config
Host *
```

```
    ServerAliveInterval 240
    AddKeysToAgent yes
    UseKeychain yes

Host github
    Hostname github.com
    IdentityFile ~/.ssh/id_rsa_github

Host myproject-nginx
    Hostname 46.101.136.102
    User root
    IdentityFile ~/.ssh/id_rsa_django_cookbook
```

通过下列命令以及 ssh，作为根用户连接至专用服务器。

```
$ ssh myproject-nginx
```

在域配置中，将域的 DNS A record 指向专用服务器的 IP 地址。在当前示例中，我们仅使用了 myproject.46.101.136.102.xip.io 以展示如何利用 SSL 证书设置 Django 网站的服务器。

12.7.2　实现方式

执行下列步骤创建产品环境下的部署脚本。

（1）确保 deployment/ansible_templates 目录中安装了服务配置使用的 Jinja 模板。

（2）针对 Ansible 脚本，生成 deployment/production 和 deployment/production/ansible 目录。

（3）创建 hosts 目录，并包含 remote 文件，如下所示。

```
# deployment/production/ansible/hosts/remote
[servers]
myproject-nginx

[servers:vars]
ansible_python_interpreter=/usr/bin/python3
```

（4）创建一个 vars.yml 文件，其中包含用于安装脚本的变量和配置用的 Jinja 模板。

```
# deployment/production/ansible/vars.yml
---
# a unix path-friendly name (IE, no spaces or special characters)
project_name: myproject
```

```
user_username: "{{ project_name }}"

# the base path to install to. You should not need to change this.
install_root: /home

project_root: "{{ install_root }}/{{ project_name }}"

# the python module path to your project's wsgi file
wsgi_module: myproject.wsgi

# any directories that need to be added to the PYTHONPATH.
python_path: "{{ project_root }}/src/{{ project_name }}"

# the git repository URL for the project
project_repo: git@github.com:archatas/django-myproject.git

# The value of your django project's STATIC_ROOT settings.
static_root: "{{ python_path }}/static"
media_root: "{{ python_path }}/media"

locale: en_US.UTF-8
timezone: Europe/Berlin

domain_name: myproject.46.101.136.102.xip.io
django_settings: myproject.settings.production

# letsencrypt settings
letsencrypt_email: hello@myproject.com
```

（5）此外还需要一个包含密码值（如密码和身份验证密钥）的 secrets.yml 文件。对此，首先创建一个不包含敏感信息的 sample_secrets.yml 文件，且仅包含变量名，随后将其复制至 secrets.yml 并填写密码。这里，前一个文件处于版本控制下，而后一个文件则被忽略。

```
# deployment/production/ansible/sample_secrets.yml
# Django Secret Key
django_secret_key: "change-this-to-50-characters-long-random-string"

# PostgreSQL database settings
db_name: "myproject"
db_user: "myproject"
db_password: "change-this-to-a-secret-password"
db_host: "localhost"
db_port: "5432"
```

```
# Email SMTP settings
email_host: "localhost"
email_port: "25"
email_host_user: ""
email_host_password: ""

# a private key that has access to the repository URL
ssh_github_key: ~/.ssh/id_rsa_github
```

（6）在 deployment/production/ansible/setup.yml 处创建一个 Ansible 脚本，用于安装所有的依赖项和配置服务。从 https://raw.githubusercontent.com/PacktPublishing/Django-3-Web-Development-Cookbook-Fourth-Edition/master/ch12/myproject_virtualenv/src/django-myproject/deployment-nginx/production/ansible/setup.yml 处复制该文件的内容。

（7）在 deployment/production/ansible/deploy.yml 处创建另一个 Ansible 脚本，用于处理 Django 项目。从 https://raw.githubusercontent.com/PacktPublishing/Django-3-Web-Development-Cookbook-Fourth-Edition/master/ch12/myproject_virtualenv/src/djangomyproject/deployment-nginx/production/ansible/deploy.yml 处复制该文件的内容。

（8）创建一个 bash 脚本，用以启动部署。

```
# deployment/production/ansible/setup_remotely.sh
#!/usr/bin/env bash
echo "=== Setting up the production server ==="
date

cd "$(dirname "$0")"
ansible-playbook setup.yml -i hosts/remote
```

（9）添加 bash 脚本的执行许可并运行该脚本。

```
$ chmod +x setup_remotely.sh
$ ./setup_remotely.sh
```

（10）如果脚本出现任何错误，一般情况下是内容变化导致需要重启专用服务器。对此，可通过 ssh 连接至服务器并重启，如下所示。

```
$ ssh myproject-nginx
Welcome to Ubuntu 18.04.3 LTS (GNU/Linux 4.15.0-74-generic x86_64)

* Documentation: https://help.ubuntu.com
* Management: https://landscape.canonical.com
* Support: https://ubuntu.com/advantage
```

```
System information as of Wed Jan 15 11:39:51 CET 2020

System load: 0.08 Processes: 104
Usage of /: 8.7% of 24.06GB Users logged in: 0
Memory usage: 35% IP address for eth0: 142.93.167.30
Swap usage: 0%

* Canonical Livepatch is available for installation.
- Reduce system reboots and improve kernel security. Activate at:
https://ubuntu.com/livepatch

0 packages can be updated.
0 updates are security updates.

*** System restart required ***

Last login: Sun Jan 12 12:23:35 2020 from 178.12.115.146
root@myproject:~# reboot
Connection to 142.93.167.30 closed by remote host.
Connection to 142.93.167.30 closed.
```

（11）创建另一个 bash 脚本用于更新 Django 项目。

```
# deployment/production/ansible/deploy_remotely.sh
#!/usr/bin/env bash
echo "=== Deploying project to production server ==="
date

cd "$(dirname "$0")"
ansible-playbook deploy.yml -i hosts/remote
```

（12）添加 bash 脚本的执行许可。

```
$ chmod +x deploy_remotely.sh
```

12.7.3 工作方式

Ansible 脚本是幂等的，这意味着，我们可多次执行该脚本，并且总是会得到相同的结果，即一个最新的专用服务器，同时安装并运行 Django 网站。如果服务器存在任何技术硬件问题，并且持有数据库和媒体文件备份，那么我们可较快地在另一台专用服务器上安装相同的配置。

产品部署脚本执行下列操作。

❑ 设置虚拟机的主机名。

- 更新 Linux 包。
- 设置服务器的本地化设置项。
- 安装 Linux 依赖项，包括 Python、Nginx、PostgreSQL、Postfix、Postfix、Memcached 等。
- 创建 Linux 用户和 Django 项目的 home 目录。
- 创建 Django 项目的虚拟环境。
- 创建 PostgreSQL 数据库用户和数据库。
- 配置 Nginx Web 服务器。
- 安装 Let's Encrypt SSL 证书。
- 配置 Memcached 缓存服务。
- 配置 Postfix 电子邮件服务。
- 复制 Django 项目储存库。
- 安装 Python 依赖项。
- 设置 Gunicorn。
- 创建 secrets.json 文件。
- 迁移数据库。
- 收集静态文件。
- 重启 Nginx。

可以看到，安装过程与预发布服务器十分类似。为了保持灵活性和可调整性，我们可将其单独存储在 deployment/production 目录中。

从理论上讲，可省略预发布环境，但可尝试在虚拟机上实现部署过程而非直接在远程服务器上安装则更具实际意义。

12.7.4　延伸阅读

- 第 1 章中的“创建一个项目文件结构”示例。
- 第 1 章中的“利用 pip 处理项目依赖项”示例。
- 第 1 章中的“以动态方式设置 STATIC_URL”示例。
- “针对预发布环境利用 mod_wsgi 在 Apache 上部署”示例。
- “针对产品环境利用 mod_wsgi 在 Apache 上部署”示例。
- “针对预发布环境在 Nginx 和 Gunicorn 上部署”示例。
- 第 13 章中的“创建和恢复 PostgreSQL 数据库备份”示例。
- 第 13 章中的“设置常规作业的定时任务”示例。

第 13 章 维　　护

本章主要涉及下列主题。

- ❑ 创建和恢复 MySQL 数据库备份。
- ❑ 创建和恢复 PostgreSQL 数据库备份。
- ❑ 设置常规作业的定时任务。
- ❑ 日志事件，以便进一步检查。
- ❑ 通过电子邮件获取详细的错误报告。

13.1 简　　介

此时，我们应持有一个或多个开发、发布完毕的 Django 项目。在开发环节的最后一步中，我们将讨论如何维护项目和监测项目。

13.2 技术需求

当与本章代码协同工作时，我们需要安装最新稳定版本的 Python、MySQL 或 PostgreSQL 数据库，以及虚拟环境下的 Django 项目。

读者可访问 GitHub 储存库的 ch13 目录以查看本章的全部代码，对应网址为 https://github.com/PacktPublishing/Django-3-Web-Development-Cookbook-Fourth-Edition。

13.3 创建和恢复 MySQL 数据库备份

考虑到网站的稳定性，能够从硬件故障和黑客攻击中恢复是十分重要的。因此，我们应始终进行备份以确保一切正常工作。相应地，代码和静态文件通常处于版本控制下，因而可以此实现恢复，但数据库和媒体文件则应定期进行备份。

在当前示例中，我们将讨论如何生成 MySQL 数据库的备份。

13.3.1　准备工作

确保 Django 项目可运行 MySQL 数据库，并将该项目部署至远程产品（或预发布）服务器。

13.3.2　实现方式

下列步骤将备份和恢复 MySQL 数据库。

（1）在项目的 home 目录的 commands 目录下，创建一个 bash 脚本 backup_mysql_db.sh。通过变量和函数定义启动该脚本，如下所示。

```
/home/myproject/commands/backup_mysql_db.sh
#!/usr/bin/env bash
SECONDS=0
export DJANGO_SETTINGS_MODULE=myproject.settings.production
PROJECT_PATH=/home/myproject
REPOSITORY_PATH=${PROJECT_PATH}/src/myproject
LOG_FILE=${PROJECT_PATH}/logs/backup_mysql_db.log
DAY_OF_THE_WEEK=$(LC_ALL=en_US.UTF-8 date +"%w-%A")
DAILY_BACKUP_PATH=${PROJECT_PATH}/db_backups/${DAY_OF_THE_WEEK}.sql
LATEST_BACKUP_PATH=${PROJECT_PATH}/db_backups/latest.sql
error_counter=0

echoerr() { echo "$@" 1>&2; }

cd ${PROJECT_PATH}
mkdir -p logs
mkdir -p db_backups

source env/bin/activate
cd ${REPOSITORY_PATH}

DATABASE=$(echo "from django.conf import settings;
print(settings.DATABASES['default']['NAME'])" | python manage.py
shell -i python)
USER=$(echo "from django.conf import settings;
print(settings.DATABASES['default']['USER'])" | python manage.py
shell -i python)
PASSWORD=$(echo "from django.conf import settings;
print(settings.DATABASES['default']['PASSWORD'])" | python
```

```
manage.py shell -i python)

EXCLUDED_TABLES=(
django_session
)

IGNORED_TABLES_STRING=''
for TABLE in "${EXCLUDED_TABLES[@]}"; do
   IGNORED_TABLES_STRING+=" --ignore-table=${DATABASE}.${TABLE}"
done
```

（2）添加命令来创建数据库结构和数据的转储。

```
echo "=== Creating DB Backup ===" > ${LOG_FILE}
date >> ${LOG_FILE}

echo "- Dump structure" >> ${LOG_FILE}
mysqldump -u "${USER}" -p"${PASSWORD}" --single-transaction --no-data
"${DATABASE}" > "${DAILY_BACKUP_PATH}" 2>> ${LOG_FILE}
function_exit_code=$?
if [[ $function_exit_code -ne 0 ]]; then
    {
       echoerr "Command mysqldump for dumping database structure
        failed with exit code ($function_exit_code)."
       error_counter=$((error_counter + 1))
    } >> "${LOG_FILE}" 2>&1
fi

echo "- Dump content" >> ${LOG_FILE}
# shellcheck disable=SC2086
mysqldump -u "${USER}" -p"${PASSWORD}" "${DATABASE}"
${IGNORED_TABLES_STRING} >> "${DAILY_BACKUP_PATH}" 2>> ${LOG_FILE}
function_exit_code=$?

if [[ $function_exit_code -ne 0 ]]; then
    {
       echoerr "Command mysqldump for dumping database content
        failed with exit code ($function_exit_code)."
       error_counter=$((error_counter + 1))
    } >> "${LOG_FILE}" 2>&1
fi
```

（3）添加命令压缩数据库转储并创建一个符号链接 latest.sql.gz。

```
echo "- Create a *.gz archive" >> ${LOG_FILE}
gzip --force "${DAILY_BACKUP_PATH}"
function_exit_code=$?
if [[ $function_exit_code -ne 0 ]]; then
    {
        echoerr "Command gzip failed with exit code
         ($function_exit_code)."
        error_counter=$((error_counter + 1))
    } >> "${LOG_FILE}" 2>&1
fi

echo "- Create a symlink latest.sql.gz" >> ${LOG_FILE}
if [ -e "${LATEST_BACKUP_PATH}.gz" ]; then
    rm "${LATEST_BACKUP_PATH}.gz"
fi
ln -s "${DAILY_BACKUP_PATH}.gz" "${LATEST_BACKUP_PATH}.gz"
function_exit_code=$?
if [[ $function_exit_code -ne 0 ]]; then
    {
        echoerr "Command ln failed with exit code
         ($function_exit_code)."
        error_counter=$((error_counter + 1))
    } >> "${LOG_FILE}" 2>&1
fi
```

（4）通过记录执行上述命令所花费的时间来完成脚本。

```
duration=$SECONDS
echo "------------------------------------------" >> ${LOG_FILE}
echo "The operation took $((duration / 60)) minutes and $((duration
% 60)) seconds." >> ${LOG_FILE}
exit $error_counter
```

（5）在同一目录中，创建一个 bash 脚本 restore_mysql_db.sh，如下所示。

```
# home/myproject/commands/restore_mysql_db.sh
#!/usr/bin/env bash
SECONDS=0
PROJECT_PATH=/home/myproject
REPOSITORY_PATH=${PROJECT_PATH}/src/myproject
LATEST_BACKUP_PATH=${PROJECT_PATH}/db_backups/latest.sql
export DJANGO_SETTINGS_MODULE=myproject.settings.production

cd "${PROJECT_PATH}"
```

```
source env/bin/activate

echo "=== Restoring DB from a Backup ==="

echo "- Fill the database with schema and data"
cd "${REPOSITORY_PATH}"
zcat "${LATEST_BACKUP_PATH}.gz" | python manage.py dbshell

duration=$SECONDS
echo "------------------------------------------"
echo "The operation took $((duration / 60)) minutes and $((duration
% 60)) seconds."
```

（6）运行两个脚本。

```
$ chmod +x *.sh
```

（7）运行数据库备份脚本。

```
$ ./backup_mysql_db.sh
```

（8）运行数据库恢复脚本。

```
$ ./restore_mysql_db.sh
```

13.3.3　工作方式

备份脚本将在/home/myproject/db_backups/下创建备份文件，并将日志保存在/home/myproject/logs/backup_mysql_db.log 中，如下所示。

```
=== Creating DB Backup ===
Fri Jan 17 02:12:14 CET 2020
- Dump structure
mysqldump: [Warning] Using a password on the command line interface can
be insecure.
- Dump content
mysqldump: [Warning] Using a password on the command line interface can be
insecure.
- Create a *.gz archive
- Create a symlink latest.sql.gz
------------------------------------------
The operation took 0 minutes and 2 seconds.
```

如果操作成功，脚本将以代码 0 返回并退出；否则，退出代码为执行脚本时错误的数量。同时，日志文件将显示错误消息。

在 db_backups 目录中，存在一个压缩后的 SQL 备份且标有一周中某一天的信息，如 0-Sunday.sql.gz、1-Monday.sql.gz 等。另一个文件则称作 latest.sql.gz，实际上是符号链接。基于工作日的备份可在定时作业下正确设置最近 7 天的备份；符号链接则可通过 SSH 快速或自动将最新的备份传输至另一台计算机。

注意，我们从 Django 设置项中获取数据库证书，并在 bash 脚本中使用这些证书。

除会话表外，我们转储了全部数据，因为会话是临时的且十分消耗内存空间。

当运行 restore_mysql_db.sh 脚本时，对应的输出结果如下所示。

```
=== Restoring DB from a Backup ===
- Fill the database with schema and data
mysql: [Warning] Using a password on the command line interface can be
insecure.
------------------------------------------
The operation took 0 minutes and 2 seconds.
```

13.3.4　延伸阅读

- 第 12 章中的“针对产品环境利用 mod_wsgi 在 Apache 上部署”示例。
- 第 12 章中的“针对产品环境在 Nginx 和 Gunicorn 上部署”示例。
- “创建和恢复 PostgreSQL 数据库备份”示例。
- “设置常规作业的定时任务”示例。

13.4　创建和恢复 PostgreSQL 数据库备份

在当前示例中，我们将学习如何备份 PostgreSQL 数据库，并在硬件故障或受到黑客攻击时对其进行恢复。

13.4.1　准备工作

确保 Django 项目能够运行 PostgreSQL 数据库，同时将项目部署至远程模拟或产品服务器上。

13.4.2　实现方式

下列步骤将备份和恢复 PostgreSQL 数据库。

（1）在项目 home 目录的 commands 目录下，创建一个 bash 脚本 backup_postgresql_db.sh，利用变量和函数定义启动脚本。

```
/home/myproject/commands/backup_postgresql_db.sh
#!/usr/bin/env bash
SECONDS=0
PROJECT_PATH=/home/myproject
REPOSITORY_PATH=${PROJECT_PATH}/src/myproject
LOG_FILE=${PROJECT_PATH}/logs/backup_postgres_db.log
DAY_OF_THE_WEEK=$(LC_ALL=en_US.UTF-8 date +"%w-%A")
DAILY_BACKUP_PATH=${PROJECT_PATH}/db_backups/${DAY_OF_THE_WEEK}.backup
LATEST_BACKUP_PATH=${PROJECT_PATH}/db_backups/latest.backup
error_counter=0

echoerr() { echo "$@" 1>&2; }

cd ${PROJECT_PATH}
mkdir -p logs
mkdir -p db_backups

source env/bin/activate
cd ${REPOSITORY_PATH}

DATABASE=$(echo "from django.conf import settings;
print(settings.DATABASES['default']['NAME'])" | python manage.py
shell -i python)
```

（2）添加命令创建一个数据库转储。

```
echo "=== Creating DB Backup ===" > ${LOG_FILE}
date >> ${LOG_FILE}

echo "- Dump database" >> ${LOG_FILE}
pg_dump --format=p --file="${DAILY_BACKUP_PATH}" ${DATABASE}
function_exit_code=$?
if [[ $function_exit_code -ne 0 ]]; then
    {
        echoerr "Command pg_dump failed with exit code
         ($function_exit_code)."
        error_counter=$((error_counter + 1))
    } >> "${LOG_FILE}" 2>&1
fi
```

（3）添加命令压缩数据库转储，并创建一个符号链接 latest.backup.gz。

```
echo "- Create a *.gz archive" >> ${LOG_FILE}
gzip --force "${DAILY_BACKUP_PATH}"
```

```
function_exit_code=$?
if [[ $function_exit_code -ne 0 ]]; then
    {
        echoerr "Command gzip failed with exit code
         ($function_exit_code)."
        error_counter=$((error_counter + 1))
    } >> "${LOG_FILE}" 2>&1
fi

echo "- Create a symlink latest.backup.gz" >> ${LOG_FILE}
if [ -e "${LATEST_BACKUP_PATH}.gz" ]; then
    rm "${LATEST_BACKUP_PATH}.gz"
fi
ln -s "${DAILY_BACKUP_PATH}.gz" "${LATEST_BACKUP_PATH}.gz"
function_exit_code=$?
if [[ $function_exit_code -ne 0 ]]; then
    {
        echoerr "Command ln failed with exit code
         ($function_exit_code)."
        error_counter=$((error_counter + 1))
    } >> "${LOG_FILE}" 2>&1
fi
```

（4）通过记录执行上述命令所花费的时间来完成脚本。

```
duration=$SECONDS
echo "------------------------------------------" >> ${LOG_FILE}
echo "The operation took $((duration / 60)) minutes and $((duration
% 60)) seconds." >> ${LOG_FILE}
exit $error_counter
```

（5）在同一目录中，创建一个 bash 脚本 restore_postgresql_db.sh，如下所示。

```
# /home/myproject/commands/restore_postgresql_db.sh
#!/usr/bin/env bash
SECONDS=0
PROJECT_PATH=/home/myproject
REPOSITORY_PATH=${PROJECT_PATH}/src/myproject
LATEST_BACKUP_PATH=${PROJECT_PATH}/db_backups/latest.backup
export DJANGO_SETTINGS_MODULE=myproject.settings.production

cd "${PROJECT_PATH}"
source env/bin/activate

cd "${REPOSITORY_PATH}"
```

```
DATABASE=$(echo "from django.conf import settings;
print(settings.DATABASES['default']['NAME'])" | python manage.py
shell -i python)
USER=$(echo "from django.conf import settings;
print(settings.DATABASES['default']['USER'])" | python manage.py
shell -i python)
PASSWORD=$(echo "from django.conf import settings;
print(settings.DATABASES['default']['PASSWORD'])" | python
manage.py shell -i python)

echo "=== Restoring DB from a Backup ==="

echo "- Recreate the database"
psql --dbname=$DATABASE --command='SELECT
pg_terminate_backend(pg_stat_activity.pid) FROM pg_stat_activity
WHERE datname = current_database() AND pid <> pg_backend_pid();'

dropdb $DATABASE

createdb --username=$USER $DATABASE

echo "- Fill the database with schema and data"
zcat "${LATEST_BACKUP_PATH}.gz" | python manage.py dbshell

duration=$SECONDS
echo "------------------------------------------"
echo "The operation took $((duration / 60)) minutes and $((duration
% 60)) seconds."
```

（6）执行两个脚本。

```
$ chmod +x *.sh
```

（7）运行数据库备份脚本。

```
$ ./backup_postgresql_db.sh
```

（8）运行数据库恢复脚本。

```
$ ./restore_postgresql_db.sh
```

13.4.3　工作方式

备份脚本将在/home/myproject/db_backups/下创建备份文件，并将日志保存在/home/

myproject/logs/backup_ postgresql_db.log 中，如下所示。

```
=== Creating DB Backup ===
Fri Jan 17 02:40:55 CET 2020
- Dump database
- Create a *.gz archive
- Create a symlink latest.backup.gz
------------------------------------------
The operation took 0 minutes and 1 seconds.
```

如果操作成功，脚本将以代码 0 返回并退出；否则，退出代码为执行脚本时错误的数量。同时，日志文件将显示错误消息。

在 db_backups 目录中，存在一个压缩后的 SQL 备份且标有一周中的某天的信息，如 0-Sunday.sql.gz、1-Monday.sql.gz 等。另一个文件则称作 latest.backup.gz，实际上是符号链接。基于工作日的备份可在定时作业下正确设置最近 7 天的备份；符号链接则可通过 SSH 快速或自动将最新的备份传输至另一台计算机中。

注意，我们从 Django 设置项中获取数据库证书，并在 bash 脚本中使用这些证书。

当运行 restore_postgresql_db.sh 脚本时，对应的输出结果如下所示。

```
=== Restoring DB from a Backup ===
- Recreate the database
pg_terminate_backend
----------------------
(0 rows)

- Fill the database with schema and data
SET
SET
SET
SET
SET
set_config
------------

(1 row)

SET

…

ALTER TABLE
```

```
ALTER TABLE
ALTER TABLE
-----------------------------------------
The operation took 0 minutes and 2 seconds.
```

13.4.4　延伸阅读

- 第 12 章中的“针对产品环境利用 mod_wsgi 在 Apache 上部署”示例。
- 第 12 章中的“针对产品环境在 Nginx 和 Gunicorn 上部署”示例。
- “创建和恢复 PostgreSQL 数据库备份”示例。
- “设置常规作业的定时任务”示例。

13.5　设置常规作业的定时任务

通常情况下，网站会涉及一些管理任务，并定期在后台运行，如每个星期一次、一天一次或每隔一个小时一次。这可通过调度任务（也称作定时任务）完成。这些任务可定义为脚本，并在指定时间间隔后在服务器上运行。在当前示例中，我们将创建两个定时任务，其中，第一个任务将清除数据库中的会话，第二个任务则备份数据库数据。这两项任务均在每晚运行。

13.5.1　准备工作

首先需要将 Django 项目部署至远程服务器上。随后通过 SSH 连接至服务器。这些步骤均假设我们正在使用一个虚拟环境，但类似的定时任务可针对 Docker 项目被创建，甚至还可直接运行于应用程序容器中。相应地，代码文件提供了替代语法，而其他步骤则基本相同。

13.5.2　实现方式

下列步骤将创建两个脚本并定期运行。

（1）在生产或预发布服务器上，访问项目用户的 home 目录，其中包含了 env 和 src 目录。

（2）在 env 目录旁边创建 commands、db_backups 和 logs 文件夹，如下所示。

```
(env)$ mkdir commands db_backups logs
```

（3）在 commands 目录中，创建一个 clear_sessions.sh 文件。我们可利用终端编辑器（如 vim 或 nano）编辑该文件，如下所示。

```
# /home/myproject/commands/clear_sessions.sh
#!/usr/bin/env bash
SECONDS=0
export DJANGO_SETTINGS_MODULE=myproject.settings.production
PROJECT_PATH=/home/myproject
REPOSITORY_PATH=${PROJECT_PATH}/src/myproject
LOG_FILE=${PROJECT_PATH}/logs/clear_sessions.log
error_counter=0

echoerr() { echo "$@" 1>&2; }

cd ${PROJECT_PATH}
mkdir -p logs

echo "=== Clearing up Outdated User Sessions ===" > ${LOG_FILE}
date >> ${LOG_FILE}

source env/bin/activate
cd ${REPOSITORY_PATH}
python manage.py clearsessions >> "${LOG_FILE}" 2>&1
function_exit_code=$?
if [[ $function_exit_code -ne 0 ]]; then
    {
        echoerr "Clearing sessions failed with exit code
         ($function_exit_code)."
        error_counter=$((error_counter + 1))
    } >> "${LOG_FILE}" 2>&1
fi

duration=$SECONDS
echo "------------------------------------------" >> ${LOG_FILE}
echo "The operation took $((duration / 60)) minutes and $((duration
% 60)) seconds." >> ${LOG_FILE}
exit $err
or_counter
```

（4）生成 clear_sessions.sh 可执行文件，如下所示。

```
$ chmod +x *.sh
```

（5）假设采用 PostgreSQL 作为项目的数据库。随后在同一目录中，如前所述，创建一个备份脚本。

（6）测试脚本并检查是否可正确执行，即运行脚本并在 logs 目录中检查*.log 文件，如下所示。

```
$ ./clear_sessions.sh
$ ./backup_postgresql_db.sh
```

（7）在远程服务器上的项目的 home 目录中，创建一个 crontab.txt 文件，如下所示。

```
# /home/myproject/crontab.txt
MAILTO=""
HOME=/home/myproject
PATH=/home/myproject/env/bin:/usr/local/sbin:/usr/local/bin:/usr/sb
in:/usr/bin:/sbin:/bin:/usr/games:/usr/local/games:/snap/bin
SHELL=/bin/bash
00 01 * * * /home/myproject/commands/clear_sessions.sh
00 02 * * * /home/myproject/commands/backup_postgresql_db.sh
```

（8）以 myproject 用户身份安装 crontab 任务，如下所示。

```
(env)$ crontab crontab.txt
```

13.5.3　工作方式

根据当前设置，clear_sessions.sh 将于 1:00 A.M.被运行；而 backup_postgresql_db.sh 则在 2:00 A.M.被运行。另外，运行日志被保存于~/logs/clear_sessions.sh 和~/logs/backup_postgresql_db.log 文件中，如果出现任何错误，用户可检查这些文件用于回溯。

每天，clear_sessions.sh 将执行 clearsessions 管理命令。顾名思义，这将利用默认的数据库设置清除数据库中的过期会话。

数据库备份脚本则稍显复杂。每天，该脚本利用 0-Sunday.backup.gz、1-Monday.backup.gz 等命名模式创建当日的备份文件。因此，我们可以恢复 7 天前后的备份数据。

crontab 文件遵循特定的语法，每行代码包含一个特定的时间并由一系列的数字表示，随后是既定时间运行的任务。这里，时间由 5 部分构成并以空格分隔，如下所示。

（1）分钟（0～59）。

（2）小时（0～23）。

（3）月份中的天数（1～31）。

（4）月份（1～12）。

（5）星期天数（0～7，其中，0 表示星期日，1 表示星期一等。7 再次表示为星期日）。

另外，*号表示将使用每个时间范围。因此，下列*号意味着 clear_sessions.sh 将于每月的每天、每月、每星期中的每天的 1:00 A.M 运行。

```
00 01 * * * /home/myproject/commands/clear_sessions.sh
```

关于 crontab 规范，读者可访问 https://en.wikipedia.org/wiki/以了解更多内容。

13.5.4　更多内容

前述内容定义了定期执行的命令，并且结果日志也处于激活状态，但是我们并不了解定时任务是否成功执行，除非我们登录服务器并每天以手动方式查看日志。为了解决这一枯燥的手动检查问题，我们可通过健康检查（healthchecks）服务（https://healthchecks.io/）自动监视定时任务。

当使用健康检查服务时，我们可以调整定时任务，以便在每次成功执行作业后 ping 一个特定的 URL。如果脚本失败并以非 0 代码退出，健康检查服务将了解到任务未成功执行。每天，我们可通过电子邮件的方式获取定时任务的概况及其执行状态。

13.5.5　延伸阅读

- 第 12 章中的“针对产品环境利用 mod_wsgi 在 Apache 上部署”示例。
- 第 12 章中的“针对产品环境在 Nginx 和 Gunicorn 上部署”示例。
- “创建和恢复 MySQL 数据库备份”示例。
- “创建和恢复 PostgreSQL 数据库备份”示例。

13.6　日 志 事 件

在前述示例中，我们看到了日志针对 bash 脚本的工作方式。不仅如此，我们还可以记录 Django 网站上出现的事件，如用户注册、向购物车中添加一件商品、购买门票、银行交易、发送 SMS 消息、服务器错误等。

提示：

不应对敏感信息予以记录，如用户密码或信用卡详细信息。

另外，应使用分析工具而非 Python 日志机制跟踪网站的整体应用状况。

在当前示例中，我们将学习如何将网站的结构化信息记录至日志文件中。

13.6.1　准备工作

当前示例将在 likes 应用程序的基础上完成。

在 Django 项目的虚拟环境中，安装 django-structlog，如下所示。

```
(env)$ pip install django-structlog==1.3.5
```

13.6.2　实现方式

下列步骤将设置 Django 网站中的结构化日志机制。

（1）在项目的设置项中添加 RequestMiddleware。

```
# myproject/settings/_base.py
MIDDLEWARE = [
    "django.middleware.security.SecurityMiddleware",
    "django.contrib.sessions.middleware.SessionMiddleware",
    "django.middleware.common.CommonMiddleware",
    "django.middleware.csrf.CsrfViewMiddleware",
    "django.contrib.auth.middleware.AuthenticationMiddleware",
    "django.contrib.messages.middleware.MessageMiddleware",
    "django.middleware.clickjacking.XFrameOptionsMiddleware",
    "django.middleware.locale.LocaleMiddleware",
    "django_structlog.middlewares.RequestMiddleware",
]
```

（2）在同一文件中，添加 Django 日志配置。

```
# myproject/settings/_base.py
LOGGING = {
    "version": 1,
    "disable_existing_loggers": False,
    "formatters": {
        "json_formatter": {
            "()": structlog.stdlib.ProcessorFormatter,
            "processor": structlog.processors.JSONRenderer(),
        },
        "plain_console": {
            "()": structlog.stdlib.ProcessorFormatter,
            "processor": structlog.dev.ConsoleRenderer(),
        },
```

```
        "key_value": {
            "()": structlog.stdlib.ProcessorFormatter,
            "processor":
             structlog.processors.KeyValueRenderer(key_order=
             ['timestamp', 'level', 'event', 'logger']),
        },
    },
    "handlers": {
        "console": {
            "class": "logging.StreamHandler",
            "formatter": "plain_console",
        },
        "json_file": {
            "class": "logging.handlers.WatchedFileHandler",
            "filename": os.path.join(BASE_DIR, "tmp", "json.log"),
            "formatter": "json_formatter",
        },
        "flat_line_file": {
            "class": "logging.handlers.WatchedFileHandler",
            "filename": os.path.join(BASE_DIR, "tmp", "flat_line.log"),
            "formatter": "key_value",
        },
    },
    "loggers": {
        "django_structlog": {
            "handlers": ["console", "flat_line_file", "json_file"],
            "level": "INFO",
        },
    }
}
```

（3）设置 structlog 配置。

```
# myproject/settings/_base.py
structlog.configure(
    processors=[
        structlog.stdlib.filter_by_level,
        structlog.processors.TimeStamper(fmt="iso"),
        structlog.stdlib.add_logger_name,
        structlog.stdlib.add_log_level,
        structlog.stdlib.PositionalArgumentsFormatter(),
        structlog.processors.StackInfoRenderer(),
        structlog.processors.format_exc_info,
```

```
        structlog.processors.UnicodeDecoder(),
        structlog.processors.ExceptionPrettyPrinter(),
        structlog.stdlib.ProcessorFormatter.wrap_for_formatter,
    ],
    context_class=structlog.threadlocal.wrap_dict(dict),
    logger_factory=structlog.stdlib.LoggerFactory(),
    wrapper_class=structlog.stdlib.BoundLogger,
    cache_logger_on_first_use=True,
)
```

（4）在 likes 应用程序的 views.py 文件中，记录点赞/未点赞对象。

```
# myproject/apps/likes/views.py
import structlog

from django.contrib.contenttypes.models import ContentType
from django.http import JsonResponse
from django.views.decorators.cache import never_cache
from django.views.decorators.csrf import csrf_exempt

from .models import Like
from .templatetags.likes_tags import liked_count

logger = structlog.get_logger("django_structlog")

@never_cache
@csrf_exempt
def json_set_like(request, content_type_id, object_id):
    """
    Sets the object as a favorite for the current user
    """
    result = {
        "success": False,
    }
    if request.user.is_authenticated and request.method == "POST":
        content_type = ContentType.objects.get(id=content_type_id)
        obj = content_type.get_object_for_this_type(pk=object_id)

        like, is_created = Like.objects.get_or_create(
            content_type=ContentType.objects.get_for_model(obj),
            object_id=obj.pk,
            user=request.user)
        if is_created:
```

```
            logger.info("like_created",
            content_type_id=content_type.pk,
            object_id=obj.pk)
        else:
            like.delete()
            logger.info("like_deleted",
            content_type_id=content_type.pk,
            object_id=obj.pk)

        result = {
            "success": True,
            "action": "add" if is_created else "remove",
            "count": liked_count(obj),
        }

    return JsonResponse(result)
```

13.6.3　工作方式

当访问者浏览站点时，特定的事件将记录于 tmp/json.log 和 tmp/flat_line.log 文件中。django_structlog.middlewares.RequestMiddleware 则记录了 HTTP 请求处理的开始和结束状况。除此之外，我们还记录了 Like 实例的创建和删除时间。

json.log 包含 JSON 格式的日志，这意味着，我们可从编程角度解析、查看和分析日志。

```
{"request_id": "ad0ef355-77ef-4474-a91a-2d9549a0e15d", "user_id": 1, "ip":
"127.0.0.1", "request": "<WSGIRequest: POST
'/en/likes/7/1712dfe4-2e77-405c-aa9b-bfa64a1abe98/'>", "user_agent":
"Mozilla/5.0 (Macintosh; Intel Mac OS X 10_15_2) AppleWebKit/537.36 (KHTML,
like Gecko) Chrome/79.0.3945.130 Safari/537.36", "event":
"request_started", "timestamp": "2020-01-18T04:27:00.556135Z", "logger":
"django_structlog.middlewares.request", "level": "info"}
{"request_id": "ad0ef355-77ef-4474-a91a-2d9549a0e15d", "user_id": 1, "ip":
"127.0.0.1", "content_type_id": 7, "object_id": "UUID('1712dfe4-2e77-
405c-aa9b-bfa64a1abe98')", "event": "like_created", "timestamp":
"2020-01-18T04:27:00.602640Z", "logger": "django_structlog", "level":
"info"}
{"request_id": "ad0ef355-77ef-4474-a91a-2d9549a0e15d", "user_id": 1, "ip":
"127.0.0.1", "code": 200, "request": "<WSGIRequest: POST
'/en/likes/7/1712dfe4-2e77-405c-aa9b-bfa64a1abe98/'>", "event":
"request_finished", "timestamp": "2020-01-18T04:27:00.604577Z", "logger":
"django_structlog.middlewares.request", "level": "info"}
```

flat_line.log 文件包含了简短格式的日志，且易于阅读。

```
(env)$ tail -3 tmp/flat_line.log
timestamp='2020-01-18T04:27:03.437759Z' level='info'
event='request_started' logger='django_structlog.middlewares.request'
request_id='a74808ff-c682-4336-aeb9-f043f11a7316' user_id=1 ip='127.0.0.1'
request=<WSGIRequest: POST '/en/likes/7/1712dfe4-2e77-405c-aa9b-
bfa64a1abe98/'> user_agent='Mozilla/5.0 (Macintosh; Intel Mac OS X 10_15_2)
AppleWebKit/537.36 (KHTML, like Gecko) Chrome/79.0.3945.130 Safari/537.36'
timestamp='2020-01-18T04:27:03.489198Z' level='info' event='like_deleted'
logger='django_structlog' request_id='a74808ff-c682-4336-aeb9-f043f11a7316'
user_id=1 ip='127.0.0.1' content_type_id=7
object_id=UUID('1712dfe4-2e77-405c-aa9b-bfa64a1abe98')
timestamp='2020-01-18T04:27:03.491927Z' level='info'
event='request_finished' logger='django_structlog.middlewares.request'
request_id='a74808ff-c682-4336-aeb9-f043f11a7316' user_id=1 ip='127.0.0.1'
code=200 request=<WSGIRequest: POST '/en/likes/7/1712dfe4-2e77-405c-
aa9b-bfa64a1abe98/'>
```

13.6.4 延伸阅读

- “创建和恢复 MySQL 数据库备份”示例。
- “创建和恢复 PostgreSQL 数据库备份”示例。
- “设置常规作业的定时任务”示例。

13.7 通过电子邮件获取详细的错误报告

当执行系统的日志时，Django 使用 Python 内建的日志模块或前述示例中的 structlog 模块。默认的 Django 配置看上去十分复杂。在当前示例中，我们将学习对其进行调整，并发送包含完整 HTML 内容的错误电子邮件，这类似于当出现错误时 Django 以 DEBUG 模式提供的内容。

13.7.1 准备工作

定位虚拟环境中的 Django 项目。

13.7.2　实现方式

下列步骤将发送与错误相关的详细信息。

（1）首先设置项目的 LOGGING 设置项。访问 env/lib/python3.7/site-packages/django/utils/log.py 并获取 Django 日志实用程序文件。作为 LOGGING 字典，将 DEFAULT_LOGGING 目录复制至项目的设置项中。

（2）向 mail_admins 处理程序添加至 include_html 设置项。前两个步骤的结果如下所示。

```
# myproject/settings/production.py
LOGGING = {
    'version': 1,
    'disable_existing_loggers': False,
    'filters': {
        'require_debug_false': {
            '()': 'django.utils.log.RequireDebugFalse',
        },
        'require_debug_true': {
            '()': 'django.utils.log.RequireDebugTrue',
        },
    },
    'formatters': {
        'django.server': {
            '()': 'django.utils.log.ServerFormatter',
            'format': '[{server_time}] {message}',
            'style': '{',
        }
    },
    'handlers': {
        'console': {
            'level': 'INFO',
            'filters': ['require_debug_true'],
            'class': 'logging.StreamHandler',
        },
        'django.server': {
            'level': 'INFO',
            'class': 'logging.StreamHandler',
            'formatter': 'django.server',
        },
        'mail_admins': {
```

```
            'level': 'ERROR',
            'filters': ['require_debug_false'],
            'class': 'django.utils.log.AdminEmailHandler',
            'include_html': True,
        }
    },
    'loggers': {
        'django': {
            'handlers': ['console', 'mail_admins'],
            'level': 'INFO',
        },
        'django.server': {
            'handlers': ['django.server'],
            'level': 'INFO',
            'propagate': False,
        },
    }
}
```

日志配置包含 4 部分内容，即日志器、处理程序、过滤器和格式器。

- 日志器是日志系统的入口点。每个日志器包含一个日志级别，即 DEBUG、INFO、WARNING、ERROR 或 CRITICAL。当消息写入至日志器时，消息的日志级别将和日志器的级别进行比较。如果满足或超出日志器的日志级别，则进一步被处理程序处理；否则该消息被忽略。
- 处理程序表示为引擎，并定义了日志器中每条消息发生的情况。处理程序可被写入至控制台、通过电子邮件发送至管理员、保存至一个日志文件中、发送至 Sentry 错误日志服务等。在当前示例中，我们设置了 mail_admins 处理程序的 include_html 参数，因为针对 Django 项目中出现的错误，回溯和本地变量需要完整的 HTML 内容。
- 过滤器对消息提供了额外的控制，这些消息由日志器传递至处理程序。如在当前示例中，当 DEBUG 模式设置为 False 时，电子邮件才被发送。
- 格式器用于定义如何作为一个字符串渲染一条日志消息。当前示例并未使用格式器。关于日志信息的更多内容，读者可查看官方文档，对应网址为 https://docs.djangoproject. com/en/3.0/topics/logging/。

13.7.3　更多内容

我们定义的配置将发送与服务器错误相关的电子邮件。如果流量较大，假设数据库为此而崩溃，那么我们将收到大量的电子邮件，甚至会导致电子邮件服务器死机。

为了避免这种情况，我们可使用 Sentry（https://sentry.io/for/python/）。Sentry 跟踪所有的服务器错误，且针对每种错误类型仅发送一封通知电子邮件。

13.7.4　延伸阅读

- 第 12 章中的“针对产品环境利用 mod_wsgi 在 Apache 上部署”示例。
- 第 12 章中的“针对产品环境在 Nginx 和 Gunicorn 上部署”示例。
- “日志事件”示例。